The

Renewable Energy Handbook

The
Renewable Energy Handbook

A Guide to Rural Independence, Off-Grid and Sustainable Living

William H. Kemp

ΛZTEXT PRESS

Trade distributor:

NEW SOCIETY PUBLISHERS

Aztext Press
2622 Mountain Road, Tamworth, Ontario Canada K0K 3G0
michelle@aztext.com • www.aztext.com

Canadian Cataloguing in Publication Data
Kemp, William , 1960-
    The Renewable Energy Handbook
    A Guide to Rural Independence, Off-Grid and Sustainable Living
Rev. 1
Includes index
ISBN 0-9733233-2-9

1. Renewable energy sources—Handbooks, manuals, etc. I. Title.
TJ808.3.K44 2003   621.042   C2003-904237-5

Disclaimer:
"The installation and operation of Renewable Energy systems involves a degree of risk. Ensure that all proper installation, regulatory and common sense safety rules are followed. If you are unsure of what you are doing, STOP! Seek skilled and competent help by discussing these activities with your dealer or electrician.

    Electrical systems are subject to the rules of the National Electrical Code ™ in the United States and the Canadian Electrical Code ™ in Canada. Wood and pellet stove units are also subject to national and local building code requirements.

    In addition, local utility, insurance, zoning, and many other issues must be dealt with prior to beginning any installation. In the words of Emerson, "People don't plan to fail, they just fail to plan." Nowhere is this more apparent than when installing renewable energy systems.

    The author and publishers assume no liability for personal injury, property damage, consequential damage or loss, including errors and omissions, from using the information in this book, however caused.

    The views expressed in this book are those of the author personally and do not necessarily reflect the views of contributors or others who have provided information or material used herein.

Printed and bound in Canada

This book is dedicated to our planet Earth.

We humans, blessed with Her bounty, relentlessly abuse what we should nurture. There are many who say they wish to save Her, but few who are willing to pay their share of the bill.

Acknowledgements

A book of the magnitude of the Renewable Energy Handbook requires the technical knowledge and access to information from many individuals and companies, not to mention the patience of family while writing and editing the text.

It is customary to identify those contributors who have helped with technical data required to put this book together and make sure the information is the most up to date.

I am indebted to, and would like to thank, Steve Anderson Biodiesel; Sean Twomey and the delightful staff at Arbour Environmental Shoppe; David Elzina, ARISE Technologies; Michael Reed of Array Technology Inc. for information on tracking solar mounts; Mike Bergey of Bergey Wind Power for technical publications and photographs on wind power systems; Don Bishop of Benjamin Heating Products for information regarding dual energy heating boilers; Christine Paquette, Executive Director - Biodiesel Association of Canada for assistance and review on the Biofuels Chapter; Bogart Engineering, Energy Meters; Liujin, Les and Jose from Carearth Inc. for installing a great solar thermal system on my house and providing information and photographs; Dankoff Solar Products; Vanessa Percival at Embers Wood Stoves; Kevin Pegg of Energy Alternatives; Energy Systems and Design (micro hydro); Bob Fisher, Enerworks (solar thermal); Nicole and Aidan Foss for being excellent critics and mentors of world energy policy; Paul Gipe for his assistance and mentoring; Harmen Pellet Stoves Company; Peter Talbot, Home Power Systems (microhydro); Ross and Kathryn Elliot of Homestead Building Solutions for help with the home air leakage tests; George Peroni at HydroCap Corporation for information regarding hydrogen gas reclaiming devices for batteries; Tania Glithero, Iogen Corporation; Madawaska Mill Works (hot tubs and Saunas); Steve Howell, MARC-IV Consulting; Patrice Feldman at Morning Star Corporation for her quick response regarding PV controller design and installation; Rajindar Rangi and Tony Tung of Natural Resources Canada, Energy Technology Branch for guidance and providing access to reams of technical bulletins and contacts; Mike McGahern, Ottawa Solar Power; "The Boys" from Renewable Energy of Plum Hollow; Deborah Doncaster and Melinda Zytaruk, Ontario Sustainable Energy Association (OSEA); Outback Power Systems; Tim Rurak, Ralph's Radio; Gerald Van Decker of Renewability Energy Inc. for assistance with the waste water heat recovery system; Sharp Solar Corporation; John Supp of South West Wind Power for the great site photographs and micro-wind technical data; Snorkel Hot Tub Company; Surrette Battery Company; Dr. David Suzuki, for endorsing this book; Jaeson Tanner for cover and web design; Govindh Jayaraman of Topia Energy Inc.; Ian Micklethwaite of Warren Publishing Co. Ltd.; World Energy LLC. Sean McAdam, Veggiegas.ca; Pam Carlson of Xantrex Technology Inc. for the tremendous support of technical documentation, photographs and the nice Xantrex coffee mug!; and lastly, Stefani Kuykendall of Zomeworks Corp. for information on fixed PV mounting hardware. (If I have missed anyone, please accept my apologies and understand that your assistance was truly appreciated!)

A huge thank you must also be given to Joan McKibbin my editor. Joan worked tirelessly under very demanding deadlines and also stopped me from writing too many run-on sentences. I am surprised that she only needed a cup of tea and a hot bath after pulling our last "all nighter". A lesser person would have required something a bit more potent!

I am also indebted to those who have had to work directly with me in the preparation of the text. A big thanks goes out to: Carol McGregor and Cam Mather for their excellent line drawings and graphics, not to mention having to work under Bill's demanding time lines; Lorraine Kemp, Katie Mather and Michelle McNamee for their great help behind the camera and proving the old adage that a picture is worth a thousand of my words. (All images without photo credits by them); To Karen and Jamie Wilson, Tina and Mike Hubicki, The Beevor Family, Cherie and John Graham of Church Key Brewing Company, Fairmont Resorts, Hillary Houston and Raymond Lebfrevre, The Adams, Lorraine Kemp and Michelle and Cam Mather for letting me invade their privacy to extract the photographs of R.E. systems throughout the book.

And of course thanks to Lorraine Kemp and Michelle Mather (and Katie and Nicole) for putting up with Cam Mather and me during our single-minded focus on writing and producing this book, using 100% Renewable Energy! (Aztext Electronic Publishing Ltd. is available for your needs; contact cam@aztext.com.)

Contents

Preface

The majority of the population doesn't give much thought to energy. Sure, everyone gripes a bit (maybe a lot) when the cost of electricity or gasoline rises for the tenth time this year. But do most people really ever think about energy? Where does it come from, what impact does it have on the family budget, the environment, or our children's future?

Most of us are so busy trying to stay afloat in our hectic lives that these questions only surface as polite dinner discussions. Every once in a while someone stops to ponder these issues. Perhaps that's why you've picked up The Renewable Energy Handbook.

Within the covers of this handbook, you will be provided with the background information, purchasing data, and step-by-step instructions to allow you to:

- stretch your heating and utility dollar;
- operate your boat, RV, cottage or full-time residence on clean, noise-free, renewable energy;
- create energy for domestic hot water and space heating from independent and clean sources;
- worry just a little bit less about acid rain, ozone depletion, airborne smog, nuclear waste, or world oil supply problems.

Whether you are just curious or an industry expert, if these issues are on your mind, then The Renewable Energy Handbook is for you.

We will discuss how to stretch your energy dollar, allowing you to do much more with less. Step-by-step guidance and easy-to-understand instruction will help you pave the way to better energy management and renewable energy production, whether you own a sailboat in the Caribbean, a cabin in the Ozarks, or an estate ranch in the prairies.

We will guide you to a new relationship with the environment that lets you consume energy without worrying about the effects of dumping today's pollution on tomorrow's children. Unlike fossil fuels or nuclear energy, renewable sources are just that: renewable. No matter how much energy we capture and use, there will always be more for everyone else.

And most important of all, The Renewable Energy Handbook will show you how to do it correctly, do it economically, and do it now!

1
Introduction

Energy is the life breath of modern society. Without energy, almost every event we take for granted comes to a screeching halt. If you happened to live on the eastern side of North America during the blackout of 2003, you will acutely understand how life without electricity changes everything. For a few dozen hours, modern life stopped— *completely*: people stuck in elevators; no building or traffic lights; traffic gridlocks; no air conditioning, cell phones, computers, television or email. The entire world appeared to go black.

North American dependence on electrical energy and fossil fuels is, per capita, the highest in the world. In the early 1900's the United States was the world's largest producer of oil. Over the last 100 years most of the domestic wells have been sucked almost dry, leaving the country to import the majority of its oil from domestically unstable and expensive sources. The importation of large amounts of fossil fuel causes the *exportation* of equally large numbers of dollars that could have been invested in a North American clean energy industry.

Family budgets are straining from the cost of energy to power our homes and fuel our cars. Many people are surprised to learn that they must work an average of two months per year just to pay their home energy bills. With the cost of all types of energy steadily increasing, the outlook for the future can only get worse.

What about the environment? Climate change is one of the key challenges facing worldwide sustainable development. Global warming, caused mainly by the burning of fossil fuels, deforestation, and inefficient use of energy, is pushing ecosystems to the brink of catastrophic failure.

Atmospheric concentrations of carbon dioxide (CO_2) have increased dramatically since the industrialization of society over the last 150 years and are at the highest levels measured during the past 500,000 years. Global energy use amongst industrialized countries is continuing to increase unabated. Underdeveloped countries, where more than one–third of the world's population does not have access to electricity, are starting to ask for their fair share of the energy pie. Exponential human population growth and accompanying energy consumption are expected to cause global temperatures to increase between 2 and 5.8°C over the coming century, according to the *Intergovernmental Panel on Climate Change* report issued in 2001.[1]

This temperature increase sounds innocent enough until you realize that a change of this magnitude has never previously occurred and will amplify weather effects throughout the earth's ecosystems. Increased levels of drought, flooding, and storms in areas already sensitized to these environmental stresses will result. The press and respected scientific papers are full of articles describing the effects of climate change: melting glaciers, reduced arctic pack ice, increasing ocean water levels, and the destruction of coral reefs. These ecologically sensitive issues are harbingers of far more devastating events to come.

As you are contemplating generating some or all of your personal energy requirements, you may be aware of these issues. But you may wonder if there is anything you can do at a personal level. The key to reversing climate change is action. The key to generating clean energy is education. *The Renewable Energy Handbook* will provide the tools required to empower you to meet these goals. The book will introduce you to the concept of *eco-nomics*, following a logical sequence of planning and actions that will ensure that you can meet your goals of energy independence and self-sufficiency, saving money on your personal power station installation while at the same time preventing damage to the world's ecology. Hence the **eco-nomic** path of energy efficiency and clean energy sources.

Energy Efficiency

Before you generate one watt of power, you must begin with energy efficiency:

North Americans need a new strategy, one that focuses on reducing energy use through efficiency and conservation rather than on increasing supply. Experience from across North America has proven that it is significantly cheaper to invest in energy efficiency than to build or even maintain polluting sources of electricity supply. This theory holds true whether we are describing the nuclear power plant down the road or your own home-based, renewable energy system. Or, as Amory Lovins, the energy guru from the Rocky Mountain Institute, is fond of saying, "It is far less expensive and environmentally more responsible to generate *negawatts* than megawatts."

Using less energy isn't about making drastic lifestyle changes or sacrifices. Conservation and efficiency measures can be as simple as improving insulation standards for new buildings, replacing incandescent light bulbs with compact fluorescent models, or replacing an old refrigerator with a more efficient one. In fact, energy efficiency often provides an improvement in lifestyle. A poorly insulated or drafty house may be impossible to keep comfortable no matter how much energy (and money) you use trying to keep warm.

California learned firsthand that saving energy means saving money and the environment. You may recall the rolling blackouts and severe power shortages that afflicted the state a few years ago. It was predicted that dozens of generating stations would be required on an urgent basis to solve the state's energy problems. At the time of writing, the total number of power stations built to solve the problem stands at zero. Faced with the realization that construction cycles for significant generating capacity would require several years, forward-thinking officials looked to energy efficiency instead. The state's energy–efficiency standards for appliances and buildings have helped Californians save more than $15.8 billion in electricity and natural gas costs. One–third of Californians cut their electricity use by 20% to qualify for a 20% rebate on their bill. The government introduced a renewable-energy buy-down and accompanying net-metering program that has seen thousands of clean, photovoltaic power systems installed on residential rooftops.[2]

In addition to saving electricity and reducing fossil fuel burning, California's conservation and efficiency efforts reduced greenhouse gas emissions by close to 8 million tonnes and nitrogen oxide emissions by 2,700 tonnes during 2001 and 2002.[2]

Consider the following points:
- Switching to compact fluorescent lamps will reduce lighting energy consumption by 80%.
- High efficiency appliances such as refrigerators, washing machines, and dishwashers can reduce energy consumption by a factor of 5 times.
- Using on-demand water heaters will reduce hot water energy costs by up to 50%.
- Low-flow showerheads, aerator faucets, and similar fittings will reduce water consumption and resulting heating costs by approximately one-

half as well as reducing well-pump energy consumption and septic or leaching bed stress.

- A well-sealed and insulated house can reduce home heating and cooling costs by 50%.
- Adding a solar thermal water-heating system can further reduce hot water heating costs by 50%.

Many of the items on this list are neither expensive nor difficult to implement. Best of all, implementing these products not only dramatically reduces smog and greenhouse gas emissions, but greatly reduces the size and capital cost of your off-grid power station components. Expressed in a slightly more dramatic way, the United States Department of Energy states that for every dollar you invest in energy efficiency, you can reduce the cost of generating power by three to five dollars.

Clean Energy Sources

Sheikh Zaki Yamani, a former Saudi Arabian oil minister, sums up this position very well: "The Stone Age did not end for lack of stone, and the Oil Age will end long before the world runs out of oil." North Americans are addicted to oil in a manner that is not comprehensible to Europeans or to the developing world. While the rest of the world has developed pricing signals to achieve greater fuel and energy economy, we in North America still show up at the altar of the mighty "SUV".

The United States has less than 5% of the world's population (300 million out of 6.4 billion) and consumes a whopping 30% of *all* of the world's resources. This includes not only energy but also water, steel, aluminum, timber, and just about everything else you can imagine. When we enter a Wal-Mart or local grocery store, it is difficult to imagine that there could possibly be shortages of anything we desire. Unfortunately, present levels of consumption are not sustainable in the long term. Consider for a moment that if all of the world's population were to consume sources at the same level as the United States, we would require approximately 20 times current levels. In other words, we would require an additional seven planets worth of stuff and energy.

Indeed, with China, Russia, India and many other emerging-economy nations with combined populations many times that of the United States and Canada, it will not take long before our resources dry up and ecosystems give up in frustration.

Many people simply do not believe this. I have personally heard educated people say that climate change, air pollution, and fossil fuel shortages are the fabrications of doomsayers and quacks. However, scientific evidence suggests that these environmental hazards are not a fabrication, but pose a definite threat. Although frugality or erring on the side of caution is currently not in vogue in North America, I would suggest that this generation has a responsibility to our children and our children's children to not squander our resources.

Our current way of life includes the belief that cheap energy is our God-given right. Never mind that the cost of gasoline or imported heating oil does not include the vast subsidies lavished on the oil industry. The price of a gallon of gasoline neglects the ongoing American military presence in the Middle East, depletion subsidies, cheap access to government land, as well as monies to advance drilling and exploration technologies. None of these "hidden" costs even touches on the environmental and health damage caused by the burning of fossil fuels. For example, air pollution, largely from the burning of fossil fuels, kills an estimated 5,800 Ontarian's prematurely each year and results in hundreds of thousands of incidents of illness, absenteeism and asthma attacks—costing the economy billions of dollars. [3]

If the same level of subsidization and "external costs" were to be heaped on the renewable energy industry, every home would be equipped with solar electric and thermal panels, allowing energy independence.

Fossil fuels don't just power our cars. Home heating and coal-fired electrical power plants rely on fossil fuels as well. Contrary to popular belief, it is the electrical power-generation industry and not the transportation sector that contributes the largest amounts of smog, acid rain, and green house gas emissions to the world's ecosystems. [4] With much of the developing world's population connecting to the electrical grid every day, it is obvious that world energy demand will mushroom as more (and larger) appliances are brought online, further exacerbating the problem. In the developed world, middle class families are demanding more and more appliances that were considered luxury items only one generation ago. Central air conditioning, multiple refrigerators, computers, chest freezers, hot tubs, and swimming pools, all luxury or unimaginable items in our parents' day, all consume enormous amounts of energy.

Energy efficiency and demand-side management of these electrical loads will only go so far. No matter how efficient our appliances are, energy is still required to power our homes. Reducing our reliance on fossil fuels and working towards cleaner energy supplies based on the efficient use of natural gas, biofuels, geothermal and renewable sources will go a long way toward creating a carbon-neutral energy supply.

The World Wildlife Fund (WWF) has introduced a program to get major power utilities to switch from high-carbon fuels—especially coal—to cleaner options. The key technologies and policies of their Powerswitch! program include:

- energy efficiency and demand-side management strategies
- large-scale wind energy projects—mainly offshore
- large-scale biomass and biomass cofiring at existing coal power stations
- increased support for renewable energies
- high rates of aluminum recycling
- rapid growth of combined heat and power systems (CHP)
- fuel switch from coal to natural gas as a transition fuel

On the home scale, several of these options are completely viable, reliable and plain fun, as you will learn in the following chapters.

By adopting demand-side management and efficient appliances, focusing on conservation, and shifting to cleaner fuels and renewable energy supplies, homeowners like you and me will be less vulnerable to energy price increases and security of supply problems and will help protect the world's ecosystems. After all, if we don't get involved, who will?

Footnotes
1. *Climate Change 2001 – Synthesis Report*. IPCC.
2. *Bright Future*. David Suzuki Foundation, September 2003
3. *Ontario Medical Association* http://www.oma.o rg/phealth/smogmain.htm
4. *New Scientist*, June 2003

1.2 Renewable Energy <u>*Eco*</u>-*nomics*

Now that you are motivated to "do your bit" we should take a minute to chart a path and make sure that the "bit" you "do" is the right one.

At this point you may be eager to run out and purchase a wind turbine or install solar thermal hot water heating. Not so fast! No matter how much of a "keener" you are for eco-stuff, there is right and wrong way to do everything. In fact, anyone who has had a bad experience with renewable energy systems probably didn't take the time to fully understand the technology and the economics of that decision. That is, they did not understand the *eco-nomics*.

Figure 1-1. This infrared thermal image of a house shows that the modern home is poorly constructed. With current technology, home heating and electrical consumption can be reduced by 50% without any decrease in quality of life. (Courtesy NASA/IPAC).

Figure 1-2. This home incorporates special flexible photovoltaic panels mounted in place of roofing shingles. Modern renewable energy-powered homes use advanced technologies that provide the same amenities and level of comfort as their utility-connected cousins. (Courtesy Arise Technologies Inc.)

For off-grid systems, it is very important to understand that energy efficiency is much more important that energy generation. Remember that for every dollar you invest in efficiency measures you save three to five dollars in generating equipment. It stands to reason that you **must** perform all of the updates, installations, and conversions to an energy-efficient lifestyle **before** you invest one dollar in electrical generating equipment.

The delivered price of grid-supplied electricity averages approximately 10 to 12 cents per kilowatt-hour in North America, while power produced by an off-grid system costs many times this amount. Therefore, the capital cost of energy-efficient upgrades and installations has a very short payback time as inefficient appliances require electrical generating equipment that will cost many times as much.

North Americans have an unfortunate habit of looking at "first cost" rather than "life cycle" or "true cost" when analyzing appliances and consumer goods. The everyday incandescent light bulb is an excellent example. Thomas Edison's mid-1800s invention was a technological marvel in its day. Replacing smoky, dim kerosene lamps with clean and bright electric lights can never be undervalued. However, just as we no longer use a horse and buggy or wireless radio sets from that era, the inefficient Edison lamp should now be relegated to the museum. The replacement is the modern compact fluorescent lamp which uses 4 1/2 times less energy and lasts 10 times longer.

At first glance, the initial cost of the compact fluorescent lamp appears uneconomic compared with the cost of the old-fashioned incandescent lamp. Typical compact fluorescent lamps have a first cost of approximately three dollars where the incandescent lamp is approximately 40 cents. But look closer. The compact fluorescent lamp lasts approximately 10 times longer, making first cost lower ($.40 x 10 incandescent lamps = $4.00, giving the same life as 1 compact fluorescent = $3.00). Based on first cost alone, our compact fluorescent lamp is already one dollar less expensive.

Now let's look at energy costs. Assuming you pay $0.12 for each kilowatt hour of electricity delivered to your home[5], a 23 W compact fluorescent lamp (which provides light equivalent to a 100 W incandescent lamp) will save over $92 in energy costs over the life of the bulb[6]. Now, multiply this savings times the average 25 light bulbs per house and you have just put a cool $2,300.00 *after-tax* dollars in your pocket.

Now consider these numbers from an off-grid point of view. Assume that an off-grid system installation is $25,000 based on an energy efficient homes' electrical energy consumption of 4 kWh of energy per day. The typical home using inefficient lights and appliances will easily use 7 to 10 times this amount of energy. If a typical home were to be converted to an off-grid design, the system would be much larger, driving capital costs in excess of $200,000!

This is real money. It's your choice: write a monstrous check to your renewable energy dealer or spend a few thousand dollars investing in energy-efficient appliances and products before you take the off-grid plunge. It's your money; why not keep the dollars in your pocket instead of making the dealer rich? (Chapter 2 and the accompanying worksheet in Appendix 7 will help you choose the economic alternative.)

There is no refuting economics in this case. Businesses, cruise ships, hotels, and our European neighbors have used compact fluorescent, color-corrected lighting for years without any degradation of their lifestyle. Indeed, compact fluorescent lighting is flicker free, dimmable, of equivalent brightness and quality to incandescent lights, available in a wide selection of sizes, usable outdoors, and available with a seven-year money-back guarantee.

Our single 23 W compact fluorescent light will reduce greenhouse gas emissions by 1140 pounds (518 kg) and acid rain-producing compounds by 8 pounds (3.6 kg) over the life of the bulb. For the entire house, a reduction of 28,500 pounds (12,950 kg) of greenhouse gas emissions and 200 pounds (91 kg) of acid rain-producing compounds will result. As an added bonus, there will be nine fewer dead light bulbs added to your local landfill or recycling depot.

In this example, true cost is many times lower than first cost, providing a rapid return on investment far greater than any stock market analyst could ever (legally) achieve.

This is the case for _**eco-nomics.**_

---

5 Electrical utilities are quick to point out that the cost of generation is "x" cents per kilowatt hour. What they forget to tell you is that the electricity has to be delivered to your door using transmission and distribution systems that must be paid for. Of course we can't forget the bit of tax added for the government. On average the cost of electricity delivered to your home is approximately double the cost of generation in most jurisdictions. To calculate your actual cost of electrical energy, divide your total electrical bill charges by the number of kilowatt hours of electricity used during that billing period.

6 At 12 cents per kWh a 100 W bulb requires $120 in electricity, a 23 W compact florescent (CF) bulb requires $27.60, saving you $92.40 in energy costs over the life of the 10,000 hour life of the CF bulb. Bulbs of varying wattages use more or less energy directly based on their power consumption compared to this example.
Incandescent Bulb = $0.12/kWh of electricity x 10,000 hours x 100 W/bulb = $120.00
CF Bulb = $0.12/kWh of electricity x 10,000 hours x 23 W/bulb = $27.60
Savings per bulb = $120.00 per Incandescent Bulb - $27.60 per CF bulb = $92.40

What about Off–Grid System Capital Costs?

Off-grid systems have a unique way of looking at payback criteria. In most (but certainly not all) cases, a homeowner who wishes to build a home or cottage on a desirable lot which is a distance away from the utility lines must pay for the line extension to the home. One does not have to run the wire very far before a serious amount of cash is burned up. It may only require a line extension of 0.5 miles (0.8 km) to exceed the cost of an off-grid system. In that case, the payback time is the instant the first light is turned on!

Obviously, the further away from the utility lines, the faster the payback, to the point where an off-grid system may be the only economic way to build in that location. Islands and mountaintop locations with beautiful views and limited intruding neighbors come to mind.

"Off Gridders" (that group of people who are often fond of granola and Birkenstocks) can use the cost of utility-line extensions to their advantage. A building lot that has no access to electrical power might just not be worth as much in the seller's eyes, even though the buyer has no intention of making the electrical utility connection.

> **A Word about Pricing:** Many of the products described in this handbook are either manufactured in the United States or priced in U.S. dollars. To avoid confusion, all prices are shown in U.S. dollars.

1.3 What Is Energy?

All of the earth's energy comes from the sun. In the case of renewable energy sources and how we harness that solar energy, the link is often very clear: sunlight shining through a window or on a solar heating panel creates warmth, and when it strikes a photovoltaic (or PV) panel the sunlight is converted directly into electricity; the sun's energy causes the winds to blow, which moves the blades of a wind turbine, causing a generator shaft to spin and produce electricity; the sun evaporates water and forms the clouds in the sky from which the water, in the form of raindrops, falls back to earth becomes a stream that runs downhill into a micro-hydroelectric generator.

While these energy sources are renewable, they are also variable and intermittent. The sun generally goes down at night and may not shine for several cloudy days. If we just wanted to use our energy when it is available, the various systems used to collect and distribute it would be a whole lot simpler. But we humans are just not that easily contented. I know for a fact that most people want their lights to turn on at night, even though the sun stopped shining on the PV panels hours ago. In order to ensure that heat and electricity are available when we need it, a series of cables, fuses, and a bewildering array of components are required to capture nature's energy and deliver it to us on demand.

The process of capturing and using renewable energy may seem far too complicated and expensive for the average person. However, while the components themselves are complicated, the theory and techniques required to understand, install, and live with renewable energy are not.

Figure 1-3. Most people want their lights to turn on at night. This little issue complicates the requirements of a renewable energy system.

Why Are We Discussing Math?

Although it is not absolutely necessary to be an expert in heating, wind, and electrical energy, an understanding of the basics will greatly assist you in operating your renewable energy system. This knowledge will also help you make better decisions when it comes to purchasing the most energy-efficient appliances.

Trying to save money by reducing energy costs can only be accomplished by understanding your "energy miles per gallon" quotient. It is fairly simple to determine gas fuel economy or efficiency, but it is much more difficult to determine energy efficiency for a house full of appliances and heating equipment.

The mathematical theories described in *The Renewable Energy Handbook* are simplified. Equations are used throughout the text, but they are limited to those areas that are most important. While the mathematics might be a bit light for some, for the rest of us the math and logic are fairly straightforward; a simple calculator (powered by the sun of course) will make the task that much simpler. For those so inclined, there are plenty of references to data sources that will make even the most ardent "techie" happy.

The Story of Electrons

If you can remember back to your high school days in science class, you may recall that an atom consists of a number of electrons swirling around a nucleus. When an atom has either an excess or a lack of electrons in comparison to its "normal" state, it is negatively or positively charged, respectively.

In the same way that the north and south poles of two magnets are attracted, two oppositely charged atoms are also attracted. When a negatively charged atom collides with a positively charged one, the excess electrons in the negatively charged atom flow into the positively charged atom. This phenomenon, when it occurs in far larger quantities of atoms, is called

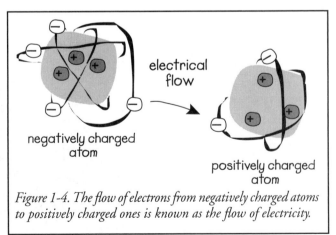

Figure 1-4. *The flow of electrons from negatively charged atoms to positively charged ones is known as the flow of electricity.*

the flow of electricity.

The force that causes electricity to flow is commonly known as *voltage* (or V for short). The actual flow of the electrons is referred to as the *current*. So where does this force come from? What makes the electrons flow in the first place? The trigger that brings about the flow of electrons can come from several energy sources. Typical sources are chemical batteries, photovoltaic cells, wind turbines, electric generators, and the up-and-coming fuel cell. Each source uses a different means to trigger the flow of electrons. Waterfalls, coal, oil, or nuclear energy are commonly used to generate commercial electricity. Fossil or nuclear fuels are used to boil water, which creates the steam that drives a turbine and generator; falling water drives a turbine and generator directly. The spinning generator shaft induces magnetic fields into the generator windings, forcing electrons to flow.

Let's use the flow of water as a visual aid to understanding the flow of electricity that is otherwise invisible.

Let's presume that a greater amount of water moving past you per second equates to a higher flow and that a river with a high flow of water has a large current. With electricity, a large number of electrons flowing from one atom to the next is similar to a large flow of water. Therefore, the greater the number of electrons flowing past a given point per second, the greater the electrical current.

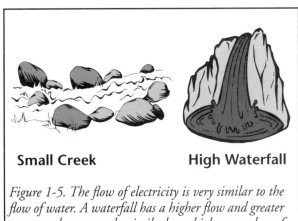

Small Creek　　　　**High Waterfall**

Figure 1-5. The flow of electricity is very similar to the flow of water. A waterfall has a higher flow and greater pressure than a creek; similarly, a higher number of electrons moving from atom to atom increases electrical current.

The flow of water is typically measured in gallons per minute or liters per second; electrical current is measured in amperes (or A for short). If the measured current of electrons in a material is said to be 2 A, we know that a certain number of electrons have passed a given point per second. If the current were increased to 4 A, there would be twice the number of electrons flowing through the material.

Gravity is a factor in considering water pressure. A meandering creek has very little water pressure, owing to the small hill or height that the water has to fall. A high waterfall has a much greater distance to fall and therefore has increased water pressure. For example, if a water-filled balloon were to fall on you from a height of 2 feet you would find this quite refreshing on a hot day. A second balloon falling from a height of 100 feet would exert a much higher pressure and probably knock you out! This increase in water pressure is similar to the electrical pressure or voltage that forces the electrons to flow through a material.

Conductors and Insulators

Electricity flows through conductors in the same manner as water flows through pipes and fittings. When electrons are flowing freely through a material we know that the material is offering little resistance to that flow. These materials are known as conductors. Typical conductors are copper and aluminum, which are the substances used to make electrical wires of varying diameters. Just as a fire hose carries more water than a garden hose, a large electrical wire carries more electrons than a thin one. A larger wire can accept a greater flow of current (i.e. handle higher amperage).

Before electricity can be put to use, we must create an electrical circuit. We do this by connecting a source of electrical voltage to a conductor and causing electrons to flow through a load and back to the source of voltage.

The drawing in Figure 1-6 shows how a simple flashlight works. Electrons stored in the battery (we will cover that one later) are forced by the voltage (pressure) to flow into the conductor from the battery's negative contact (-) through the light bulb and back to the battery's positive contact (+). Current flowing in this manner is called *direct current* (or DC for short). The light will stay lit until the

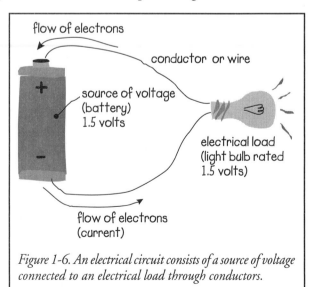

Figure 1-6. An electrical circuit consists of a source of voltage connected to an electrical load through conductors.

battery dies (runs out of electrons) or until we turn off the switch.

Oh yes, a switch would be a good idea. This handy little device allows us to turn off our flashlight to prevent using all of the electrons in the battery. From our description of a circuit, we can assume that if the electrical conductor path is broken, the light will go out. How do we break the path? The flow of electrons may be interrupted using a nonconductive substance wired in *series* with the conductor. Any substance that does not conduct electricity is known as an insulator.

Typical insulators include rubber, plastics, air, ceramics, and glass.

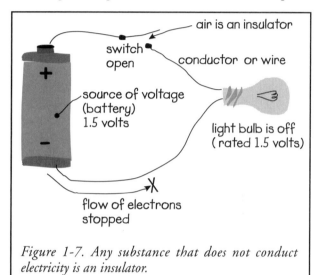

Figure 1-7. Any substance that does not conduct electricity is an insulator.

Batteries, Cells, and Voltage

You may have noticed that your brand of flashlight has two or even three cells. Placing cells in a stack or in *series* causes the voltage to increase. For example, the flashlight shown in Figure 1-7 contains a cell rated at 1.5 V. Placing two cells in series, as shown in Figure 1-8, increases the voltage to 3 V (1.5 V + 1.5 V = 3 V). The light bulb in Figure 1-8 is glowing very brightly and will quickly burn out because it is rated for 1.5 V. In this example, a higher voltage (pressure) is

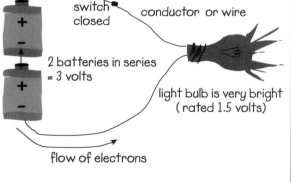

Figure 1-8. A higher voltage forces more electrons to flow (higher current) than the bulb is rated to tolerate. Always ensure that the source and load voltage are rated equally.

causing more current to flow through the circuit than the bulb is able to withstand. Likewise, if the cell voltage were lower than the rating of the bulb, insufficient current would flow and the bulb would be dim. This is what happens when your flashlight batteries are nearly dead and the light is becoming dim: the batteries are running out of electrons.

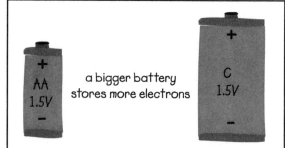

Figure 1-9. Cells of different sizes have different amounts of electron storage. Just as a 2-liter jug contains twice as much water as a 1-liter bottle, so a larger battery stores more electrons and works for longer periods than a smaller one.

Next time you are waiting in the grocery store lineup, put down that copy of *The National Enquirer* and take a look at the battery display. The selection will include AA, C, and D sizes of cell, which all have a rating of 1.5 V. Can you guess the difference between them?

A larger cell holds more electrons than a smaller one. With more electrons, a larger battery cell can power an electrical load longer than a smaller one. The circuit shown in Figure 1-10 shows a set of jumper wires connecting the cell terminals in *parallel*. This parallel arrangement creates a battery bank of 4 AA-size cells, which has the same number of electrons as and the capacity of the C-size cell. Any grouping of cells, whether connected in series, in parallel, or both, is called a battery bank or battery.

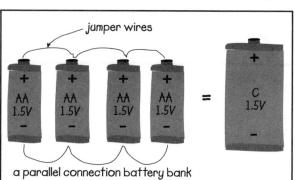

Figure 1-10. Wiring cells in parallel increases the capacity of the battery bank.

Obviously, a house requires far more electricity than a simple flashlight does. Off-grid homes generate electricity from renewable sources (more on that later) and may store it in battery banks such as the one shown in Figure 1-11.

Figure 1-11. Off-grid homes or homes equipped with emergency backup systems require large amounts of electricity, which is stored in a battery bank such as this one.

Each battery cell is numbered 1 through 12 and has a nominal voltage of 2 V. If you look carefully, you can see that each cell has a (+) and (-) terminal. Each terminal is wired in *series* to the next battery in the manner illustrated in Figure 1-8. Therefore a series string of 12 cells rated at 2 V each creates a battery bank rated at 24 V DC (Vdc).

You will also note that there are two such banks of batteries. The bank on the left is identical to the bank on the right. By wiring the two banks in *parallel* we create a total battery capacity that is twice as large as a single bank. This is exactly the same as wiring the 4 AA cells in parallel to make the equivalent of the large C cell. Obviously, the more electricity we use the greater the battery size and cost.

Off-Grid Houses Don't Usually Run on Batteries

You are probably aware that your house works on 120/240 V and not 12 V, 24 V, or 48 V from a battery. Early off-grid houses, small cottages, boats, and many recreational vehicles can and do use 12 Vdc systems. But don't consider using low-voltage DC for anything but the smallest of systems. With the limited selection of appliances and the difficulty of wiring a low-voltage, full-time home, low-voltage DC systems are not a viable option.

The modern off-grid home is supplied with electricity in the form of 120/240 V *alternating current* (Vac). In a DC circuit, as defined earlier, current flows from the negative terminal of the battery through the load and back to the positive terminal of the battery. Since the current is always flowing one way, in a direct route, this is called direct current.

In an AC circuit, the current flow starts at a first terminal and flows through the load to the second terminal, similar to the flow in a battery circuit. However, a fraction of a second later the current stops flowing and then reverses direction, flowing from the second terminal through the load and back to the first terminal, as shown in Figure 1-12.

Alternating Current in the Home

Generating electricity in the modern, grid-connected world is accomplished by using various mechanical turbines to turn an electrical generator. In the early years of the electrical system, there was considerable debate as to whether the generator output should be transmitted as AC or DC. For a time, both AC and DC were generated and transmitted throughout a city. However, as time passed, it became clear for safety, transmission, and other practical reasons that

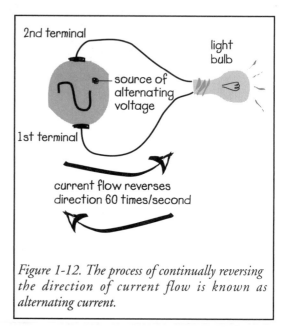

Figure 1-12. The process of continually reversing the direction of current flow is known as alternating current.

AC was more desirable. As the old saying goes, "The rest is history." All modern houses and electrical appliances are standardized in North America to operate on either 120 Vac or 240 Vac. For this reason, it is advisable for off-grid homes to convert the low-voltage electrical energy stored in the battery bank to 120/240 Vac. The device used for this electrical conversion is the inverter, which will be discussed in more detail in Chapter 11.

Power, Energy, and Conservation

Conserving energy and doing more with less is not only good for the planet but also helps keep the size and cost of your renewable-energy power station within reasonable limits. We discussed earlier how a bigger battery could run a light bulb longer than a smaller battery. This is fairly obvious. What might not be so obvious is that if we replace the "ordinary" light bulb with a more efficient one we might not need the larger battery in the first place.

Let's review how any typical electrical circuit operates: A source of electrons under pressure (i.e. voltage) flows through a conductor to an electrical load and back to the source. For household circuits, the voltage (pressure) is usually fixed at 120 Vac or 240 Vac. For battery-supplied circuits, the voltage is usually fixed at either 12 Vdc, 24 Vdc, or 48 Vdc, depending on the size of the load and the amount of electron flow (i.e. current) that is required to make the load operate. Let's say that an ordinary light bulb requires 12 V

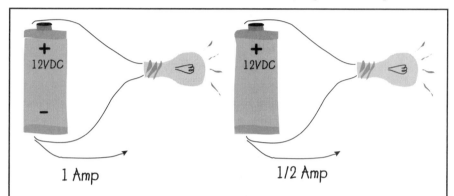

Figure 1-13. The lamp on the right uses 50% of the electrical energy of the one on the left and both have equal brightness. The lamp on the right is said to be twice as efficient.

of pressure and 1 A of current flow to make it light. Now suppose that we can find a light bulb that uses 12 V and only 0.5 A of current flow. Assuming that both lights are the same brightness, we infer that the second light is twice as efficient as the first. Stated another way, we would need only half the battery-bank size (at lower cost) to run the second lamp for the same period of time, or the same size of battery bank would run the second lamp for twice as long.

The relationship between the pressure (voltage) required to push the electrons to flow in a circuit and the number of electrons actually flowing to make the load operate (current or amps) is the power (commonly expressed as *watts*, or W) consumed by the load. Using the above example, let's compare the power of the two circuits:

Ordinary Lamp:
   12 V x 1 A = 12 W
More Efficient Lamp:
   12 V x 0.5 A = 6 W

The second lamp uses half the number of electrons to operate. The power of a circuit is simply the voltage multiplied by the current in amps, giving us the instantaneous flow of electrons in the circuit measured in watts.

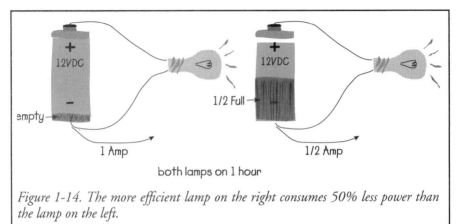

Figure 1-14. The more efficient lamp on the right consumes 50% less power than the lamp on the left.

Remember that the more efficient bulb operates twice as long on a battery as the other bulb. How does time factor into this equation? If a battery has a known number of electrons stored in it and we use them up at a given rate, the battery will become empty over time. The use of electrons (power) over a period of time is known as *energy*.

Energy is *power* multiplied by the *time* the load is turned on:

Ordinary Lamp:

12 V x 1 A x 1 hour = 12 Watt hours; or
12 W x 1 hour = 12 Watt hours

More Efficient Lamp:

12 V x 0.5 A x 1 hour = 6 Watt hours; or
6 W x 1 hour = 6 Watt hours

Your electrical utility charges you for energy, not power. You run around turning off unused lights to cut down on the amount of time the lights are left on. If we think about it long enough, we can also use these calculations to figure out how much energy is stored in a battery bank. For example, if a 12 V battery bank can run a 10 A load for 30 hours, how much energy is stored in the battery?

Battery Bank Energy (in Watt Hours):

12 V x 10 A x 30 hours = 3,600 Watt hours or 3.6 kilo-Watt hours of energy

An interesting thing about batteries is that their voltage tends to be a bit "elastic." The voltage in a battery tends to dip and rise as a function of its state of charge. For example, our nominal 12 V battery bank only registers 12 V when the batteries are nearly dead. When they are under full charge, the voltage may reach nearly16 V. Because of this swing in the voltage, many batteries are not sized in *Watt hours* of energy, but *Amp hours.*

The math is similar; just drop the voltage from the energy calculation:

Battery Bank Energy (in Amp Hours):
10 A x 30 hours = 300 Amp hours of energy

To convert Amp hours of energy to Watt hours, simply multiply Amp hours by the battery voltage. Likewise, to convert Watt hours of battery bank capacity to Amp hours, divide Watt hours by the battery voltage.

That pretty well covers everything you need to know about electrical energy. With a bit of practice (and you'll get plenty of that in the chapters to follow) you will know this stuff well enough to brag at your next office party.

Heating Energy

If you happen to be sitting beside a nice warm wood stove as you read this book, you can feel the heat from the fire radiating towards you. This warmth and the effects of reading the section above on electrons may have caused you to doze off to sleep. If not, you might consider that heat is not the same thing as electricity, for if it were, it could not reach you since air is an insulator.

In fact, the idea that heat is a form of energy baffled many earlier physicists and took a long time to be understood. It took an even longer period for the theory of heat to make its effect on industry. Consider that houses built around the turn of the century gave no thought to conserving heat by means of insulating or any of a number of techniques which seem quite obvious to us now.

So what exactly is heat? Early researchers thought that heat was a substance, something you could put in a bottle—a fluid they called *caloric.* This theory persisted until the middle of the 19th century and was not nearly as silly as you might think. Consider: Heating a liquid causes it to expand, as if something is added to it; when wood is burned, a small pile of ashes is left behind, as if something has escaped or evaporated; a jar of boiling hot water placed next to a jar of cool water causes the cool water to warm up. Perhaps *caloric* flowed from the warmer jar to the colder?

Although we now understand that the notion of *caloric* is incorrect, the concept of heat flowing as if it were a fluid is not that far from the truth, as the above examples illustrate. As scientists continued scratching their heads, the current laws of heat and thermodynamics slowly replaced the theory of *caloric*.

It is now understood that heat is a form of energy caused by the motion of molecules, or groups of atoms. All matter is made up of molecules, which (like

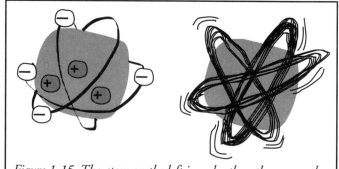

Figure 1-15. The atom on the left is cooler than the one on the right because it is moving more slowly.

teenagers) are always in motion. As a substance is warmed, its molecules move faster; as it is cooled, its molecules move less. The temperature of a substance is directly related to the motion of these tiny molecules.

One of the primary practical considerations relating to heat is that it does not like to stay still (also like teenagers). Consider any type of heat and you will notice that it always wants to move from the hotter object to a colder one. A hot pan placed in cold water will cool, while warming the water. In the same manner, in winter your expensive heated air wants to get outside as quickly as possible to help melt the snow. Although this might be a useful task, the cost in dollars and energy is just a bit beyond reach!

When we wish to stop the flow of electricity in a circuit, we use a non-conductive device called an insulator. Heat energy can also be slowed down in its relentless path into or out of our homes by the use of an insulator for heat (insulation), which is discussed in Chapter 2.

Because heat is energy, it is also possible to quantify this energy in the same way that electricity is quantified in kilowatt-hour units. As a matter of fact, if we were to heat, cool, and operate our homes completely on electricity, the energy usage charge would indeed be in kWh, as all of the energy used would be sourced from the electrical utility. However, as you know there are many sources of heating/cooling energy to choose from, including natural gas, electricity, propane, oil, wood and wood by-products, and even coal. Each of these sources of heat energy is delivered to you in a bewildering array of units, making comparison shopping very difficult. To complicate matters further, the efficiency of the heating or cooling equipment using these various sources varies greatly.

To level the playing field, we will use the English "British thermal unit," or BTU, as our standard for comparing heating and cooling energy. For readers more accustomed to the metric system, the calorie or joule measure will provide the same basis for comparison. Just make sure you don't use both systems at the same time, or you may be wondering why you have to chip ice out of your toilets next January.

One BTU is the quantity of heat required to raise, by one Fahrenheit degree, the temperature of one pound of water. Using these units allows you to quickly compare two heating sources which have different base units. For example, suppose you are trying to compare the cost of heating a house with oil or propane. Which is more economical?

Assume that the current cost to heat your home is $500.00 per year and that you require 250 gallons of oil at $2.00 per gallon. Propane costs $1.75 per gallon. You assume that propane costs 25 cents less per gallon and that it will be cheaper to use. Sorry, it doesn't work like that. The first step is to find out how much heat energy in BTUs is stored in each energy source. Refer to the cross-reference chart in Appendix 1 and look up information on both energy sources. You will see that oil contains 142,000 BTUs per gallon and propane contains 91,500 BTUs per gallon. Therefore:

250 gallons of oil x 142,000 BTU/gallon = 35,500,000 BTU per heating season is required to heat your house.

And, if we were to assume that we need 250 gallons of propane:
250 gallons of propane x 91,500 BTU/gallon = 22,875,000 BTU

But your house requires 35.5 million BTU, so we have a shortfall of:
35,500,000 required − 22,875,000 from propane = 12,625,000 BTU shortfall

Now let's take a look at the costs. The current heating charge using oil is $500.00, and our assumption that 250 gallons of propane would be sufficient produced our first estimate of:

250 gallons of propane x $1.75 per gallon = $437.50 or a savings of $62.50/year

However, in order to make up the shortfall of 12+ million BTU, we have to purchase more propane:

12,625,000 BTU shortfall ÷ 91,500 BTU/gallon propane = 138 more gallons

138 more gallons x $1.75 / gallon of propane = $241.50 additional cost

The total cost of heating your house with propane is now *$437.50 + $241.50 = $679.00.*

As you can see, your cost for using propane over heating oil went from an estimated saving of $62.50 to an increase of $179.00 per year! Of course these costs are not real, but the example shows that the amount of heat energy "stored" in a fuel is known and can be used for comparison when all fuels have the same base units of BTU or joules. This also applies to renewable sources such as firewood, wood pellets, and even solar heating systems.

You must also consider the efficiency of the heating appliance. Suppose you find two nearly identical heating devices at the same price with output ratings of 120,000 BTUs per hour. (The fuel type is not important for this calculation). One has an efficiency of 80% and the other 65%. It stands to reason that the 80%-efficiency-rated unit is the better buy. Over the lifetime of a product, this difference in efficiency can add up to tens of thousands of dollars.

Not interested in doing the math? Contact a competent home energy dealer and discuss these issues. Most dealers can show how fuel prices affect the operating cost of the heating appliances they sell. You will have to go to the fortune teller down the street to determine future energy costs. The one thing you won't need a fortune teller for is to understand that *all* energy costs are rising.

Cooling Energy

An important consideration for homeowners is how to "make" cool air in the dog days of summer. No matter how big your air conditioning unit is, you do not make cold air. A better way to understand a mechanical cooling system is to think of it as a *heat pump*. Modern air conditioning units are complex devices that move heat from one location to another. A window air conditioner pumps heat from your warm indoor room and moves it outside. This may seem a little difficult to understand until you realize that the outside part of the A/C unit is quite hot because of the indoor heat moved there by the internal mechanism. As the outside component (the condenser) is hotter than the outdoor air, the heat (guess what?) wants to move from the hot condenser to the relatively cooler outdoor air. This phenomenon of air-conditioning heat transfer explains why the unit has to work harder in hotter weather. As the outside air temperature approximates that of the condenser, heat transfer slows, making the unit run for longer periods. This is also why you want to place the condenser facing north or out of direct sunlight.

This cooling process is not efficient in even the best-designed air conditioners or more complex heat-pump systems. Regardless of the type of A/C system installed, the energy used to cool a given area is measured either by the BTUs of heat moved or by the electrical energy consumed by the unit. Commercial A/C units are usually rated both ways, with some units still carrying a "cooling tonnage" rating. This archaic rating compares the cooling

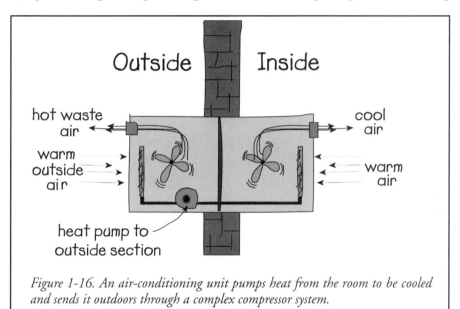

Figure 1-16. An air-conditioning unit pumps heat from the room to be cooled and sends it outdoors through a complex compressor system.

capacity of the A/C unit to a ton of ice. Just in case you were wondering, this is approximately equal to 12,000 BTU of cooling capacity.

With all of the heating and cooling energy flowing here and there it seems that just maybe the *caloric* theory really should have been given more consideration after all. At least we could measure the stuff in gallons or liters and not worry about BTUs and cooling tons.

1.4 Energy, Pollution and Climate Change

Everyone has heard about them, but do you actually know what smog, climate change, and greenhouse gas emissions really are? Ask anyone who lived in Los Angeles in the early 1980s and they will tell you about dirty skies and endless haze. For someone with asthma or other respiratory sensitivities it meant staying indoors for countless hours or depending on medical inhalers to struggle with each breath. Steve Ovett wished the air were cleaner in Los Angeles. Although he effortlessly beat his rivals during the 800- and 1,500-meter Olympic men's races, going 3 years and 45 races undefeated, the air in Los Angeles was his undoing. During the 1984 Olympics he collapsed during the 1500 metre race due to smog-induced asthma and spent two nights in hospital.

Smog and soot pollution is created when the emissions from burning fossil fuels combine with atmospheric oxygen and ultraviolet light from the sun. Estimates indicate that over half a million deaths worldwide are directly attributed each year to excessive levels of airborne pollution. When smog and soot particles combine with raindrops, acid rain is formed.

Clean rain water is neither acidic nor base, meaning that this life-giving element is non-corrosive and restorative to everything it touches. Acid rain, on the other hand, is precisely that—water which has become acidic. When it comes into contact with metal, rock, or plants the result is a corrosive action and the decay and destruction of a wide array of living and lifeless objects.

Smog has been greatly reduced in the developed world in recent years as a result of decreased sulfur concentrations in gasoline and diesel fuels. Improvements in automotive and coal-fired power-station emissions have also gone a long way in reducing smog over the last 20 years; however, the problem is far from over. Increasing demand for coal-fired electrical energy and ever-increasing numbers of automobiles on the world's highways continue to exacerbate the problem.

Carbon dioxide and methane are two greenhouse gases that are increasing in concentration in the atmosphere and are directly linked to global warming. Carbon dioxide concentrations are higher than at any time in the last 420,000 years and have directly contributed to a rise in the Earth's surface temperature of 0.6 degrees over the last century. It should also be noted that the 1990s were the warmest decade on record, with 1998 recorded as the single warmest year of the last 1,000 years.

Scientists predict that if the earth's average temperature were to rise by approximately 2°C, numerous devastating effects could occur:
- increased magnitude and frequency of weather events
- rising ocean levels causing massive flooding and devastation of low-lying areas
- increased incidents of drought affecting food production
- rapid spread of non-native diseases

So where do these gases come from? Carbon dioxide is a byproduct of the burning of any carbon-based fuel: gasoline, wood, oil, coal, or natural gas. With few exceptions, the world's current energy economy is fueled by carbon.

We have to go back half a million years and more to learn how this carbon fuel came to be. All carbon fuel sources began as living things in prehistoric times. Peat growing in bogs absorbed carbon dioxide from the air as part of the photosynthetic process of plant life. The rolling, heaving crust of the earth entombed the dead plant material. Sealed and deprived of oxygen, the plant matter could not rot. Over the millennia, shifting soils and ground heating compressed the organic material into soft coal that we retrieve today from shallow open-pit mines. Allowed to simmer and churn longer, under higher heat and pressure, oil and hard coal located in deep underground mines are created.

Provided these prehistoric fuel sources remain trapped underground there is no net increase in atmospheric carbon dioxide. However, the process of burning any of these fuels reduces the carbon stored in the coal or oil and drives off the trapped CO_2. Conversely, the burning of renewable carbon-based fuels such as wood, biomass, and oilseed-based fuels does not contribute to net greenhouse gas emissions. These plant materials "recycle" carbon dioxide from the atmosphere into carbon during the growing cycle. Burning these materials releases the same CO_2 that was originally absorbed. Although the burning of prehistoric fossil fuels such as coal and oil are simply giving back carbon dioxide the atmosphere lost millennia ago, these "new" emissions increase concentration levels, causing the affects of global warming.

It is not necessary for society to make a wholesale switch to a non-carbon-based fuel such as hydrogen in order to reduce the effects of global warming. By selecting a more fuel-efficient car, switching to compact fluorescent lamps, or improving the energy efficiency of our home we can drastically reduce greenhouse gas emissions and enjoy significant savings along the way.

From an *eco-nomic* point of view, doing more with less is the path to sustainable development.

2
Energy Efficiency

When the price of gasoline jumps for the tenth time in a month, most people tend to cut back on their driving—or at least talk about it. Some people might even consider trading their vehicle in for something with better gas mileage than a Hummer.

Family budgets are straining from the rising cost of living: taxes, mortgages, car payments, and energy bills for homes and vehicles. As a society we rarely consider the source of these costs. People hop on the bandwagon of the middle-class dream fueled by the two incomes that make it happen. Large homes in the suburbs require enormous amounts of heat, light, and air conditioning, and many people have two cars to get to distant jobs. All of this eats away at our precious discretionary income—and our free time.

Figure 2-1 An energy-conserving lifestyle does not mean a spartan lifestyle. Choosing the most energy-efficient appliances and products that "do more with less" is how to make it happen. The hybrid 2004 Toyota Prius doubles gas mileage efficiency and reduces pollutants by 90%. Courtesy Toyota Corporation.

Doesn't energy conservation mean giving things up, not living the middle-class dream North Americans have come to expect, or living a spartan life of near poverty? Well, yes and no. Yes, energy conservation does mean giving things up, but they are mostly wasteful, inefficient things. And no, it does not mean that you must compromise the quality of your lifestyle.

How can we reduce our heating and electrical loads or transportation costs to embrace energy conservation without giving anything up? The answer is efficiency. As the current corporate downsizing mantra goes: "Do more with less."

I recently heard a politician asking us all to do our part to conserve energy by restricting ourselves to five-minute showers. This is a misdirected effort. People may do their part by conserving energy or resources in times of severe famine, war, or national disaster, but it is never sustainable. As soon as things are back to normal people immediately return to their old habits. Conservation is not sustainable if there is a perception that it will interfere with quality of life.

The better approach is to take a ten-minute shower but use a low-flow showerhead. Installing this five-dollar device will not reduce your quality of life and in fact might improve it, as many models incorporate a massage feature. By reducing energy and water consumption by 50% or more, they will put real money in your pocket.

This is the *eco-nomic* approach: doing more with less; adopting an energy efficient, sustainable lifestyle.

It may sound like nickel-and-dime stuff, but in fact the opposite is true. The constant "leakage" of energy dollars here and there can become quite significant. Refer back to Chapter 1 and you will recall that the simple act of switching common incandescent lamps to high-efficiency compact fluorescent models will put over $2,300 *after-tax* in your pocket over the operating life of the lamps. If you are in a 35% tax bracket this translates into a savings of over $3,100.

This is just one example of dozens of tips that, added together, translate into serious energy and financial savings. What you do with the savings is up to you. Have you been trying to scrape together enough cash to take the kids on vacation or top up your retirement savings plan? If so, this is one of the easiest ways to do it.

For those of you considering an off-grid lifestyle, energy efficiency is doubly important. One of the most interesting statistics being floated about is from the U.S. Department of Energy, which states that for every dollar you spend on energy efficiency you will save three to five dollars on the cost of generating equipment. This rule applies to both the big nuclear plant built by the mega-utility and your own personal renewable energy power station at home.

I am well acquainted with explaining this logic; on Sunday afternoons, tourists often stop by to have a chat about the "big fan on the pole over there." Enjoying a cup of solar-power-brewed coffee (which somehow always tastes better), they mull over why they don't consider using renewable energy as well.

At some point in our conversation, I am asked the usual question: "Our hydro bill is a couple of hundred bucks a month. How much to install a rig like yours to power our house?"

The answer is always the same: "Too much." Applying the Department of Energy rule discussed above, it is much less expensive to optimize the energy consumption of the house than to build a bigger power station.

And no, you don't have to resort to coal-oil lamps and a black and white television with a rabbit ear antenna. You will be introduced to homes in Chapter 4 that will surprise you with the number and type of appliances that can be used in an off-grid home. No one has to suffer in making the transition.

Let's start our quest for home-energy efficiency with an understanding of the house itself. First, we'll look at some of the most obvious design features that can be incorporated into the building. Next, we'll take a look around inside. Lighting, heating, electronics, and major appliances all consume energy and must be considered in an energy-efficient lifestyle.

2.1

New Home-Design Considerations

By far the largest energy usage in North American homes is heating and cooling. As we learned in Chapter 1.3, heat has a nasty habit of wanting to escape outside in winter. In summer, the opposite is true. Heat from the hot summer sun just can't wait to come inside and enjoy the air conditioning with you.

It is not necessary to design your own home in order to achieve satisfactory energy-efficiency results. Many builders now recognize that energy-efficient homes can almost compete with granite countertops and hardwood floors in attracting the buying public. Look for builders that are certified for R-2000 or other high-quality construction methods.

Heat Loss and Insulation

Heat loss and gain in your home is linked to the level and quality of insulation in the ceilings, walls, and basement. Not so obvious are the losses associated with windows, doors, and the sealing in various joints and holes in the structural cavity. The key to designing an energy-efficient house is to ensure that it is well constructed and airtight and contains adequate levels of insulation.

The national and local building codes in your area set the *minimum* standards that your building must meet, but it is fairly easy to incorporate upgrades into your design that will pay for themselves many times over. With the world's current political climate and uncertainty over fossil fuel supplies in the coming years, these upgrades will provide a safety net against increasing energy prices.

Insulation in the walls and ceilings of our homes slows the transfer of heat into or out of our house (in much the same manner that electrical energy cannot flow through a non-conductive material). The

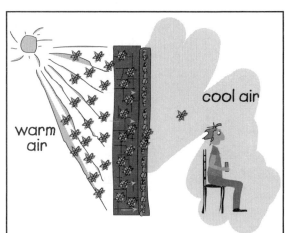

Figure 2-2. Heat loves to move from the hot summer outdoors to enjoy the air conditioning inside with you. Some materials such as brick and stone allow heat to travel fairly easily. Proper insulation slows the flow of heat.

higher the quality and thickness of the insulation, the harder time the heat has getting through. Many people believe that because hot air rises, most of the heat loss in the house will be up through the ceiling. Not so. Heat moves wherever it can, shifting from warm areas to colder ones, whether that be upwards, downwards, or sideways. Keep this in mind as we visit all areas of the home during our insulation spree.

Table 2-1 lists typical recommended insulation values for a very well insulated home. The "R" value (English system) and "RSI" value (metric system) indicate quality levels of insulation. The higher the value, the better the insulation level. If you live in colder climates where the number of heating days is high, it may be in your best interests to increase these thermal resistance values.

As the price of heating fuel continues to rise over the coming years, any effort you make now to create a more energy-efficient home will be paid back several times over. I recall my parents' first home, built at a time when fuel costs weren't considered at all. This home was constructed using 2 x 4" lumber (63 x 125 mm) and minimal levels of insulation, with no wind barrier and no vapor barrier. But with heating-fuel costs in the $0.25/gallon range, no one cared.

How times have changed. With oil prices now ten times this price, no one can afford to let their heating dollars escape because of poor home planning or construction. Although no one has a crystal ball, I suspect that energy prices will rise considerably faster in the coming years. Our parents had little or no concern about supply, environmental issues, security, or hostile foreign governments. We do not have this luxury.

Table 2-1: Minimum Recommended Insulation Values

Insulation Quality (over unheated space)	Walls	Basement Wall	Roof	Floor
R Value	23	13	40	30
RSI Value	4.1	2.2	7.1	5

Moisture Barriers

A vapor barrier consisting of 6-mil-thick (0.006") polyethylene plastic can be attached to the wall structure on the warm side of the insulation. The vapor barrier completely surrounds the inside of the house and must be well sealed at the joints and overlapped edges. During construction, care must be taken to ensure that this barrier extends without a break from one floor to the next.

The function of the vapor barrier is to seal the house in a plastic bag, controlling air intake and leakage. It also stops warm, moist air from penetrating the insulation and contacting cooler air, condensing, and causing mold and rot problems in the wall. Additionally, a vapor barrier ensures that the air inside the insulation remains still. These conditions are absolutely essential in making the insulation system function properly.

Uncontrolled moisture can cause wood rot, peeled paint, damaged plaster, and ruined carpets. Moisture can also directly influence the formation of molds, which are allergens for many people. Moisture control is not to be taken lightly. Left uncontrolled, it can cause damage to building components.

Some insulation materials such as urethane foam spray or sheet styrofoam are fabricated with millions of trapped air or nitrogen bubbles in the plastic material. Although more expensive than traditional fiberglass, inch-for-inch they provide higher insulation ratings and also form their own vapor barrier.

Wind Barriers

A properly taped and well-jointed Tyvek®-style wind and water barrier on the outside of the house, just under the siding material, helps to keep winter winds from whistling right through the insulation. Controlling wind pressure infiltration into the building structure is an area most people overlook. Even if the breeze isn't blowing through your hair while the windows are closed, wind leakage is important.

Insulation works by keeping dry air very still. Even the slightest movement will greatly reduce the insulation value of the system. Think about how much force you exert wrestling with an umbrella in even a light wind. Now imagine the same effect wind has on the entire surface area of a house. Quite frankly, it's amazing more homes don't blow down! The force of the wind on the house structure can penetrate the insulation. This upsets our still air requirement and lowers insulation values.

Provided all of the above insulation techniques are followed carefully, you will have a very efficient home-insulation system.

Examples of these techniques are shown in the cross-section or side view of the bungalow in Figure 2-3. This design strengthens your local building code by increasing insulation in areas that you may not have considered.

Basement Floor

This floor comprises a 3" (75 mm)-thick sheet of extruded foam-board insulation installed directly on top of packed crushed stone. Over this can be applied an 8-mil layer of polyethylene moisture barrier, with all joints taped

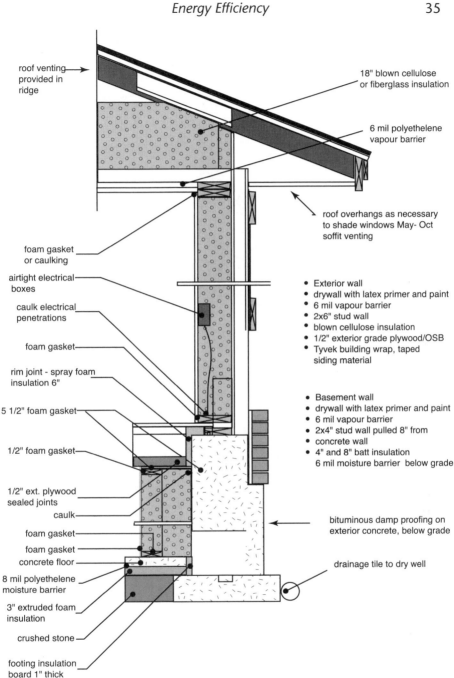

roof venting provided in ridge

18" blown cellulose or fiberglass insulation

6 mil polyethelene vapour barrier

roof overhangs as necessary to shade windows May- Oct soffit venting

foam gasket or caulking

airtight electrical boxes

caulk electrical penetrations

foam gasket

rim joint - spray foam insulation 6"

5 1/2" foam gasket

1/2" foam gasket

1/2" ext. plywood sealed joints

caulk

foam gasket

foam gasket

concrete floor

8 mil polyethelene moisture barrier

3" extruded foam insulation

crushed stone

footing insulation board 1" thick

- Exterior wall
- drywall with latex primer and paint
- 6 mil vapour barrier
- 2x6" stud wall
- blown cellulose insulation
- 1/2" exterior grade plywood/OSB
- Tyvek building wrap, taped siding material

- Basement wall
- drywall with latex primer and paint
- 6 mil vapour barrier
- 2x4" stud wall pulled 8" from concrete wall
- 4" and 8" batt insulation 6 mil moisture barrier below grade

bituminous damp proofing on exterior concrete, below grade

drainage tile to dry well

Figure 2-3. This cross-section of a high-efficiency house illustrates additional insulation and sealing techniques that will greatly reduce your heating and cooling energy usage.

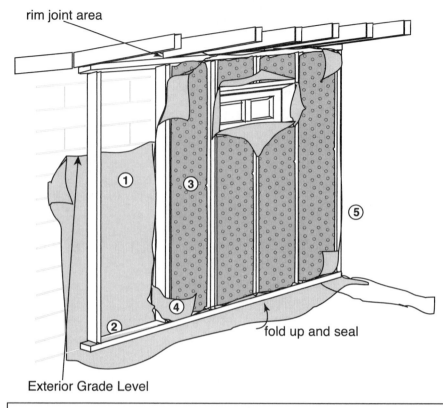

rim joint area

① ③ ⑤ ④ ②

fold up and seal

Exterior Grade Level

Interior insulation involves: 1) a moisture barrier; 2) new frame wall; 3) insulation; 4) air and vapor barrier; 5) finishing

Figure 2-4. The technique illustrated provides excellent insulation value and ensures that your basement will also remain dry and mold free.

or overlapping seams caulked with acoustical sealant. Such a membrane prevents ground-based moisture from entering through the floor area and thus reduces mold potential. It will also stop ground gases including radon from entering. In areas where foundation footings are not subject to frost heaving, a footing insulation board can be added between the basement wall and the concrete slab floor.

Basement Walls

The basement walls can be "finished" using standard framing techniques, with the exception that the 2 x 4" stud wall is set away from the concrete basement wall by 4" (100 mm). This increases the wall cavity space for extra insulation without requiring additional framing material. Instead of traditional batt insulation, blown-in fiberglass or cellulose is suggested. These materials ensure complete coverage and packing density, especially in the hard-to-insulate areas around plumbing lines, electrical boxes, and wiring. Figure 2-4 shows the use of a tarpaper-style moisture barrier glued to the concrete wall from just below grade. It is then folded up and glued to the bottom of the interior vapor barrier.

Rim Joints

Rim joints are not an obscure arthritic condition, but rather an obscure area of your home where the floor joists meet with the exterior support walls. (They are also known as rim joists or header joints.) They are known to be notoriously difficult to insulate and vapor barrier correctly, given the huge amount of surface area accorded this space. Many insulation contractors give only passing effort to this area, stuffing a piece of insulation in place and then stapling a swath of plastic on top, as shown in Figure 2-5a.

A much better method of insulating the rim joists is to use a spray foam material. Although more expensive than traditional methods, this process actually works. As an added benefit, the foam also provides its own vapor barrier, as shown in Figure 2-5b.

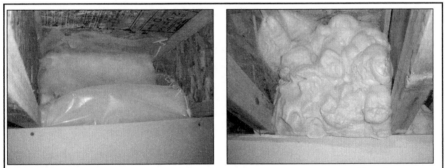

The photograph on the left (Figure 2-5a) details a too typical "stuff and run" rim joist insulation job. The photograph on the right (Figure 2-5b) shows the best way to complete rim joist insulation using sprayed urethane foam, which provides an integral vapor barrier.

Exterior Walls

Typical framed construction now employs 2 x 6 " (50 x 150 mm) stud walls. This wall-cavity thickness is suitable in all construction areas provided quality workmanship is assured. The major problem with exterior wall construction is air leakage. (Have you ever felt the breeze blowing from the electrical boxes of older homes?) The use of blown-in fiberglass, cellulose, or rock wool in the

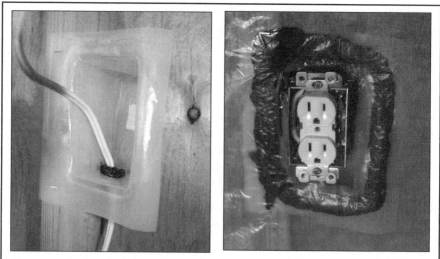

Figure 2-6 Electrical boxes and other wall penetrations must be carefully sealed with the vapor barrier. You can purchase electrical boxes that are integrally sealed or fabricate your own using a piece of 6-mil polyethylene sheet.

Figure 2-7 The vapor barrier must create a "plastic bag" effect for the entire home. Ensure that this membrane is wrapped from one floor to the next and caulked at each joint.

wall cavity will ensure complete coverage over the entire surface area. This technique also ensures that insulation works its way around electrical boxes and other obstructions in the cavity. Make sure the contractor fills the wall with a sufficient density of insulation. The material should be tightly packed and not sagging at the top of the frame wall.

The next areas to consider are electrical, phone, and cable boxes, which penetrate the exterior structure. Ensure that airtight boxes are used and caulk between them with a 6-mil layer of polyethylene vapor barrier. The vapor barrier must be overlapped and caulked tightly at all seams and with the top and bottom plates making up the wall section.

Air can easily blow between the top and bottom plates of the wall and the floor components. For this reason, these areas should have foam gaskets or caulking compound to ensure air tightness.

Windows

It would be quite easy to write an entire book on the design and treatment of windows. For our purposes, we will consider windows just from an energy standpoint. A single pane of glass is a very poor insulator. During the last thirty years, there have been many advances in window design including double- and triple-pane versions, low-emissivity coatings, argon and krypton gas fillings, and a bewildering array of styles and construction.

At the risk of oversimplifying this issue, follow these basic rules when choosing your windows:

- Purchase the best-quality window you can afford. Ensure that it is from a reputable supplier and has a good warranty. Ask for references.

- If you don't require solar gain, purchase at least double-pane windows with low-emissivity glass (also known as "low-e"). This type of glass reduces heat gain and loss. Note, though, that low-e glass and coatings should not be used with south-facing windows where winter solar gain is desired.

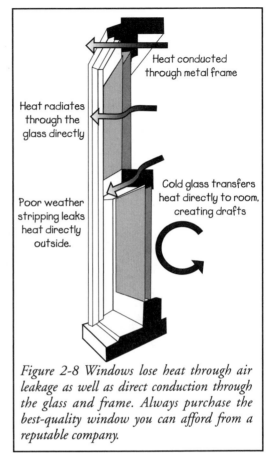

Figure 2-8 Windows lose heat through air leakage as well as direct conduction through the glass and frame. Always purchase the best-quality window you can afford from a reputable company.

- Ensure that the cavity between window panes uses argon gas, or better still krypton gas. Triple-glazed windows using low-e glass and filled with krypton gas have an insulation value 3.5 times better than that of traditional double-pane windows.

- All framing materials used in the construction of the window should have wood or vinyl cladding to prevent heat transfer through a metallic structure.

- When installing windows be sure to adequately insulate and caulk, preferably with spray foam, between the window and the house frame.

- Make certain that windows are shaded with roof overhangs, awnings, or leaves of deciduous trees from early May to mid-October to prevent undesired solar heating. In turn, ensure that windows are free to capture the winter sun for the remainder of the year.

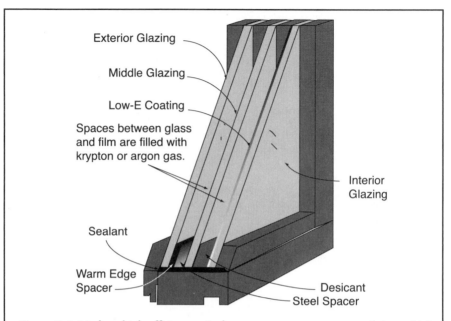

Exterior Glazing

Middle Glazing

Low-E Coating

Spaces between glass
and film are filled with
krypton or argon gas.

Interior
Glazing

Sealant

Warm Edge
Spacer

Desicant
Steel Spacer

Figure 2-9 Modern high-efficiency windows use two or more panes of glass which are separated by thermal insulation and filled with argon or krypton gas which increases the insulation value. When solar gain is not a factor (such as on the north side of the home), purchase windows with low-emissivity coatings.

Figure 2-10 Solar blinds such as these sunshades block 97% of the daylight heat energy and glare while allowing you to see outside (www.shade-o-matic.com). Quilted blockout curtains or bamboo shading blinds will also work.

Attic and Ceiling Treatment

Provided the attic is not finished and has a suitable vapor barrier, the best way to insulate this area is to blow in 18" of cellulose or fiberglass. To ensure that the area above the insulation remains as cool and dry as possible during the year, install adequate roof and soffit vents to provide proper air circulation: warm air rises and exits through the roof peak vents while cooler outside air is drawn in through the soffit vents. This upgrade will eliminate moisture damage and ice damming in winter.

If you live in an area where summer air-conditioning load is a higher concern than winter heating, you may wish to consider radiant insulation. This material is similar to aluminum foil stapled to the underside of your roof rafters. Working like a mirror, the film reflects heat back through the roof surface before it gets a chance to hit the attic insulation.

The ceiling vapor barrier should continue down the wall surface a few inches and overlap the wall vapor barrier. Where these overlap, ensure that adequate caulking is applied. The best vapor barrier job is wasted if the finishing trades rip, cut, or damage the sealed home. We are striving to have you live inside a plastic bag, so take extra care to seal and repair any cuts. Good workmanship will guarantee an energy-efficient home.

Lastly, gasket and seal the attic hatch door in the closed position. This reduces air leakage around its perimeter.

If your home has more complex design features, be sure to discuss these general guidelines with your architect during the planning stage.

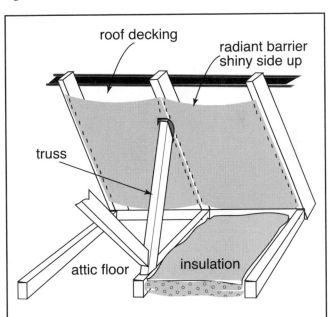

Figure 2-11 Summer air-conditioning loads can be reduced by the use of radiant insulation, which reflects heat back through the roof.

Ventilation Systems

Living in an energy-efficient house is like living in a plastic bag. With all of the air leaks sealed and the vapor barrier extending from the basement floor to the ceiling, we really have created a sealed, airtight home. Contrast this with the typical older house with leaky kitchen and bathroom fans and other cracks and leaks in the building structure. While these provide ventilation, they are completely uncontrolled and cause tremendous losses in heating and cooling energy. In a typical thirty-year-old house, it is estimated that these leaks collectively equate to a one-square-foot hole in the wall!

To ventilate your sealed home, you will have to resort to alternative measures. The methods you use will depend to a large extent on where you live.

Most high-efficiency homes utilize a device known as an air-to-air exchanger or heat recovery ventilator (HRV)—a marvel of simplicity which is brilliant in its execution. Stale, moist air from the kitchen, bathrooms, and other areas is drawn into the HRV unit and passes over a membrane before being exhausted from the house. At the same time, colder incoming fresh air is pulled into the unit and passes over the opposite side of the membrane.

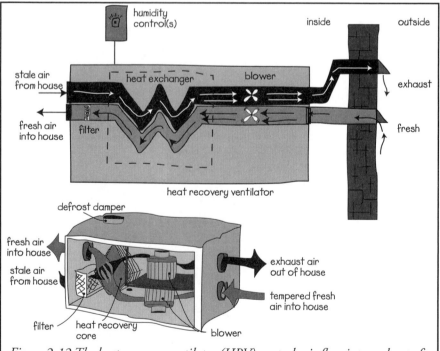

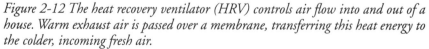

Figure 2-12 The heat recovery ventilator (HRV) controls air flow into and out of a house. Warm exhaust air is passed over a membrane, transferring this heat energy to the colder, incoming fresh air.

This causes waste heat to be transferred to the colder, dry incoming air, which is then warmed and distributed in the central furnace air duct.

Controls and timers located throughout the house can be programmed to monitor humidity and smoke. Even carbon dioxide levels can be monitored, indicating that the house is occupied and adjusting ventilation accordingly. Based on rules programmed into the control system, the HRV automatically adjusts air-exchange flow to current conditions, while at the same time saving plenty of heating dollars.

Homes in areas that have a low heating load in winter do not fully benefit from the heat-recovery aspect of the HRV. However, these homes would benefit from the ability of the HRV to filter dust and pollen as well as provide proper ventilation.

If there is a disadvantage to HRV units, it is that their fan and electrical circuits can cause difficulty in off-grid homes. Although the blower and control circuits may appear to be fairly small loads, in actual fact they consume a relatively large amount of electrical energy. (A typical unit operating 12 hours per day can consume 2,000 Watt-hours of energy, which may equate to 50% or more of a total winter day's production).

Unfortunately, there are only a few options that can substitute for a standard HRV:

- Remove the 120 Volt fans in the HRV unit and replace them with lower-flow, high-efficiency 15 Watt computer-style fans. Two 15 Watt fans operating for 10 hours per day lower HRV energy consumption to 300 Watt-hours, which is manageable. This modification does require someone who is handy with wire cutters and duct tape.
- Install an air loop intake in conjunction with a bathroom-style vent fan.
- Install an air loop intake in conjunction with a non-direct venting gas, wood, or pellet stove fireplace.

Figure 2-13 demonstrates an alternative installation to the HRV which requires no electrical energy. A fresh air intake is placed in the basement wall near grade level and at least 6 feet (2 meters) away from any exhaust gas vents. An insulated, flexible pipe is connected to the intake, looped, and placed just above the basement floor. A damper is installed in the pipe. Typically the vent outlet is connected in a wall cavity with the outlet facing into a mechanical or laundry room. Such an arrangement will limit drafts and ensure that cold air will not flow into the house until required.

When a kitchen or bathroom vent fan is activated, or a wood stove draws in room combustion air, stale air is drawn outside. This creates a partial vacuum in the house that causes fresh, dry, colder air to be drawn into the house. As the intake is placed in a little-used area, the fresh air has time to mix with warmer room air before contacting the living area. Although this system is not as efficient as an HRV, intake airflow is controlled. This is a viable alternative to relying on building leakage to make up fresh air, causing uncomfortable drafts. Adjusting the intake damper will control building humidity through the influx of drier winter air.

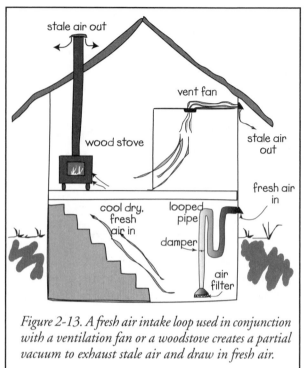

Figure 2-13. A fresh air intake loop used in conjunction with a ventilation fan or a woodstove creates a partial vacuum to exhaust stale air and draw in fresh air.

It is important to discuss this or any airflow plan with building officials to ensure that they are onside with the chosen design. It is also important to make them aware that an off-grid electrical system should not be used to power an HRV unit.

New Construction Summary

By following the tips provided above, you can achieve a satisfactory balance between capital cost and energy conservation. In contrast to a home built to standard building codes, a home designed using the principles provided in this (and the next) section can reduce your heating and cooling energy requirements and costs by 50% or more.

2.2
Updating an Older Home

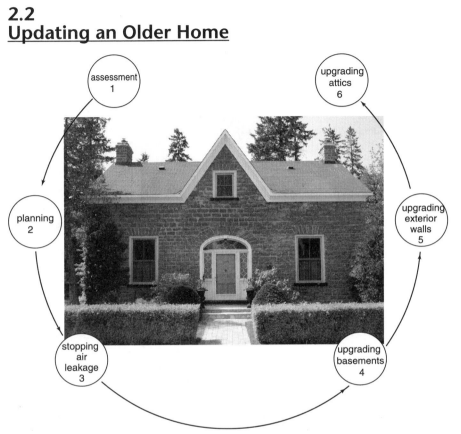

Figure 2-14 Older homes have incredible character and appeal. What they lack is energy efficiency. Proper assessments, planning, and retrofit work always make energy sense.

If your home is more than ten years old, it is likely that many of the energy-saving features described in the previous section are not incorporated into its design. Don't despair; there is no need to tear down the old place or move. Instead, we can review the entire home systematically and determine where to put your renovation dollars to make energy sense.

Where to Start?
Every home is unique and, whether it's one, ten or a hundred years old, what works for one house may not work for the next. As with any project, an assessment is the place to start. We need to check the general condition of the house and test what areas need to be beefed up. Over the last few decades, home designs have improved and architects, engineers, and building

contractors have all learned from past practices. With this in mind, there is a general checklist we can use to determine what energy shape our house is in right now.

Step 1 – Assessment

- Consider having a professional inspection of your home. Programs such as the Canadian *EnerGuide for Homes* assessment will not only identify problem areas but also provide suggested corrective action and even precalculate government rebates for implementing the suggested work.
- Air leakage is the most common problem with all older homes. Poorly fitting window frames and leaks around doors and chimneys abound. Simple corrective actions such as applying sealants and weather stripping stop breezes from blowing into the house. Correcting air leakage also helps to regulate humidity and reduce condensation problems in existing insulation.
- Many older homes treat the basement like an outcast relative: it's there, but leave it alone. Often dark, damp holes which are not properly insulated, basements can eat an enormous number of heating dollars. Moisture problems tend to be left to a dehumidifier, which is about as useful as putting a bucket under a leaky roof. Let's try to correct these issues with damp-proofing, moisture barriers, drain gutters, and proper insulation systems.

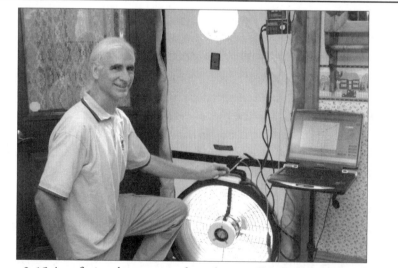

Figure 2-15 A professional assessment of your home will include a pressure door test. A computer-controlled fan will pressurize a home, calculate air leakage, and allow the assessor to identify problem areas in the home and suggest corrective action.

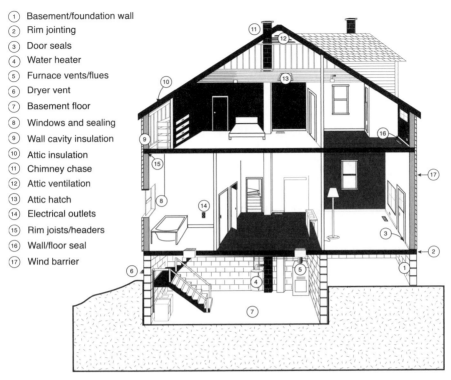

1. Basement/foundation wall
2. Rim jointing
3. Door seals
4. Water heater
5. Furnace vents/flues
6. Dryer vent
7. Basement floor
8. Windows and sealing
9. Wall cavity insulation
10. Attic insulation
11. Chimney chase
12. Attic ventilation
13. Attic hatch
14. Electrical outlets
15. Rim joists/headers
16. Wall/floor seal
17. Wind barrier

Figure 2-16 A thousand little air leaks and a poorly insulated area add up to big energy costs. A thorough assessment of these areas will let you know where you are wasting your money.

- Cavities between walls in older homes are woefully underinsulated or completely uninsulated. Stone walls look great on the outside, but are murder on the heating bill if not properly dressed inside. There are many ways of tackling these problems, from both the inside as well as the outside of the home.
- Everyone knows that heated air rises. Let's try to stop it before it escapes through underinsulated attics, leaky joints, and chimney and plumbing areas.

Air Leakage

Air leakage control is the most important step that can be taken in upgrading any home. Controlling air leakage provides many benefits:
- Heating/cooling costs are reduced as the infiltration of outside air is decreased.
- Insulation efficiency is increased, since still air allows insulation to work properly.

- Humidity and condensation levels can be controlled.
- Home comfort is improved as drafts and cold spots are eliminated.

It is possible to hire professional contractors to assess the quality of the air barrier system in your house. Alternatively, a simple yet effective method is to conduct the test yourself using several incense sticks left over from your psychedelic days in college.

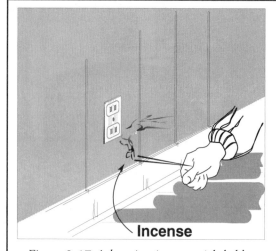

Figure 2-17 A burning incense stick held near suspected areas of leakage on a windy day will cause the smoke to move towards or away from the leak.

Hold the incense near the suspected leakage area on a windy day and observe what happens to the smoke. Smoke drawn towards or blown away from a suspect area indicates an air leakage path. Mark the location down on your list. Continue testing in this manner, paying attention to:

- electrical outlets, including switches and light fixtures
- plumbing penetrations that include the attic vent stack, plumbing lines to taps and dishwashers, etc.
- floor-to-ceiling joints and other building areas that are storied
- baseboards, crown molding, and doorway molding

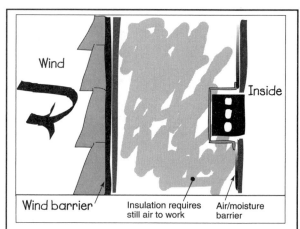

Figure 2-18 Proper wind and air barriers combined with adequate insulation keep the heat in. If any one element is missing or faulty, you may be throwing money away.

- fireplace damper area and chimney exit through the attic or wall
- attic hatch
- windows and doors (Ensure that glass fits tightly and casing area is sealed.)
- ventilation for appliances (kitchen and bathroom fans, dryer vents, gas stoves, water heaters, etc.)
- pipes, vents, wiring, and plumbing lines in the basement and attic (Move the insulation if necessary to access these areas.)

In an older home, it is quite likely that the smoke from the incense will blow just standing in the middle of the room. Don't despair. Just mark all of the areas down on your sheet and perhaps record a "severity rating" from 1 to 10. This way you can tackle the tough problem areas first.

The Basement

Most homeowners don't even consider unfinished basements as a source of heat loss. Part of this mentality comes from the mistaken idea that heat *only* rises and that earth is a good insulator. Both are wrong. Heat travels in any direction it chooses, but it always travels from a warm area to a colder one. Additionally, older basements have large surface areas of uninsulated walls and flooring that act as a heat sink, drawing warm air to these cooler surfaces.

As we discussed in the *Air Leakage* section above, there is also a lot of heat loss through crevices in the walls, around and through windows, and at the top of the foundation wall where it meets the first floor. An uninsulated basement can account for up to 35% of the total heat loss in a home.

It is not possible to simply add insulation and air barriers to a damp, leaky basement without first correcting any underlying problems. Any areas that accumulate water in the spring or after a heavy rain must be repaired, as wet insulation has no energy-efficiency value and will ultimately contribute to mold and air quality issues. Check the basement for dampness, water leaks, and puddles in wet periods, and look for major or moving cracks in the foundation wall. Also ensure that a sump pit and pump have been installed in areas where persistent water accumulation occurs.

Some basement wetness problems are corrected by sloping the landscaping away from the foundation wall or adding rain gutters to the house eaves. If you are not sure that these measures will correct the basement water problems, it will be necessary to excavate around the perimeter of the house to provide supplementary drainage and to apply damp-proofing and external, waterproof insulation treatments to the foundation wall.

Wood-Framed Walls

Wood-framed walls are the easiest to insulate, as they are very similar to new construction. The major concerns relate to wall-cavity thickness and access. When assessing these wall structures, attempt to determine if the wall is empty or if there is already some form of insulation in place.

The simplest means of checking these walls for insulation is to remove the cover plate of an exterior-wall electrical plug or switch and, using a flashlight and a thin probe, check for insulation behind the electrical box. CAUTION! Before completing this test, turn off the circuit breaker or remove the fuse for that circuit and test the outlet or switch to verify that it is off. This test will have to be conducted at a few points around the house to ensure that your sample investigation is accurate.

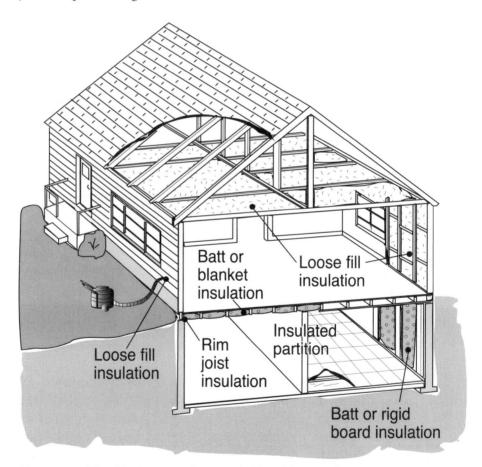

Batt or blanket insulation

Loose fill insulation

Loose fill insulation

Rim joist insulation

Insulated partition

Batt or rigid board insulation

Figure 2-19 The older home can be upgraded by adding insulation to the structure as shown above. Air and vapor barriers may also be added.

Brick Walls

Brick walls of homes are almost always constructed of veneer with a frame wall on the interior side. Usually there is a small air gap between the brick and the frame wall to allow air to circulate and prevent moisture on the brick from rotting the framing members.

This air gap must not be filled with insulation. However, the frame wall might allow for insulation to be added. Use the same tests as those used for a standard frame wall to determine the depth and area that may be insulated.

Stone or Other Solid Walls

Stone, concrete block, cut stone, and other solid-wall treatments are similar in nature to brick walls. Solid wall treatments are not suitable for insulating and sealing and will require extensive reframing on the inside to add an insulation cavity and vapor barrier. Alternatively, special insulating siding and coverings may be added to the exterior.

Attics and Roof Areas

Most homeowners love to dump their home insulation upgrade dollars in the attic. Perhaps this results from the mistaken idea that all heat rises and the losses must therefore be highest in the attic? Could it also be because attics are one of the easiest places to insulate first? Just dump a few bags of insulation in the old attic and, *voilá!*, your heating bill goes down by 50%?

Sorry, not so fast. The attic does lose heat, but it actually loses less than an uninsulated basement or exterior wall. Most homeowners, no matter how old their house, may have had a passing thought about attic insulation and may have even added a few inches of something up there. Adding more insulation on top of old is not a problem unless the air leakage tests discussed earlier confirm that problems exist. If this is the case, it may be necessary to remove or move aside existing insulation to gain access to the leaking area. The necessity of ensuring quality air sealing in the attic area cannot be understated.

Another potential problem in the attic is moisture, which may originate from different sources such as a leaky roof, ice damming, or frost. Moisture in the attic can also come from within the house through leaky ventilation fan outlets from the bathroom and kitchen areas.

Ventilation of the attic itself is important to provide summer cooling and winter dryness, but many older homes do not have adequate (or any) vents. Ventilation is provided by air intake vents in the soffit and roof vents in the gable or peak where hot air can exit. There should be a ratio of 1 unit of roof vent area for every 300 units of attic floor area. For example, a 1200-

square-foot attic should have 4 square feet of vent area.

A common upgrading practice is to add electric vent fans to the attic area, usually in the gable or roof peak. This is not required, nor is it recommended. An electric exhaust fan increases airflow in the attic and may exceed the intake capacity of the soffit vents. If this should occur, additional air will be drawn from the main part of the house, which is exactly the opposite of what is desired.

Check the attic several times during the year, for example after a heavy rain or on a very cold day. Look for wet areas, mold, and rot or small "drip holes" on the insulation or attic floor surfaces.

Attics come in all manner of shapes, sizes and designs. The trickier it is getting access to the existing insulation, (or locating where new insulation should be placed) the more difficult it will be to upgrade, and therefore professional assistance may be required.

Step 2 – Planning the Work

You may wish to tackle some of the upgrading work yourself. Most of the tools required are pretty common household items and the few specialized tools can be rented from your local rental depot or borrowed from a friend. One couple recently completed a blown-in cellulose cathedral ceiling while balancing on a scaffold and using an insulation blower loaned to them by the material supplier. The do-it-yourself approach might not be for everyone, but if you do it correctly, it can result in considerable savings.

Some upgrading work is best left to the professionals: urethane foam spray insulation, for example, requires a truckload of specialized equipment; and if excavating around the foundation of the house is a job that just doesn't make the top of your "list of things to do in life," a backhoe and an experienced operator will really work wonders.

Building Codes

The various national, state or provincial, and local building codes as well as their variations are enough to frustrate any do-it-yourselfer. But don't cut corners. The reason the codes are there in the first place is for your health and well-being. Get to know your local building official. Although there are plenty of horror stories circulating about these officials, they are generally started by people who began the upgrade work before completing the planning and getting a building permit. You will find that most officials are very helpful, providing guidance on technical issues and referrals to qualified contractors or suppliers in your area. Work *with* them.

Safety

Working around the house climbing ladders or working with insulation and chemicals can be dangerous. Make sure you have the proper safety equipment including work boots, dust masks, rubber or latex gloves, and eye protection. Attic and basement areas may not be well illuminated; ensure that you have a suitable light. Use caution on the attic "floor." Often this floor is nothing more than the drywall or lath and plaster finish material on the ceiling below. It will barely hold a cat, let alone your body weight.

Step 3 – Stopping Air Leakage

Now that we are armed with our "list of air leaks" from the assessment phase, we can start getting down to business:

Caulking

Seal up small cracks, leaks, and penetrations on the inside (warm side) of exterior walls, ceilings, and floors. Sealant applied on the inside lasts longer as the material is not exposed to the elements outside.

Caulking is done using an inexpensive gun with a tube of appropriate material. There are literally hundreds of caulking materials available. Discuss with a building supply store the type best suited to your project. After a caulking job, many people are dissatisfied with the brand they used and/or the job they did. Avoid the tendency to purchase poor-quality materials which are difficult to apply and do not last.

Remember to purchase high-temperature silicone or polysulfide compounds for areas around wood-stove chimneys or hot water heater flue vents.

- Identify the area to be caulked. Determine the compound type appropriate for the job.
- Never caulk in cold weather. Caulking should be applied as close to room temperature as possible.
- Clean the area to be caulked. Large cracks and holes greater than 1/4" will require a filler of oakum or foam rope sold for this purpose.
- The nozzle of the compound should be cut just large enough to overlap the crack. Insert a piece of coat hanger wire or a long nail into the nozzle to break the thin metal seal.
- Pull the caulking gun along at right angles to the crack, ensuring that sufficient compound is dispensed to cover both sides of the crack. Remember that caulking shrinks, so it's better to go a bit overboard.
- The finished caulking "bead" should be smooth and clean. The surface of the bead may be smoothed with a finger dipped in water.

- Some compounds require paint thinner or other chemicals to clean up. Check the label of the compound before using it to determine the cleanup method.

Figure 2-20 It takes a bit of practice to make a good caulking joint. Go slowly and don't cut the nozzle too large; otherwise you'll end up with hard-to-clean goop!

Electrical Boxes

Air leakage around electrical boxes is so common that there is an off-the-shelf solution available. Special fireproof foam gaskets and pads may be added just behind the decorative plate as shown in Figure 2-21. To ensure a superior seal, use an indoor latex caulking compound on the gasket face before applying the decorative cover.

If the room is being renovated at the same time, install plastic "hats" around the entire electrical box. These hats have an opening for the supply wires and small flaps that can be sealed to the vapor barrier. Ensure that the area where the supply wire enters the hat is well sealed with acoustic sealant.

Windows

Older homes often have single panes of glass puttied into wooden frames. When the putty dries out, air leakage occurs. Remove old putty and replace it with glazing compound to ensure a high degree of flexibility and a long life.

Figure 2-21 Air leakage is so common in older homes that off-the-shelf sealing pads have been designed to aid in stopping leaks.

Old-fashioned putty is not recommended.

The area between the window and the frame is another area prone to leakage. Access to this area usually requires the removal of the window casing trim. Use oakum, foam rope or, better still, urethane foam spray to air/vapor seal and insulate this area in one application.

Baseboards, Moldings, and Doorway Trim

Trim pieces such as baseboards, moldings, and door casing are used to cover the gap between one framing section and another. For example, a premanufactured door is placed in a framed section of the wall called the "rough opening." Obviously the door must be smaller than the opening in order to fit into it. If this opening is not properly sealed, considerable air leakage will result.

The best way to seal these areas is to use methods similar to those described for window framing (see Figure 2-22). For smaller gaps or areas where urethane foam may be too messy (near carpets or finished wood), oakum or foam rope may be jammed into the gap.

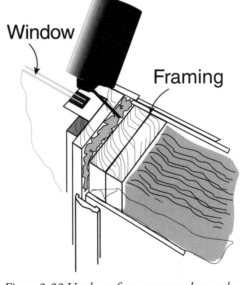

Figure 2-22 Urethane foam spray works wonders in small nooks and crannies such as those between the window and house framing.

Fireplaces

Fireplaces warm the heart, but rob you blind. Air leakage up and down the chimney when the unit is not in use is enormous. When a fire is burning, the suction it creates draws large volumes of expensive, heated room air up the chimney, drawing in cold outside air to replace the heated air that literally went up in smoke!

What to do? Simple. Replace the fireplace with an airtight wood-burning stove or a similar controlled combustion unit. We will discuss these items in greater detail in Chapter 3.

If you really must keep the fireplace, make a removable flue plug that properly seals the chimney when it is not in use. Check to make sure the damper closes as it should. If you detected air leakage around the chimney and framing as part of your assessment in Step 1, seal any cracks with heat-resistant sealant and mineral wool or fiberglass batting.

A quick word on glass fireplace doors; don't waste your money. If you like the look of them or are worried about flying sparks, so be it, but be warned that the majority of these door units are cheaply made and will not provide any air sealing capacity.

Attic Hatch

Seal the attic hatch in the same manner as any exterior doorway trim. Use hook and eye screws to ensure that the hatch fits snugly against the weather stripping (see Figure 2-23).

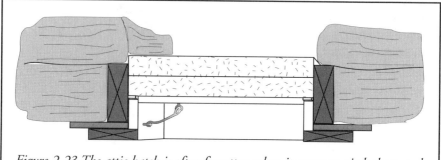

Figure 2-23 The attic hatch is often forgotten when it comes to air leakage and insulation. Use exterior doorway sealing gaskets and hook and eye screws to ensure that the hatch door fits snugly, preventing air leakage.

Air and Vapor Barrier

So far in our discussion of air leakage we have been concerned with the worst but most easily corrected areas. If you are undertaking more extensive renovations, it may be possible to improve large sections of the air and vapor barrier. Before we continue with how to perform this type of major upgrade, let's quickly review the function of each component to be sure we understand how this work can be implemented (see Figure 2-18).

On the exterior of the house, directly under the siding material, is a layer of spun-bonded olefin, which is often sold under the trade name Tyvek®. This material stops wind from penetrating into the insulation area and repels rain-water. Interestingly, it is also permeable to water vapor, which allows trapped moisture in the insulation cavity to escape outdoors, preventing rot.

On the warm side of the insulation is the vapor barrier. In newer homes, this barrier consists of 0.006" (6-mil) polyethylene sheet (think food wrap on steroids) affixed to the wall studs and carefully sealed.

The vapor barrier acts as a secondary wind barrier, stilling the air trapped in the insulation. It also prevents warm, moist household air from penetrating into the insulation. (Remember—heat moves toward colder areas.) Moist air contacting cold insulation and wood siding on the exterior side of the insulation will quickly condense, forming water. Should this water be present for extended periods of time, it will cause structural rot and mold problems. Many a house has been seriously damaged due to high inside humidity coupled with an inadequate vapor barrier.

Your Options

During renovation, it may be possible to install both an air barrier and a vapor barrier. This type of upgrade can only be completed when the framing members are accessible, which typically occurs when a major rebuilding effort is in progress or an addition is being constructed. Major renovation involving more than just a few sheets of vapor barrier is discussed in *Step 5 – Upgrading Walls*. It doesn't matter if your renovation work is concentrated on the exterior or interior side of the wall; in either case there are ways to complete the work.

Exterior Wall Upgrade

If you are upgrading, removing, or adding to existing siding, it is possible to increase the air barrier with little difficulty. Removal of the old siding reveals the wood sheeting covering the framing members. A layer of Tyvek® air barrier can be taped to the exposed wall sheeting or over existing smooth siding

Figure 2-24 A home wrapped with a suitable air barrier decreases air leakage and increases insulation values.

materials. The air barrier is similar to the vapor barrier in that all joints should be carefully taped to ensure a continuous barrier to wind penetration.

If the interior wall is not being upgraded, follow *Step 3 – Stopping Air Leakage*. Although a polyethylene film is not being applied, a reasonably good vapor barrier can be made using multiple layers of latex paint over well-sealed drywall. When an external air barrier is mixed with air-leakage sealing and latex paint, you can achieve the next-best thing to new construction.

Interior Wall Upgrade

Your choices for upgrading air and vapor barriers increase if you are updating an interior wall. We will discuss this work further in *Step 5 – Upgrading Walls*.

Step 4 – Upgrading Basements

In the section on basements in *Step 1*, we discussed the need to ensure complete dryness before insulation and/or vapor barriers can be added to the inside. Persistent moisture or water leakage problems will ruin even the best-quality work.

Dampness

Minor dampness causes staining or mold growth, blistering and peeling of paint, moldy smells, and efflorescence (whitish deposits) on concrete. These problems can be corrected from the inside by cleaning up any mold and then applying damp-proofing to the foundation wall. More serious problems will have to be corrected from the outside.

Figure 2-25 Installing rain gutters and sloping the grade away from the house are two simple but effective means of keeping the basement dry.

Major Cracks

If your foundation wall has large cracks or cracks that are getting bigger, seek professional help prior to upgrading to determine if structural repairs are necessary.

What Are My Choices?

There is no doubt that insulating from the inside of the basement is the easiest and least costly method, provided that the basement is dry. Insulating from the outside will do the best job from a technical point of view. What to do? Let's begin by weighing the pros and cons of each method.

Insulating Inside

Insulating inside usually involves the addition of a wood-framed wall and adding some form of insulation material, just like a standard wall. There are several advantages of insulating inside:

- The work can be done at any time of the year.
- A completed job will increase the value of your home.
- Indoor finished space will be increased.
- It is the lowest cost way to update the basement insulation and vapor barrier.
- The landscaping, porches, walkways, and other obstructions outside will not be disturbed.

There are also some disadvantages:

- Persistently damp or wet basements cannot be upgraded.
- Furnaces, electrical panels, vent pipes, and plumbing obstructions make do-it-yourself framing more difficult.

Insulating Outside

Insulating outside the home involves more complex and expensive excavation but at the same time provides an opportunity to correct existing foundation defects. There are several advantages of insulating outside:

- The outside wall is generally straighter and simpler to insulate once the excavation has been completed.
- Most moisture and water leakage problems can be corrected at the same time.
- Foundation cracking and other damage can be inspected and repaired.
- There is no lost space inside the home as a result of the added wall thickness.
- The weight or mass of the foundation is on the warm side. This mass absorbs heating and cooling energy, helping to balance temperature fluctuations.

There are also some disadvantages of insulating outside the home:

- Excavation work is costly and may be difficult if there are porches, finished walkways, or decks abutting the foundation wall.

Insulation	Type	R-Value (approx.)	R.S.I Value
Batts	Fiberglass or Rock Wool	3 1/2" (R-11) 5 1/2" (R-19) 9 1/2" (R-30)	90mm 1.9 140mm 3.3 240mm 5.0
Loose Fill	Fiberglass or Rock Wool	R-2.7 per inch	1.9 per 100mm
	Cellulose	R-3.7 per inch	2.6 per 100mm
Rigid Board	Expanded Polystrene (Beadboard)	R-4 per inch	2.8 per 100mm
	Extruded Polystrene	R-5 per inch	3.5 per 100mm
	Polyurethane or Polyisocyanurate	R-7 to R-8 per inch	4.4 - 5.6 per 100mm
	Polyurethane	R-7 to R-8 per inch	4.9 - 5.6 per 100mm

Table 2-2 Common Insulation Materials and Their Heat-Resistance Values in R/RSI.

- Storing the excavated dirt will damage lawns and bring mud and sand into the house.
- Work cannot be done economically in the winter season.

  Once you have determined which system to use, the next step is to examine how the work should be done.

How to Insulate Outside
Step 4 - 1 - Preparation
- Prior to beginning the insulation work, remove any outside features that will get in the way. This includes decks, stairs, trees, walkways, and so forth.

- Determine where power, water, sewer, septic tank lines, gas, telephone, and other services enter the building. Contact your utility to locate unknown pipes and wires; this is usually a free service.
- Determine where the excavated dirt will go. Placing a polyethylene sheet or tarp on the grass to hold the dirt will make the cleanup easier.

Step 4 - 2 – Excavation
- USE CAUTION! The soil may be unstable and fall back into the excavation, causing injury or death.
- The excavation must extend down to the top of the footing
- Never dig below or near the base of the footing (see Figure 2-26, Item 8), as this will cause the foot to sink and the house to drop.
- USE CAUTION! Older rubble stone walls may collapse without the support of the surrounding soil. Seek expert help if you are in doubt.

Step 4 - 3 – Preparing the Foundation Wall
- Brush and fully clean the foundation wall. Scrape any loose concrete or rubblework from the wall (Item 2).
- If concrete is missing or damaged in places, it is best to apply a coating of parging (waterproof masonry cement) to the damaged or missing sections. Allow this to dry.
- Apply damp-proofing compound to the foundation wall (Item 7). Damp-proofing is a tar-like substance that is painted on the foundation wall from the top of the footing to grade level.
- Inspect the footing drain system (Item 4). If there is no footing drain, determine whether one can be added. The footing drain is made by installing a flexible pipe that has been manufactured with thousands of holes along its length. The pipe is normally "socked," with a pantyhose-like liner to prevent sand and dirt from entering.
- This pipe is laid down adjacent to the footing without affecting the undisturbed soil in this area.
- A 12" layer of "clear stone" is spread on top of the footing drain.
- A strip of filter fabric or "gardeners' cloth" should be applied on top of the clear stone to prevent sand and dirt from plugging the footing drain.
- The footing drain should be routed downhill away from the house to a drainage ditch or dry well. A dry well can made simply by excavating a hole 3-4' in diameter to a depth lower than the house footings. Place the end of the footing drainpipe in this hole and fill with crushed "clear stone" to within 12" of the top. Cover the top of the dry well and supply ditch with soil to finish.

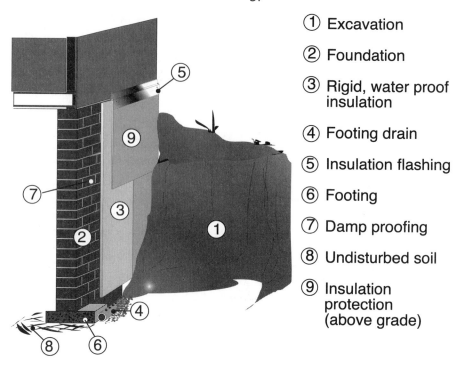

① Excavation

② Foundation

③ Rigid, water proof insulation

④ Footing drain

⑤ Insulation flashing

⑥ Footing

⑦ Damp proofing

⑧ Undisturbed soil

⑨ Insulation protection (above grade)

Figure 2-26 Excavation of the outside foundation wall is the best way to solve water leakage and foundation structural problems while upgrading insulation.

• Check that all services penetrating the foundation wall are well sealed with a suitable caulking compound. Remember to allow room for the insulation.

Step 4 - 4 – Installing the Insulation

• There are many types of insulation in use today, but the most common and easiest to work with is polystyrene rigid board. This material is supplied in sheets that are typically 2' x 8' long and have interlocking grooves running the full length. As indicated in Table 2-1, the minimum recommended insulation level of rigid board is R13/RSI 2.0. Your building supply dealer will be able to recommend which brand of insulation is available in your area. Just make sure you explain your R/RSI rating requirements.

• Rigid board insulation is very fragile and will break when exposed to wind or undue flexing. Make sure the board is protected prior to and during the installation phase of the work. A special flashing trim is installed at the top of the foundation wall that clips the insulation board in place.

Alternatively, pressure-treated plywood may be applied on-site to hold the insulation to the header joist at the top of the foundation wall (Item 5).

- Make sure that all insulation joints are well sealed and clipped together. It is important that you discuss your specific needs for a flashing and insulation-clip system with your material supplier as there are many variations available.
- Ensure that insulation overlaps at the wall corners, preventing areas of concrete from being exposed.
- A covering is required to protect the insulation from sunlight and damage from traffic and animals where it protrudes above grade (Item 9). Coverings may be purchased for the application or fabricated from any of the following:
- pressure-treated plywood
- vinyl or aluminum siding to match the house
- metal lath and cement parging

Step 4 - 5 – Backfilling the Excavation

- After backfilling the drainage pipe as discussed in Step 3, it is time to refill the excavation. If the soil that was removed earlier is heavy clay or drains poorly, it is better to remove it and use clear-running "pit run" sand as the backfill material. This will greatly assist in encouraging water to drain away from the foundation wall, further ensuring a dry basement.
- When the excavation is refilled, ensure that the finished grade slopes away from the house. This promotes drainage and allows runoff to move away from the foundation wall. This is a good time to remind you to install eavestroughs with downspout pipes that lead away from the house.

How to Insulate Inside the Basement

Insulating a basement that is known to be dry is not much different than insulating a new house. The major differences relate to the type of wall structure that is used. A fairly new poured-concrete or block wall should present few problems. Older rubble or cut stone walls tend to be "wavy" and vary in height somewhat. This makes framing more difficult, but does not change the methods involved. The dry-basement insulation method is just a repeat of new-home construction:

- A tarpaper-style moisture barrier is glued to the concrete wall starting from just below grade and then folded up and glued to the bottom of the interior vapor barrier as detailed in Figure 2-4.
- The basement wall is built using standard framing techniques except

that the 2 x 4" stud wall is "pulled" away from the foundation wall by 4" (100 mm). This increases the wall-cavity space for additional insulation without using additional framing material.

- Instead of traditional batt insulation, blown-in fiberglass or cellulose is suggested. These materials ensure complete coverage and packing density, especially in the hard-to-insulate areas around plumbing lines, electrical boxes, and wiring.

- All other finishing details are exactly the same as in a new home.

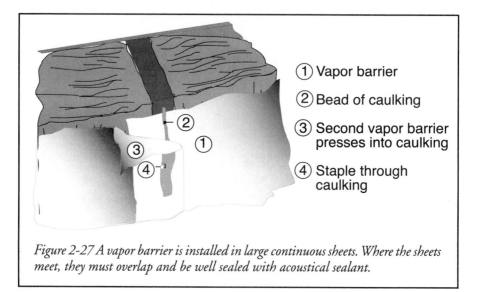

① Vapor barrier

② Bead of caulking

③ Second vapor barrier presses into caulking

④ Staple through caulking

Figure 2-27 A vapor barrier is installed in large continuous sheets. Where the sheets meet, they must overlap and be well sealed with acoustical sealant.

Problem Basements

There are a number of obstacles that can be encountered in older buildings, including:

- packed–stone or dirt floors;
- crawl spaces or basements with very low clearance heights;
- no-basement, slab-on-grade construction;
- building is on piers or blocks.

While it is possible to insulate problem basements, you should seek a professional contractor to review your specific requirements. There are too many variations in climate and construction type to generalize on how to deal with each scenario.

Step 5 - Upgrading Exterior Walls

Because of the large surface area they cover, walls account for a sizeable percentage of the heat loss in houses. Older homes, constructed of solid material such as stone, brick, or log, gave little thought to interior insulation. Many of these designs have a small air space between the exterior covering and the small inner frame wall. This area must not be insulated, as it is used as a drainage cavity for water leakage and condensation.

Hollow concrete-block walls should not be filled with insulation. The quality of the insulation and the "thermal bridging" effect of heat and cold passing through the block does not warrant the trouble or expense.

Traditional frame walls are easily insulated as there is usually an accessible cavity. Using various construction techniques, you can determine if there are cables, duct work, or other obstructions inside the wall cavity that may interfere with the application of insulation.

Solid walls of stone, brick, or log may be insulated from either the inside or outside, depending on several factors:

- The building may have heritage appeal. Homes built of traditional stone, log, or brick may be too beautiful to cover. Insulating from the inside may make the most sense in these situations.

- The outside may need refinishing. If your exterior siding or building material is looking a little tired, there are a number of ways to insulate from the outside and refinish the exterior siding at the same time.

- Interior walls may need new lath and plaster or general updating. When the inside wall surface is cracking and the wallpaper is starting to get to you, perhaps insulating from the inside is the right choice.

- Some exterior and interior finishing may be required. If energy efficiency is the ultimate quest, why not consider doing both? It is possible to add insulation to both the outside and inside

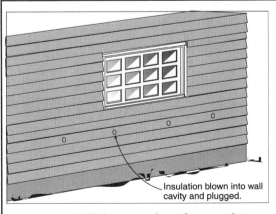

Insulation blown into wall cavity and plugged.

Figure 2-28 Cellulose or urethane foam insulation may be blown into the wall cavity from either the interior or the exterior side of the wall.

exterior walls, creating a "good as new" insulated home.

Construction type, material costs, and your skill level will determine the path you take on this upgrade.

Frame-Wall Cavities

Frame-wall cavities are by far the easiest to upgrade provided they are empty. A wall that is half-filled with insulation from an earlier job makes it almost impossible to do a good job and ultimately will not be worth the expense.

During the assessment stage, we determined how much insulation was in the wall by using a flashlight and poking around through electrical outlets (with the power off!) and other access points. If there is little or no existing insulation, a contractor will apply either cellulose fiber or polyurethane foam into the cavity. These materials are applied through small holes drilled into either the exterior siding or the interior drywall finish.

The holes drilled into exterior siding are plugged using wood dowels. Once the siding is repainted, the work is almost invisible. Brick homes that are suitable for blown insulation have selected bricks removed. The insulation is sprayed and the bricks are replaced.

If the work is being done from the inside, a hole is drilled into the drywall or lath and plaster surface at strategic locations. After the insulation is applied, the holes are filled, primed, and repainted.

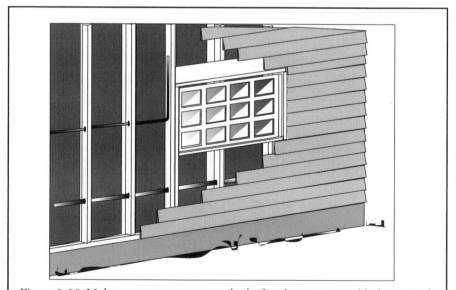

Figure 2-29 Make sure your contractor checks for obstructions or blockages in the wall, including window and door frames.

If the work is being done inside and the interior drywall or lathwork is poor, consider drilling holes into the existing surface and then covering it with a full vapor and air barrier. New drywall can then be placed directly on top of the existing surface. This technique uses up very little space and is cheaper than a fully framed interior wall.

Upgrading Insulation on the Exterior

If it's time to upgrade the old siding, this is an excellent opportunity to upgrade exterior insulation as well. There are several methods available that allow you to add insulation under the new siding, significantly increasing the overall "R/RSI" value of the home. If you are applying new siding, using blown-in insulation will eliminate the need for filling and repainting access holes. Consider these points as well:

- It is possible to add a generous amount of insulation by using high-density rigid insulation sheets or creating a new wall cavity on top of the existing siding.
- If the house is poorly insulated or made of stone, solid brick, or masonry, add a vapor barrier directly to the inside of this surface, under the new insulation. Follow the rules outlined above regarding proper vapor barrier installation and sealing.
- Remember that doors and window openings will have to be extended to allow for the additional insulation thickness.
- Ensure that water runoff from the roof will not drip between the old and new wall, ruining the insulation and causing structural rot damage. An eaves extension or flashing will prevent this.
- Make sure that the new insulation is well air-sealed. There is not much sense in doing a great insulation job with the winter winds howling between the old and new work.

Applying Exterior Insulation

There are dozens of ways exterior insulation can be applied. We will review the most popular methods and then point you to your local building supply contractor for more detailed information about materials for your specific application. Regardless of the application method you choose, be sure to wrap the entire upgrade work in spun-bonded olefin (Tyvek® brand) air barrier and ensure that it is carefully wrapped and taped at all joints.

1. It is possible to add exterior insulation by simply purchasing insulated siding materials. This is the easiest approach, although the insulation values are somewhat limited owing to the small thickness. Check with

your building supplier on the many finishes available. These preinsulated siding products install in the same manner as their uninsulated counterparts.

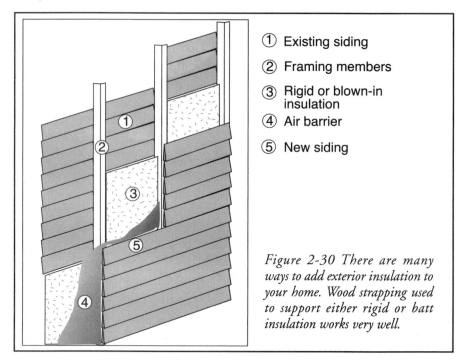

1. Existing siding
2. Framing members
3. Rigid or blown-in insulation
4. Air barrier
5. New siding

Figure 2-30 There are many ways to add exterior insulation to your home. Wood strapping used to support either rigid or batt insulation works very well.

2. Rigid board insulation can be applied directly over existing surfaces using appropriate fasteners and adhesives. Your building supplier or contractor can review which materials will work best for your application. If rigid board insulation is used, make sure that all joints are tight and well taped to reduce air leakage.

3. Rigid board or batt insulation can be added to a new wall framed on top of the existing siding structure. Batt insulation requires a thicker wall to achieve the same insulation value as rigid materials.

   a. Frame the desired wall thickness on top of the existing wall siding.
   b. Add the insulation, ensuring complete coverage between studs.
   c. Apply an air barrier of spun-bonded olefin, using well-taped edges and seams.
   d. Ensure that water runoff from the roof will not drip directly onto the top of the new wall extension.
   e. Make sure that wind cannot enter the wall cavity from the top or bottom framing plates.

Upgrading Insulation on the Interior

Determine the best application method by assessing your specific requirements. Your choices include:

- Upgrading an existing wall. This is often done when the existing wall material is damaged or lath and plaster is falling off.
- Adding rigid board insulation directly to an existing wall surface. Once the wallboard is added, a new drywall finish surface is commonly added.
- Building a new frame wall. The new wall takes more space from the interior, but provides lovely window boxes. It also allows you to increase home insulation values and apply a proper vapor barrier.

Upgrading an Existing Wall

A common upgrade for older homes is removing the lath and plaster wall (a messy job at best) and replacing it with modern drywall:

- With the internal wall studs exposed, you can easily upgrade wiring, plumbing, ductwork, central vacuum pipes, and, of course, your insulation.
- Use additional horizontal strapping as shown in Figure 2-31 to increase the wall-cavity depth from a typical 2 x 4" to 2 x 6".
- Electrical boxes and window frames must be extended to allow for the increased wall-cavity depth. Building supply stores carry box extenders for such work.
- Use batt insulation layered vertically for the existing wall and horizontally for the new wall section. Alternatively, blow in cellulose insulation after the vapor barrier has been installed.

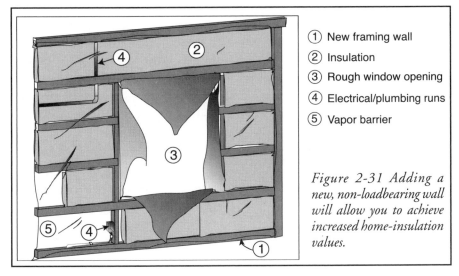

1. New framing wall
2. Insulation
3. Rough window opening
4. Electrical/plumbing runs
5. Vapor barrier

Figure 2-31 Adding a new, non-loadbearing wall will allow you to achieve increased home-insulation values.

- Install a 6-mil vapor barrier over the new framing studs.
- Apply the drywall or surface finish.

Adding Rigid Board Insulation to an Existing Wall

Rigid board insulation may be applied directly to an existing wall surface:
- As with upgrading an existing wall, windows and doorframes must be extended and electrical box extenders added.
- Your building supply store will be able to provide suitable fasteners to hold the rigid board insulation to the existing wall surface.
- Vapor and air barriers are not required when rigid board insulation is installed provided the seams are snug and well taped.
- Drywall or other finish materials may be applied directly to the rigid board insulation.

Building a New Frame Wall

You can build a new frame wall by using exactly the same techniques described earlier in *Upgrading an Existing Wall* and illustrated in Figure 2-4.

Insulating Both Sides of an Exterior Wall

A final word is required on adding insulation to both the inside and the exterior of the home. If you wish to add some of the insulation on one side of the wall and additional insulation on the other, be careful to limit the ratio to two-thirds inside and one-third outside. This mix is required to ensure that condensation does not damage the insulation system during the heating season.

Step 6 – Upgrading Attics

Air Leakage (a quick review)

Earlier, in *Step 3 – Stopping Air Leakage*, we discussed the need to prevent air from leaking into the attic space from the warm areas below. Eliminating air leakage stops the warm, moist air from condensing when it reaches colder winter air inside the attic. Condensing moisture may cause mold and structural rot. In minor cases, moisture buildup greatly reduces insulation effectiveness.

A proper vapor and air barrier **MUST** be installed on the warm side of the insulation, NOT ON THE COLD SIDE! From a practical point of view, adding a large polyethylene film to the attic-floor of the insulation is very difficult. It requires moving large amounts of existing insulation and installing the barrier around obstructions such as duct work, plumbing pipes, and wiring.

Should the interior ceiling of the house also be under renovation, you have the opportunity of adding the air and vapor barrier directly to the attic-framing members, under the drywall.

In the majority of cases, it is just not possible to add an effective air and vapor barrier into a retrofit attic. In these cases, a well-sealed and caulked attic must suffice.

Attic Ventilation (a reminder)

Attic ventilation is mandatory. A home will typically have a row of vents under the eaves called soffit vents (see Figure 2-32). These vents form the air intake for the attic ventilation system. Roof vents are installed along the roof peak (see Figure 2-33) or at the gable ends and allow hot attic air to exit. Air exiting the roof or gable vents causes suction within the attic, drawing in cooler air through the soffit vents. You will recall that a ratio of 1 unit of roof vent area to 300 units of attic floor area is required (approximately 3 square feet of roof vent space for every 900 square feet of attic floor).

Figure 2-32 Soffit vents form the air-intake section of the attic ventilation system.

Figure 2-33 Roof vents along the roof peak vent hot air out. A minimum of 1 square foot of roof venting is required for every 300 square feet of attic floor space.

Before starting any insulation work in the attic, ensure that the vents described above are present and of adequate size. If in doubt, ask your building contractor to assess their effectiveness.

Installing Attic Insulation

Once the attic has been well sealed against air leakage and the soffit and roof vents checked, you are ready to begin insulating. You have a number of choices as to which material to use. Rigid board insulation is almost never used, due to the difficulty of working in confined spaces. Blown or poured

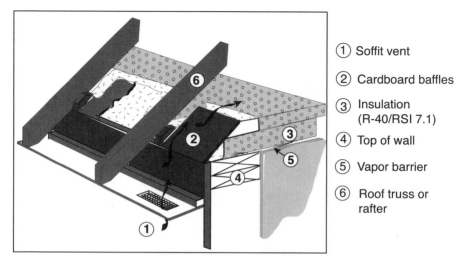

1. Soffit vent
2. Cardboard baffles
3. Insulation (R-40/RSI 7.1)
4. Top of wall
5. Vapor barrier
6. Roof truss or rafter

Figure 2-34 A well-sealed attic with plenty of ventilation and insulation is the final step in keeping your house warm in winter and cool in summer.

fiberglass and cellulose are perhaps the easiest insulation materials to use. Most building supply stores will even provide a blower unit free of charge when you purchase the insulation from them. However, many homeowners simply carry bags of material to the attic, pour it, and rake it flat.

Adding insulation to the attic is not difficult. The major concern is to ensure that attic vents are not plugged with the stuff. Figure 2-34 illustrates how cardboard baffles can be stapled or pressed into the space between the rafters or joists. These baffles prevent insulation from rolling down into the soffit vent area.

Ensure that insulation is not applied around electrical light fixtures or bathroom vent fans that are not specifically approved for insulation coverage. Overheating light fixtures are a sure way to start a house fire.

Figure 2-35 This eighty-year-old home boasts the world's largest icicle, indicating severe heat loss. A professional assessment of this home demonstrated to the owners how to reduce their heating bill by at least one-third while at the same time making their home more comfortable.

2.3
Appliance Selection

What's All the Fuss About?

A home without major appliances, computers, and entertainment systems is a lifestyle that few people would embrace. There is nothing fundamentally wrong with having air conditioning or a large-screen TV (other than watching it 18 hours a day). The use of energy to power these devices is not in itself a problem. The difficulty occurs when homeowners purchase inefficient models and use polluting, non-renewable resources to power them. There is no free lunch. At some point in the future (some would say right now), these non-renewable resources will start giving up (remember the blackout of 2003?) or kill us with their exhaust.

The good news is that governments and appliance manufacturers are beginning to understand these issues. Manufacturers are actively working to lower the energy requirements of their products. For renewable-energy-system users, these improvements have exponentially increased the number of appliances available to them. For energy conservers on the grid, these same appliances are helping put money in the bank and enabling them to live just a little bit lighter on the planet.

The average 25-year-old refrigerator (the one keeping a six-pack of beer cold in the basement) uses approximately 2,200 kilowatt-hours of energy per year to operate. Electrical rates vary greatly around North America, but the average daily rate in California (when there is

Figure 2-36 A standard Sears 18.5 cubic foot, 2-door refrigerator like this one will save thousands of dollars and kilowatt-hours of energy over its twenty-plus years of operating life.

power) is approximately 15 cents per kilowatt-hour or $330 per year operating cost [1]. A new Sears 18.5 cubic foot, 2-door unit uses 435 kilowatt-hours per year or $65.25 to operate, at a savings of $265 per year. At that rate, it would only take three or four years to pay for itself (assuming that rates don't climb any higher). Put another way, if you converted the savings into beer you could probably supply suds for the whole block on the savings alone! If that isn't incentive to switch, what is?

How to Select Energy-Efficient Appliances

It's not necessary to carry around an energy meter like the one shown in Figure 2-37 to measure the energy consumption of each appliance you wish to purchase, although it wouldn't hurt.

These meters (which are available from many sources, including www.theenergyalternative.com) plug into a wall outlet, and the appliance plugs into a receptacle on the front of the meter. The meter display indicates the electrical power consumed. Remember that power (watts or W) is a measure of the flow of electricity (amps or A) multiplied by the pressure of electricity (volts or V). Most major appliances such as washing machines, refrigerators, dishwashers, and food processors plug into a standard 120 V outlet (pressure). The flow of electricity in amps multiplied by the pressure (volts) results in the wattage. The calculation for a typical food processor drawing 2.4 amps is:

Figure 2-37 An electronic energy meter such as this one will tell you how much power an appliance requires to operate. Most models can even be programmed with your utility rates to tell you cost over a period of time.

$$120\,V \times 2.4\,A = 288\,W$$

Compare this with the power used by an electric kettle drawing 12.5 A:

$$120\,V \times 12.5\,A = 1,500\,W$$

What does this mean? Let's suppose that you and your partner are shopping for a new television set. You compare all of the models and find two

1.   Rates fluctuate from a low of 3.3 cents per kilowatt-hour in Ontario to peak daytime rates in California of 30 cents (all prices in US$). Check your utility rate and delivery charge (per kilowatt-hour) and multiply it by the yearly consumption rating of your current and proposed appliance models to calculate the savings.

that are about equal on your list of requirements. You whip out the power meter, or, if you are more conservative, authoritatively inspect the electrical ratings label, and find that model "A" uses 162 W of power while model "B" requires 1.9 A.

At this point most people would run screaming out the door having flashbacks to those grade nine "A car is moving west…"-type problems. You, on the other hand, have studied *Chapter 1.3 – What is Energy?* and recall that power (in watts) is the voltage multiplied by the current.

120 V x 1.9 A = 228 W

You close the deal by smoothly informing the salesperson that model "A" is the better TV as it will save you loads of dough over its operating life, reduce greenhouse gas emissions, or require less electrical-generation equipment if you are using off-grid energy sources. You might even save money on the purchase with the salesperson trying to get rid of you.

But power is only half of the equation. If for some strange reason you were to watch model "A" television 2 hours per day and model "B" for 1-1/4 hours per day, the energy consumption for model "B" would be lower:

Model "A" energy consumption = 162 W x 2 hours = 324 W-hours
Model "B" energy consumption = 228 W x 1.25 hours = 285 W-hours

Time factors into the energy cost. This is why you should turn off lights in empty rooms.

There Has to Be a Better Way

Most people are not so fanatical as to carry an energy meter with them when they go shopping for appliances. Some people may take a look at the electrical ratings label, provided they can keep the watts and volts straight without having to carry a copy of *The Renewable Energy Handbook*. For the rest of us, the government has made life a little easier, at least for the larger appliances. Figure 2-38 shows a Canadian "EnerGuide" label that is affixed to a high-efficiency washing machine. The EPA uses a similar label for its "EnergyGuide" program. Both programs require that labels be affixed to appliances, providing comparison data for appliance models of similar size with similar features and indicating the appliance's total energy consumption per year.

A closer look at an EnerGuide label reveals the following features:
- The bar graph running from left to right represents the energy consumption of all models of similar appliances.
- At the left side of the graph is the energy consumption of the most efficient appliance in kilowatt-hours per year.

- At the right side of the graph is the energy consumption of the least energy-efficient appliance in the same class.

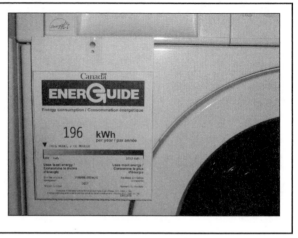

Figure 2-38 This high-efficiency washing machine carries both the "EnerGuide" and "Energy Star" program logos, allowing consumers to quickly compare the energy consumption of different models. The Energy Star logo indicates to the consumer that the appliance meets minimum energy efficiency ratings.

The bar graph on the EnerGuide label indicates energy consumption of 189 kWh per year for the most efficient appliance and 1032 kWh per year for the least efficient. Think about that for a moment. Two appliances of the same size and class, one consuming almost five-and-a-half times the amount of energy to do the same job!

Efficient washing machine = 189 kWh per year x 10 cents per kWh = $19 per yr.
Inefficient washing machine = 1,032 kWh per year x 10 cent per kWh = $103 per yr.

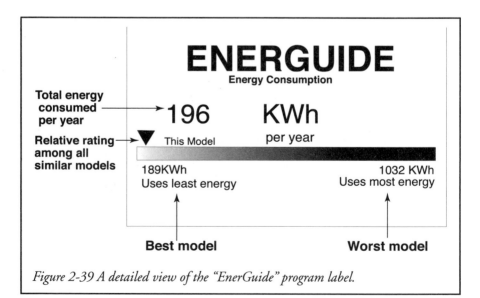

Figure 2-39 A detailed view of the "EnerGuide" program label.

Figure 2-40 This Sears high-efficiency clothes washer uses five-and-a-half times less energy than a similar-sized unit. Also consider the savings on soap and reduced waste water being pumped into the sewer or septic tank.

Assuming a delivered price of 10 cents per kilowatt-hour of electricity, the difference in operating expense is $84 per year. Over the life of the machine (say twenty years), that's a whopping $1,700!

In Figure 2-39, the triangle-shaped pointer over the bar graph on the EnerGuide label shows the energy consumption of this particular model in relation to that of the most and least efficient models in the same class. The closer the pointer is to the left side of the bar graph, the lower the operating costs.

The sample label used in Figure 2-39 is from a Sears front-loading washing machine, straight from the

Figure 2-41 Not only is the solar-powered clothes dryer energy efficient; clothes also last longer, smell better, and don't require yet another artificial fragrance sheet.

catalog pages. The most efficient model is the Staber horizontal-axis machine. The Staber is almost twice as expensive as the Sears model but only 1% more efficient. On the other hand, the least efficient model is marginally more expensive than the Sears high-efficiency model.

Small Appliances

The EnerGuide program was designed to take care of major "white goods" appliances. What about smaller appliances and electronics that don't carry the program label?

Our choices at this point become a little bit more difficult. We have to revert to using an energy meter or reading the electrical-ratings label on the device. An electrical meter such as the one shown in Figure 2-37 is a great device provided that you have access to the appliance for a sufficient amount of time to conduct a test analysis. This is important because energy consumption requires time to evaluate (energy = power multiplied by time).

Let's take a look at a conventional coffee maker to see how time factors into our assessment. A typical coffee maker such as the one shown in Figure 2-42 has an electrical rating label showing 120 V and 10.5 A. Whipping out your calculator you correctly arrive at a power consumption of 1,260 W.

Figure 2-42 A coffee maker uses a lot of energy to boil water, but considerably less to keep the pot warm. In the absence of an EnerGuide or EnergyGuide label, use care when calculating energy consumption based on the manufacturer's label.

120 V x 10.5 A = 1,260 W of power

At 7:00 o'clock on Saturday morning you stumble down the stairs and get the brew going. By 11:30 you have slugged back your third cup, draining the machine and shutting it off. Applying your caffeine-honed mathematical skills, you determine that the coffee maker was on for 4 1/2 hours, giving you an energy calculation of 5,670W.

1,260 W of power x 4.5 hours = 5,670 watt-hours of energy or 5.7 kWh

You realize that at 10 cents per kilowatt-hour for energy your morning coffee has just cost you 57 cents worth of energy, right?

$5.7 \, kWh \times \$0.10 \, per \, kWh = \$0.57$

Wrong. The math is correct, but the assumptions are wrong. You have to be very careful when calculating the energy consumption of an appliance, as it may in fact change with time. Yes, the coffee maker label does say that it requires 10.5 A or 1,260 W of power. What it does not tell you is that it only needs that much power to boil the water. Once the coffee is brewed, it uses less power to keep the pot warm.

So what is the correct answer? Actually, I don't have a clue. Without access to the coffee pot and an energy meter, it's anybody's guess. When trying to calculate the energy consumption of any small appliance with no electrical rating label, follow these general guidelines:

- Appliances that draw a lot of wattage, typically over 300 W, should not be used unless it is only for a short period of time. This applies to coffee makers, hair dryers, curling irons, electric kettles, clothes irons, car block heaters, and space heaters. Crock pots which slowly simmer food for hours at a time are actually more efficient than ovens.

- Coffee makers such as the model described above can be used, but transfer the fresh coffee to a thermos. The coffee stays warm and tastes better without using unnecessary energy.

- Consider boiling water using an electric kettle (instead of the stove) and a drip coffee basket. Transfer the fresh coffee to a thermos to keep it warm.

- Refer to *Consumer Guide to Home Energy Savings* for electrical consumption ratings of desired appliance models before you buy. (The book is available through Real Goods (www.realgoods.com), part number 82-399.)

- Refer to Appendix 2 of *The Renewable Energy Handbook* for a list of various electrical appliances and tools with their power and electrical ratings.

- All major heating appliances such as cook stoves, ovens, electric water heaters, electric clothes dryers, furnaces, central air conditioners, etc. draw an enormous amount of electrical energy. Always purchase the most efficient model and consider the addition of timers or other appropriate controls to regulate their operating time.

- Microwave ovens are much more efficient than regular electric models. This is an especially important consideration in the summertime when waste heat from the oven must be removed with expensive air conditioning.

- Try to purchase appliances that are not equipped with an electronic clock or "instant-on" anything. These devices are considered phantom loads and consume a large amount of power without doing anything for you. Cell phones, PDAs, MP3 devices, and other electronic items that require battery charging can be plugged into a power bar. Turn the power bar off once the device has been fully charged.

- Convert as much lighting as possible to high-efficiency compact fluorescent lamps (see below).

- Use high-efficiency front-loading washing machines. The Staber HXW-2304, for example, is so efficient that you require just one ounce of soap per wash load. You will use less electrical energy to operate the washer, less water per cycle, and less heating energy to heat the smaller amount of wash water.

- If you are not sure about an appliance or tool, borrow one from a friend and plug it into an energy meter like the one shown in Figure 2-37. Even if the appliance model is not exactly the same, this will give you a general idea of the power and energy consumption.

Computers and Home-Office Equipment

Many people are finding that self-employment is the way of the future, especially after the last round of corporate downsizing. An essential part of working and playing at home now revolves around computers and related home-office equipment.

With your computer, desk lamp, printer, modem, and monitor working overtime (maybe all the time if you have teenagers), how will your electrical energy consumption be affected?

Modern electronics are a marvel of efficiency and they keep getting better all the time. For example, let's compare a 7-year-old, 15-inch color computer monitor with a new flat-screen model. The older unit may use as much as 120 W of power, while the new one draws about 30 W. That's a 75% reduction in power usage.

You should also consider that few people use a computer for only a minute or two and then turn it off. The average user may have the computer and support system operating for many hours, which increases energy consumption (power x time). It makes sense to purchase the most efficient products because of this longer operating time.

Don't just think about the computer. You will almost certainly have a printer, monitor, desk lamp, room lights, fax machine, and possibly a photocopier to fully equip your home office. Teenagers (and aging rockers) can add

Figure 2-43 A wall full of home-theater equipment may not seem very energy efficient, but in fact it is. Modern electronics are a wonder of efficiency and work well with any renewable energy system. Just remember to switch off the power input using a special plug or power bar to eliminate phantom loads.

to this list a 400 Watt surround-sound system. Running an energy-efficient system for forty hours a week will still burn up a lot of juice.

Fortunately, the government has come to our rescue once more. The Energy Star program has been adopted across North America and is the equivalent of EnerGuide and EnergyGuide labeling. Energy Star provides guidelines and requirements to manufacturers who seek compliance for their electronic stereo, TV or data-processing products. Most compliant products use high-efficiency electronic components and specialized power-saving software. For example, a compliant monitor will enter a low-power sleep mode if the image has not changed within a given time frame. Another example is a laser printer which adjusts its heater temperature when idle. A laptop computer will always be more energy efficient than the equivalent desktop model, owing to the limited energy stored in the batteries. Likewise, a bubble jet printer is more efficient than a laser printer.

As with everything else in life, there are choices to be made. If you want lower per-print costs, then laser-printer toner is cheaper than bubble-jet cartridges. Laptops are more expensive than a desktop system. Just remember to shop the energy labels and look for the Energy Star logo.

Figure 2-44 Purchasing computers and home-office equipment with the Energy Star label ensures the lowest energy usage.

2.4
Energy-Efficient Lighting

Possibly the single most important invention to touch our lives is the incandescent light bulb. Prior to Mr. Edison's 1879 discovery, it was almost impossible to stay up past sundown. (My wife still has this problem.) Life before the modern light bulb meant filling a kerosene lamp and enduring poor lighting and air quality. In the early days of electrical-power production, little thought was given to energy efficiency or environmental concerns. As a result, light bulbs were—and are—inefficient and wasteful. The bulb in your floor lamp has remained essentially unchanged for over a hundred years. However, times have changed and so have lighting technologies.

The incandescent lamp and its many variations create light by heating a small coil or filament of wire inside a glass bulb. Anyone who has watched a welder working with metal knows that hot metal glows a dull red color. If the metal is heated further, it glows brighter and with an increasingly whiter light. At some point, the metal will vaporize with a brilliant shower of white sparks. The

Figure 2-45 We owe a nod of thanks to Mr. Edison and an entire industry that developed a replacement for the kerosene lamp. As technology marches forward, however, it is time to say goodbye to our old but wasteful friend—the incandescent lamp.

various stages of glowing metal are known as incandescence.

In a similar manner, electrical power applied to a light bulb causes the metal filament to glow white-hot. If you decrease the amount of power reaching the bulb (by using a dimmer switch, for example) the bulb dims and glows with a progressively redder color.

Making an incandescent lamp glow requires a large amount of energy to heat the filament. In a typical light bulb, 90% of the energy applied to the filament is wasted in the form of heat. Therefore, only 10% of the energy you are paying for is making light. What a waste!

To make matters worse, we need to think about the heat component for a moment. If it is summertime, this waste heat contributes to warming the house and increasing the air-conditioner load. An incandescent bulb hits you twice in the wallet: poor efficiency resulting in high operating cost and waste heat as well as the expense of getting rid of the waste heat using air conditioning.

Let's put this into perspective. Assume it's a warm summer evening. You have a total of fifteen 100 Watt lights on in the house. The waste heat is:

15 bulbs x 100 W x 90% heat loss = 1,350 W of heat

This is about the same amount of heat output as you get from a large electrical space heater in the winter.

You may argue that this waste heat can be used in the wintertime to help warm the house. True, except that electricity is the most expensive means of heating a home anywhere in North America and this byproduct of lighting cannot be controlled unless you want to place your house lights on a thermostat. If you are generating your own electricity off-grid, this waste is not even manageable. So what are the alternatives and how do they save energy and money?

There are numerous alternative lighting technologies, including semi-conductor-LED, mercury vapor, halogen, fluorescent, halide, and sodium. These technologies have a range of applications and different efficiencies. The most common and efficient technologies for home applications are discussed below.

Before we continue it would be prudent to say a few words about semi-conductor or LED lighting. Most people have seen LED lights installed in automotive applications or high-efficiency flashlights. Currently, LED lighting is only one-half as efficient as compact fluorescent technology. Additionally, the color quality of these bulbs is very poor and would not be acceptable for home lighting. Long life and low energy consumption compared with standard incandescent lamps gives LED lights their passing grade.

Compact Fluorescent Lamps

Mention fluorescent lamps to my wife and all she can think of are the pasty-faced girls applying makeup in the washroom at the high school prom. In the 60s and 70s some companies used the term "cool white" on their fluorescent lamps, as if this were a good thing.

Enter the compact fluorescent lamp (CF). These marvels of efficiency may look a little odd, yet they offer many advantages over standard incandescent lamps. A CF lamp is designed to last ten times longer than an incandescent lamp. Shop around when looking for these lamps. Stores such as Wal-Mart carry the electronic CF lamps for about $3.00 each. Many stores still treat CF lamps as "special," which includes a "special" price of about four times this amount.

Light output and energy efficiency are further considerations. Contrary to popular belief, light output is not measured in watts, but in *lumens*. The wattage rating of a bulb is the amount of electrical power required to make it operate. The standard 75 Watt light bulb gives off approximately 1,200 lumens of light. A 20 Watt CF lamp provides the same intensity but uses one-quarter the energy. Translate this to cost savings and the CF bulb will save $55.00 over its life span, assuming delivered energy costs of 10 cents per kilowatt-hour. This does not even take into account the savings in air-conditioning load and environmental pollution.

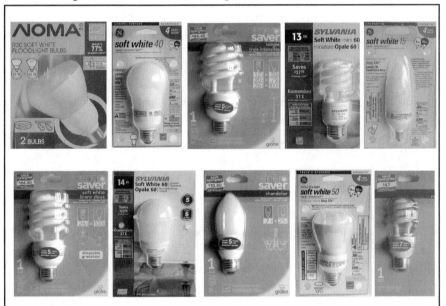

Figure 2-46 The compact fluorescent (CF) lamp is the bulb of choice in modern energy-efficient homes. The choice has never been better, as indicated by the variety available at a typical hardware store. Clockwise from top left: 30 W floodlights (available for indoors and outdoors), an 11 W CF which looks like a regular incandescent (handy for lamp shades), a 12, 20 and 26 W Trilight, two standard 13 W CF bulbs (with the frozen ice cream swirl design), a 15 W chandelier (narrow base), a 13 W outdoor bug light (yellow in colour to discourage flying insects at night), a track and recessed flood light, a chandelier (large base), another brand of 14 W CF that approximates the traditional bulb shape, and a dimmable 25 W bulb (also available in the traditional light bulb shape). If your hardware store doesn't offer this selection, it's time to ask them to, or to find another hardware store.

Lastly, CF lamps won't give you headaches or the Bride of Frankenstein look first thing in the morning. Advances in phosphor coatings and electronic ballasting eliminate ghastly color and flicker. The lighting industry applies a rating system called the Color Rendering Index (CRI) to all bulbs, although it can be a bit hard to locate the information. The closer a light source's CRI rating is to 100, the more natural and comfortable it will be to the eyes. The typical incandescent lamp has a CRI rating of 90 to 95 compared with a CF-lamp rating of 82. Compare this with a Frankenstein fluorescent bulb rating of 51. Still not convinced? Buy a couple of CF bulbs and try them out. It will be almost impossible to tell the difference in color or operation when they're in operation.

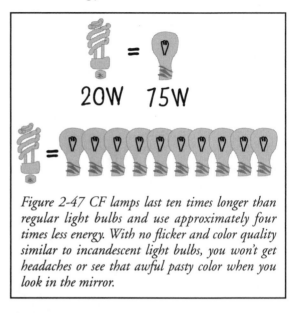

Figure 2-47 CF lamps last ten times longer than regular light bulbs and use approximately four times less energy. With no flicker and color quality similar to incandescent light bulbs, you won't get headaches or see that awful pasty color when you look in the mirror.

The Down Side

The shape and size of many CF lamps are not identical to that of standard bulbs. This may cause problems fitting the CF lamp into a conventional socket. Take note that shapes and sizes vary by manufacturer. Check several different brands to see if one will fit your application. Before you resort to changing the light fixture, see if your local hardware store is able to suggest an alternate base, harp, or socket extension.

Dining rooms often have overhead fixtures that are connected to a dimmer switch. CF lamps must never be placed in these sockets unless they are specifically designed for dimming. Dimmable CF lamps are available, but at a premium price.

CF lamps can be used in outdoor lighting systems even in winter. In this application, the lamps typically require one to two minutes to warm up and produce useful levels of light. This is not a problem for security or perimeter lighting on timers. Garage-door-opener lights or those applications where the light is required quickly should not use CF lamps.

T8 Fluorescent Lamps

Large-area or kitchen-cabinet indirect lighting often uses standard fluorescent lamps with magnetic ballasts. Although more efficient than incandescent lamps, these models are still only half as efficient as CF lamps or new "T8" fluorescent lamps with electronic ballasts. T8 lamps are more expensive than CF units but they are excellent replacements for older 4' and 8' standard fluorescent tubes. It's easy to recognize T8 lamps, as they are approximately half the diameter of conventional tubes. Likewise, electronic ballasts are very light for their size, especially when compared with older, less efficient magnetic ballasts.

Track Lighting and Specialized Lighting

There are places where even a die-hard energy conserver would never put a CF lamp. Lighting your Rembrandt or Picasso collection is one area that comes to mind. For these applications, low-voltage, high-intensity MR-16 lamps are ideal. Jewelry and watch stores use these miniature spotlights for clear, intense lighting of small objects, artwork, and display cabinets.

Figure 2-48 MR-16 flood lamps are excellent for lighting your priceless jewelry or baseball cap collection. Although they are incandescent lamps, their low wattage and high intensity make them acceptable for these applications.

MR-16 halogen gas incandescent lamps are fabricated with specialized internal reflectors that direct light in a unidirectional flood pattern. MR-16 bulbs operate on 12 V, which in turn requires a transformer (supplied with the light fixture) to drop the household 120 V supply. The low wattage of MR-16 lamps is similar to that of a larger CF lamp, so they can be considered efficient in these special applications.

Large-Area Exterior Lighting

Large outdoor areas require major lighting muscle. While it is possible to use a few million CF lamps to light your yard, an easier and even more energy-efficient method is to install high- or low-pressure sodium lamps. These have a characteristic yellow color that makes them suitable for perimeter and security lighting. They also work well in horse-riding arenas and for street lighting. Watt-for-watt, a sodium lamp produces twice as much light with the same power as a CF lamp and is up to ten times more efficient than an incandescent lamp.

Free Light

Perhaps the best lighting is free—100% energy efficient. No, don't just open the drapes. Even the best-designed house may have an area or two that suffer from low light levels during the daytime and require lighting even on the sunniest days. If you're thinking skylights, remember that typical skylights are not energy efficient, are prone to condensation problems, and often require reframing and finishing from the inside. A product available from www.sunpipe.com offers an excellent alternative. The Sunpipe is similar to a fiber-optic device on steroids. An intake piece is mounted on the roof in the same fashion as the metal chimney shown in Figure 2-49a. A reflective supply pipe is fitted down through the roof opening into the living area. A trim kit is installed and that's it.

The Sunpipe directs outside light into the immediate area, even on cloudy days. An example installation

Figure 2-49a This view details the "intake" portion of the Sunpipe unit.

Figure 2-49b The hallway in this picture is completely illuminated with a Sunpipe. Covering the intake to show the difference without the light from the Sunpipe created an almost-black photograph.

is shown in Figure 2-49b with the Sunpipe illuminating a narrow second-floor hallway. If the Sunpipe were to be covered, this hallway would be almost completely dark. A 9"-diameter pipe provides illumination equivalent to that of a 400 Watt incandescent lamp on a sunny day.

2.5
Water Supply and Energy Conservation

Introduction

Turning on a tap for a glass of water is a luxury that few people give any thought to. If you live in the city, all it takes is writing a check to your local utility a few times a year and presto! Clean, clear, life-sustaining water.

Society's lack of concern for or ignorance about this natural resource is often beyond comprehension. One only has to drop by a golf course in the desert to see firsthand the waste of a precious resource. Many people counter with: "Why not use water as we wish? Isn't water a renewable resource, like the solar energy we are capturing?"

Water is a plentiful and renewable resource, within limits. The majority of the earth's water is stored as moisture in the air, salty seawater, or ice and snow. The rest is in lakes and streams or in wells drilled to reach the underground water table.

Lake- and ground-water levels have been dropping for many years; hardly a day goes by without some newspaper

Figure 2-50 An average person in North America consumes 100 gallons (382 liters) of water per day. The conserver can easily reduce this consumption by 50%.

headline screaming about the subject. Consider that the Colorado River does not always flow into the ocean, partly because of over-extraction in the upstream watershed. The reduction in water levels is largely the result of increased human population and water usage. Industrial and climatic changes are also being blamed for an ever-increasing reduction in our second most important natural resource after air.

The average person in North America consumes 100 gallons (382 liters) of water per day. If we include the water required for industrial and agricultural processing, consumption rises rapidly to over 1200 gallons (4543 liters) per day. North Americans consume water at a rate 5-1/2 times greater than that of people in Sweden. As with gasoline, our wasteful habits can be traced back to costs. A 1988 survey of OECD countries showed that efficiency and price were interrelated, with Germany charging $2.10 per cubic meter of water and Canada practically giving it away at 31 cents.

Besides saving money on water bills and energy bills for hot water, reducing our consumption will lighten the load (electrically and from a societal viewpoint) on water- and sewage-treatment facilities. This is a significant consideration when you factor in upgrades in capacity to aging infrastructure that would almost certainly be adequate if conservation of water and sewage were undertaken. The savings to these municipalities (and property taxpayers like you and me) would be enormous.

Water is also very heavy, requiring large amounts of energy for the operation of the municipal supply and sewage pumps.

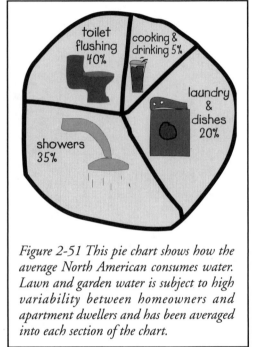

Figure 2-51 This pie chart shows how the average North American consumes water. Lawn and garden water is subject to high variability between homeowners and apartment dwellers and has been averaged into each section of the chart.

The Conserver Approach

It is estimated that 75% of our water consumption occurs in the bathroom and 20% for laundry and dishes. Figure 2-51 indicates that the remaining 5% is used for cooking and drinking. Conserving water and the energy used to heat and move it requires the same approach as with lighting and appliances: do more with less. This does not mean that the whole family has to get in the shower at the same time. It means getting the same results with more efficient appliances and methods.

Harvesting Our Water Supply

The best place to start is with the collection of the water supply. Those of you connected to municipal water supplies might want to jump ahead to the next section, "How to Conserve."

The two most common methods of collecting water are pumping from a well and pumping from a surface supply such as a lake or cistern. Let's start with the well, as this is the most common method in use.

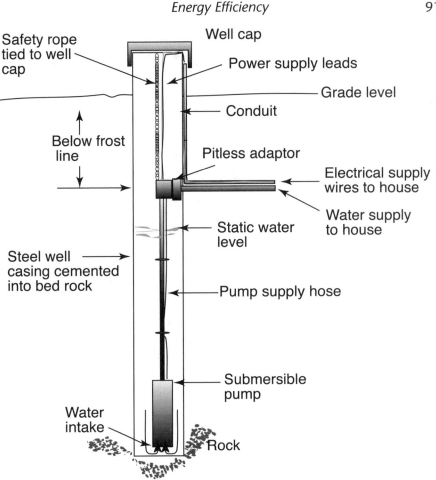

Figure 2-52. This diagram details the installation of a typical deep well submersible pump.

Deep Well Submersible Pumping

Referring to Figure 2-52, you will see an installation diagram of a typical deep well submersible pump. For any new or retrofit application the deep well submersible pump is the simplest, most reliable, and easiest-to-service design. The deep well submersible pump is available as a standard 120/240 Volt AC centrifugal unit or in high-efficiency designs such as the Dankoff ETA series pump shown in Figure 2-53 (www.dankoffsolar.com). Both designs install in the same manner, the major difference being the technology used in the pump design and the resulting efficiency. The Dankoff submersible pump is so efficient that it can be used as a remote well pump for livestock with only one small photovoltaic panel powering the unit, as shown in Figure 2-54.

Figure 2-54. The Dankoff model ETA090 can be used for stand-alone pumping applications such as this remote livestock watering well. Very efficient, but not nearly as picturesque as the old western windmill. (Dankoff Solar)

Figure 2-53. The Dankoff model ETA090 submersible deep well pump installs using the same techniques as regular AC-powered well pumps. (Dankoff Solar)

For off-grid applications, efficiency is important. However, many homeowners (including the author) use standard AC submersible pumps. Although the efficiency is lower and the power requirements higher, they are less expensive and it is easier to find service parts in the middle of the night. Couple this with low-water-usage appliances and large storage/accumulator tanks and the resulting pump usage is reduced to 10 minutes per day.

Slow-Pumping Techniques

The deep well submersible pump is the brute force method of pumping water for a home pressure system. The "do more with less" method involves a different approach that is not nearly as popular as deep well pumping but works just as well.

A slow pump is simply an ultra-high efficiency, low-Voltage pump that has low-flow characteristics compared to a standard deep well submersible pump. This lower-flow characteristic requires a means of buffering water consumption, usually by the addition of a storage tank. Figure 2-55 shows a typical

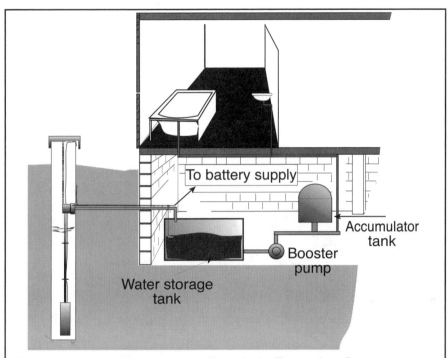

Figure 2-55. A slow-pumping system relies on high efficiency, low–flow pumping into a storage tank to increase efficiency. This design will reduce water pumping energy requirements by 50%. (Dankoff Solar)

slow-pumping submersible pump supplying water to a storage tank. A float located in the storage tank turns the submersible pump off when the tank is full and back on when the level drops.

A second pressure-booster pump is installed between the water storage tank and the household supply. This pump pressurizes an accumulator (see below) and turns off when the water pressure reaches 40 to 50 pounds per square inch (psi). This is the

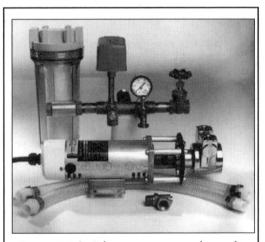

Figure 2-56. A booster pump can be used to pressurize water from a storage tank or from a lake or stream, provided freezing is not an issue.

normal maximum pressure for domestic water systems. The switch is also designed to turn the pump back on when the pressure drops below the 20 psi cut-in pressure.

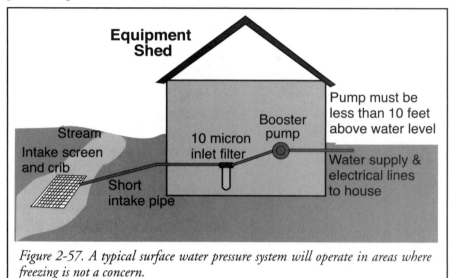

Figure 2-57. A typical surface water pressure system will operate in areas where freezing is not a concern.

Surface Water Systems

Surface water systems do not require a submersible pump unless the installation is subject to freezing. For temperate or seasonal use the booster pump can supply the water and pressurization at the same time. Placement of the pump becomes of primary importance, as pumps do not "suck" water, they push it. A typical installation is shown in Figure 2-57. An intake pipe is provided with a screened inlet (to keep the fish out) and supported inside a simple crib of rock or wood. The intake is then routed to a 10 micron sediment filter to stop grit and sand, but not bacteria, from entering the pump.

Figure 2-58. An intake sediment filter is required for all slow and booster pump applications. This filter needs to be changed, as sediment has accumulated in the filter housing.

The pump inlet must be positioned not higher than 10 feet above the water line to ensure proper operation. The booster pump then provides pressurized water to the accumulator (see below). A built-in pressure switch turns the pump on and off at the desired operating pressures in the same manner as the slow-pumping method described above.

Installations such as the one outlined in Figure 2-57 can vary greatly. One area to be aware of is the vertical height or "head" distance that water must be pushed uphill. Assuming that your cottage or home is at approximately the same level as the pump, there is no problem. For houses with solar panels that are located above the pump, water pressure cut-in and cut-out settings must be increased by 1 psi for every 29 inches (74 cm) in vertical lift, assuming the pressure gage is located at the pump.

This additional pressure is due to the weight of water. For example, if the house is located 13 feet above the level of the pump, the pressure switch setting will have to change by:

(13 feet above pump x 12 inches per foot) ÷ 29 inches per psi ≈ 5.4 psi increase

If the pressure switch is located in the house this change is not required, although you will have to be sure the pump can work with the increased pressure requirement. In addition, be aware of the maximum vertical height a pump can push water and still supply adequate water pressure. It's fine for the manufacturer to say that the pump delivers 50 psi, but if the cottage is 120 feet above the lake surface, the pressure at the top won't be enough to water your pet goldfish.

(120 feet above pump x 12 inches per foot) ÷ 29 inches per psi ≈ 49.7 psi used

Pressure Accumulation

Anyone who has been to a rural home or a cottage with an improperly installed water pressure system knows the rub. The shower pressure varies from a trickle to your being blasted against the wall. Getting a drink of water sometimes means waiting two minutes to fill the glass while the next time the whole kitchen gets flooded from the spray. During this time the well pump is kicking on and off like a wild mustang being broken.

An improvement on this comical water supply system is to even out the flow and pressure by installing a proper water accumulator tank. When your plumber offers you a coke-bottle sized accumulator, take the refund and purchase one that works. The accumulator shown in Figure 2-59 is enormous compared to the laughable, tiny units normally supplied. This model (the water accumulator, not Lorraine) holds approximately 40 gallons (152 liters) of water and will provide an even flow throughout the longest shower (teenagers notwithstanding).

In addition to smoothing out the water supply, the large tank will greatly reduce the pump cycling by storing most of your daily water usage. After applying all of the water conservation rules outlined below, Lorraine and I often manage with the pump only cycling once per day.   A further advantage of the large accumulator is water temperature buffering. Well water is very cold, often entering the house at 48°F (9°C). This water has to be heated, often with propane. The accumulator acts

Figure 2-59. A large water accumulator will provide city-quality water pressure and flow.

as a large heat sink, absorbing room air heat. (Remember from Energy 101 that heat likes to go to cold things.) This free heating will warm the water to near room temperature, saving more money by lowering the amount of fuel needed by the water heater.

Water Supply Summary

There are dozens of methods available for installing a domestic water supply system, many of which require specialized design support. If simplicity and ease of service are primary concerns, then stick to standard deep well submersible AC pumps with large accumulator tanks. This is pretty standard stuff that any plumber can understand, and when used with proper water conservation techniques this method is efficient enough for all but the smallest of renewable energy off-grid systems.

How to Conserve

Now that we have a supply of water, let's see how to make the most of it. Referring back to our pie chart in figure 2-51 will show us what the highest consumers of domestic water are.

Toilet Flushing

Conserving water when flushing a toilet is not only simple, it is actually the law in many areas wracked with inadequate water supplies. Simply remove the old clunker and replace it with a certified low-flush model (Figure 2-60). This conversion will reduce water usage by 66% or more. Conventional toilets can consume over 4.7 gallons (18 liters) per flush while low-flush models use only 1.6 gallons (6 liters).

Low-flush toilet models are available in a mind-numbing array of styles, colors, and shapes to suit even the most discriminat-

Figure 2-60 Low-flush toilets require only 33% of the water required by earlier models and come in a variety of styles to suit even the most discerning derriere.

ing derriere. Prices keep coming down, with economy models available in the $50 range. Another good reason to consider these models is the lack of condensation dripping from the tank. If your old clunker drips every time the humidity rises, that's the excuse you need to make the swap.

No matter which brand of toilet you own, if it leaks it's a water waster. A toilet that leaks four gallons of water per hour (fifteen liters) would fill a large swimming pool if allowed to continue for a year. Not sure if the toilet leaks? Place a couple of drops of food coloring in the water reservoir. Check the bowl after fifteen or twenty minutes and see if the color shows up. If it does, the next book you need is *An Introduction to Plumbing* to fix it.

For those of you considering an alternative to the pressure-water toilet and septic system, refer to Appendix 12.

Bathtubs

If you and your family prefer baths rather than showers, there is not much you can do except change an older bathtub to a new fiberglass model. Old cast iron or metal tubs that are six feet long and very deep require a lot of water and a large amount of fuel to heat it.

A fiberglass soaker bath can be ordered in a 5-foot model that is fine as long as you aren't related to Magic Johnson. A shorter bath requires less water volume and the fiberglass model resists heat loss through the sides.

If you are having a bath during the heating season, leave the water in the tub until it cools down, and then pull the plug. This little trick will transfer the heat energy from the water into the room, providing you with free warmth and humidity to boot.

Showers

Aside from turning off the water when the teenagers decide to live in the shower, the best solution is to add a low-flow shower-head. These units, such as the one shown in Figure 2-61, have built-in water-flow restrictions and special "needle orifices" to reduce flow without making you feel as if you're just standing under a leaky roof. Look for models that carry a label from a certification agency displaying a flow rate of less than 2.6 gallons (10 liters) per minute. Many models come equipped with a built-in massage feature, providing an additional incentive for making the switch.

Figure 2-61 Showerheads such as this model reduce water consumption and hot water heating costs by 50%.

The Kitchen

A very quick upgrade of the kitchen sink may be in order. Try replacing a standard-flow faucet for one with an "aerator" nozzle, such as the one shown in Figure 2-50. These devices mix air with the water stream, giving the appearance and feel of more water flow. This apparent increase in water flow offers more coverage when washing dishes and hands.

Dishwashers

Like every other major appliance in the house, dishwashers are subject to the government's EnerGuide and EnergyGuide programs. When purchasing a new model or updating a tired machine, review the EnerGuide or EnergyGuide label and purchase the most efficient model available. Better yet, try washing your dishes by hand. The energy savings provided by "manual" dishwashers is quite high, and you might find this a good time to catch up on family gossip.

Clothes Washers

Washing machines are in the same boat as dishwashers. Check the EnerGuide or EnergyGuide labels and purchase the most efficient machine you can afford. The washing machine shown in Figure 2-40 consumes five-and-a-half times less energy than a similar new machine. Shop carefully and save big!

Water Heaters

Besides changing your water heater completely, there are a few things that can be done to help conserve energy:

- Turn down the thermostat of the water heater to the lowest temperature you can accept. The thermostat on most gas units is visible and well marked. Electric units often have two thermostats underneath a removable cover. When adjusting them, ensure that the power to the unit is turned off. If you are unsure, contact your plumber for assistance.

- Install an insulation blanket as shown in Figure 2-62. These blankets are available from most building supply stores or plumbing contractors. The insulation batting helps to stop heat transfer from the hot water inside the unit. If you create a homemade blanket, make sure it is fireproof and that it does not block any air intakes or vents on a gas water heater.

- Flush the water heater at least twice per year. Built-up sediment in the bottom of the tank reduces heater efficiency. Draining a few gallons from the bottom of the tank will reduce calcium and mineral buildup.

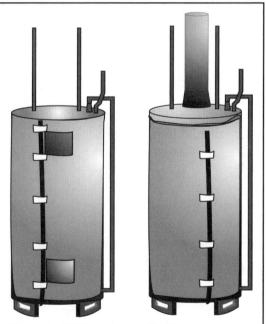

Figure 2-62 Water heater blankets save a great deal of the heat that is typically lost from hot water storage tanks. Make sure the blanket does not block any air intakes or vents on gas water heaters.

- Install pipe insulation on all hot water lines. This material is inexpensive and very easy to install (provided that the pipes are exposed).
- Add an active solar heating system that captures the sun's rays to warm the water in your tank. Solar thermal systems can easily provide 50% of your annual hot water heating requirement and provide financial returns on investment of 10% or better. Chapter 5-3 covers this equipment in detail.

Water heaters have an average life span of ten years. Before you discover a new indoor swimming pool courtesy of your leaking unit, install an upgraded model. A water heater that is reaching the end of its useful life is a bit like a refrigerator: the savings resulting from the energy efficiency of newer models will easily offset the installation cost.

New to North America are the instantaneous water heaters such as the Bosch model shown in Figure 2-63. These units have long been used in Europe where energy costs prohibit wastefulness. Your plumber can easily replace your old storage unit and give you back a bit of floor space at the same time. Do not be misled by the old wives' tale that on-demand water heaters cannot supply sufficient hot water for the entire house. This is a fallacy of days gone by which is propagated by ignorance of the technology.

Figure 2-63 Instant water heaters such as this model from Bosch do not waste energy by storing hot water all day long. They work by rapidly heating the incoming cold water, supplying enough hot water for the entire house. Do not be misled by the old wives' tale that on-demand water heaters cannot supply sufficient hot water for the entire house. This is a fallacy of days gone by which is propagated by ignorance of the technology.

Solar Hot Water Heating

Solar water heating systems are a natural extension of efficient water heating systems and are discussed in Chapter 5-3.

Landscaping
Lawns

Possibly the greatest waste of water (and energy) is using it to maintain the perfect lawn. If there was ever a sinkhole for energy, this is it. We start by trucking in tons of topsoil and adding fertilizers, pesticides, and lime to get it growing. Then we use more fuel to cut, trim, and edge our lawns, all the while wasting energy. Added to this is our need to use purified drinking water when rain would almost certainly suffice. Why water in the middle of a hot summer day and have 50% of the water evaporate or run along the sidewalk? If we must have perfect lawns, then we should try to water them more efficiently.

Figure 2-64 An automatic sprinkler system supplied by nature in the form of rain will save a considerable amount of purified drinking water that most people use to water their lawns. Try switching to drought-resistant white clover which has the added benefit of slow growth, reducing the amount of cutting required.

Watering at dusk or early in the morning is best. Apply only as much water as your local conditions require. Check with a nursery or garden center to find out how much water is appropriate. Over-spraying the lawn and watering the sidewalk is just plain wasteful. Spend the $20 on a better sprinkler unit and maybe a few dollars more on a timer. Program the sprinkler to suit the size and area of your lawn. If you are planting a new lawn, consider low-maintenance

Figure 2-65 Clover (left) vs regular lawn grass (right). This illustrates why white clover is a better choice for ground cover; it is able to withstand severe drought conditions as its root system penetrates the ground to a considerable depth providing water and nourishment.

ground covers or drought-resistant white clover grasses instead.

Are you in the market for a lawnmower? Consider a rechargeable electric model: virtually no noise, minimal maintenance compared to gas models, and no pollution. It is also a negligible burden on your electrical system, as the battery charging and usage occur coincidentally with peak summer sun hours. You might also wish to consider a modern reel mower. It is easy to use and you get to burn a few calories at the same time.

Flower Gardens

The roof of your house captures a lot of rain during a downpour. Try containing some of this water in a series of rain barrels that are interconnected to supply flowerbeds. A few 50-gallon (200-liter) barrels interconnected with poly-pipe as shown in Figure 2-66 makes a nostalgic-looking storage system. Connect the rain barrels with weeping hose and you have an energy-efficient and labor-saving watering can. Remember that mulch doesn't just look good and suppress weeds; it also keeps the soil from drying out, further conserving water and energy—and your time.

Although garden lighting does not help with water conservation, it helps to showcase your pretty flowers and hard work. This lighting will better harmonize with nature if it is solar powered. The price of solar-powered lights has been dropping while their quality has steadily increased. The model shown in Figure 2-67 was purchased for under $5.00 and will continue to run long after you should be in bed.

Figure 2-66
Capturing rainwater is a time-honored tradition. Automate the process by connecting the barrels to a weeping hose for an automatic and labor-saving watering system.

Water Conservation Summary

The tasks described above will easily reduce your water consumption by more than 50%. Add to this the decreased electrical energy required to pump the water, reduced water heating costs, decreased soap consumption, and diminished load on the municipal supply and sewage system, and you simply cannot justify putting off these upgrades.

Figure 2-67 Solar-power garden lights won't help with water conservation, but they certainly accentuate the beauty of the flowers.

Figure 2-68. Although the urban lawn is ubiquitous, it consumes an enormous amount of energy, water and your time trying while trying to make it look better than your neighbours. Xeriscaping (reducing resource requirements in landscaping) offers a beautiful and low-maintenance alternative as shown above.

Summary

While some of the items listed in this chapter may seem excessive (some would say neurotic) you cannot deny the simple fact that every step taken to reduce energy consumption will **greatly** reduce the cost of capital installing your off-grid system.

You do not have to reduce energy consumption below the level required to maintain your desired lifestyle. Just bear in mind the wise words of the U.S. Department of Energy: for every dollar you invest in efficiency, you will save three to five dollars in the cost of energy-generation equipment.

Before we start exploring the nuts, bolts, and minute details of off-grid equipment, we will digress for the next two chapters to understand how energy efficiency and off-grid generation interact. In Chapter 3, we will take a quick peek at how things work and interconnect in "A Primer for Off-Grid Living." Chapter 4 will continue the theme by offering a tour of "A Showcase of Selected Off-Grid Homes."

3
A Renewable Energy Primer

I remember the day I popped the "big question" to Lorraine. The anticipation and excitement were reaching their peak; she was clearly caught off guard. After a moment's pause, the reply: "But can I still use my hairdryer?"

That was her only concern. The rest, I was sure, would be easy....

Eleven years have passed and yes, Lorraine can still use her hairdryer. Living off the grid is both comfortable and satisfying. Our motivation for leaving "life on the grid" was simple. Lorraine wanted to move closer to her family and still have the room and privacy to support her "addiction" to animals. The lot at the back of the family farm fit the bill (and the wallet). There was only one downside. It was about $13,000 from the nearest hydro lines.

My work as an electrical/electronics designer made me think. Why not try to make our own juice? Surely I could whip up something for around $13,000. In hindsight, that was naïve, but ultimately our system has grown to the point where it supports our desired lifestyle and is far more reliable than the power utility. (Our neighbours suffered through a 16-day outage during the ice storm of 1998 as well as enduring the blackout of 2003 and numerous other "miscellaneous" outages.)

What is the trick? A willingness to give it a try, a sense of adventure, sure, but most of all it's about energy conversation. The previous chapters have been drilling this into your mind. Just in case you skipped all of that, I will take a moment to reiterate its importance once more.

Energy conservation on or off the electrical grid is not Spartan living. A big screen TV, computers, a treadmill, and a cappuccino maker are examples of devices that can be found in any home. Add in the lights and stereo in the horse stable, a hot tub on the deck, and a garage with electric door openers and you might think that this house is a large electrical consumer. In fact, the opposite is true. We operate our house on between 3 and 6 kWh per day, depending on the season. Contrast this with other North American homes that use 6 to 10 times this energy intensity.

An energy-efficient house has much of the same "stuff" as a regular house, but it can use 10 times less electricity than the average home. Before you say, "Yes, but all the expensive things to operate are on propane," remember that most people use natural gas for the majority of their heavy heating loads as well.

Off-Grid System Overview

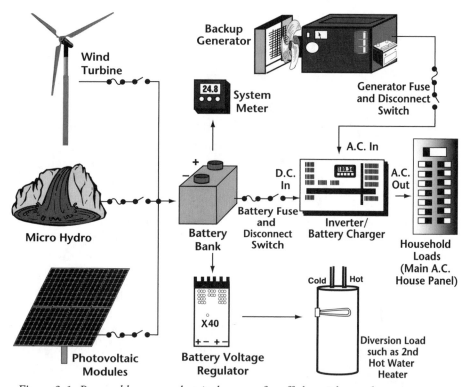

Figure 3-1. Renewable-energy electrical systems for off-the-grid use rely on one or more renewable energy sources charging a storage battery bank. An inverter converts the energy stored in the battery to regular household power to operate your appliances. On-the-grid systems are similar, except that battery storage is optional and the house remains connected to the grid.

This is not to say that an off-the-grid home is automatically the most energy efficient or environmentally friendly. It is important to remember that all forms of energy must be used wisely to make economic and environmental sense.

The Fun Stuff – Making Electricity

We now know how to stretch our electrical energy dollar further and live within the means of an off-the-grid system. Let's see how to make the energy that we need to run a typical household. It should be obvious that if we consume 3 to 6 kWh per day, we need to produce about that much too.

All off-the-grid systems work in much the same manner: collect energy from a renewable source (wind, sun, stream), convert it to electricity, and store the energy in a battery bank. When you need this energy to power your appliances, take some of the energy from the battery, convert it to alternating current (just like the power utility), and feed it to your unsuspecting appliances. Most off-grid systems also contain a backup power source, usually in the form of a propane, gasoline, or diesel generator (genset for short). These units supply power to charge the battery bank during dark, rainy periods when the renewable system is not able to support the necessary electrical loads.

Figure 3-2. Unlike fossil fuels or nuclear energy, renewable energy sources are just that: renewable. No matter how much sunlight we capture with our photovoltaic panels, there will always be more for everyone else, and we are not dumping today's pollution on tomorrow's children. This photo shows a 1200 Watt peak output PV array on a sun tracker mount.

On-the-grid or grid-interconnected systems are very similar to off-the-grid designs. The renewable energy is fed directly to the inverter and converted to alternating current, which in turn supplies the grid. If you produce more energy than you require, the surplus is sold to the grid so that your neighbor down the street can also use some. Selling energy causes your electrical meter to run "backwards," providing you with a credit in your electrical energy "bank account." When you require more energy than you are producing, the electrical grid supplies you with energy, causing your meter to debit your account in the normal manner. Every once in a while your utility will send you a statement indicating whether you owe money or the present balance in your energy account is positive. The selling of electrical energy back and forth is called net metering. It is the law in many North American jurisdictions, requiring the utility to purchase your excess energy at the same retail price you pay for theirs. This really is a great deal for people who are energy efficient; that is, people who generate more electricity than they consume.

Figure 3-3. With advances in materials technology, you will find wind power used in every location on earth. The Bergey 1500 small wind turbine is specifically designed for home-size off-grid and grid-interconnected applications.

Some folks want the ability to live off the grid but are nervous about cutting the lines to their electrical utility. Connecting an off-the-grid system to the electrical grid is known as grid- interactive mode and allows you the best of both options. During periods of excess power production, the home is fully powered by the renewable sources and excess power is sold to the grid, supplying neighbouring homes However, should the grid fail, the system will revert to full off-grid operation, allowing you to operate electrical loads connected to a separate "essential or dedicated off-grid load" panel. (Most grid-connected homes are not sufficiently energy efficient to be fully powered by an economical grid interactive system, necessitating a separate electrical panel connected to essential loads).

Grid-*dependent* systems do not require a battery bank and voltage regulation equipment or backup generation. This lowers the installed cost of the equipment compared to that of off-the-grid systems. On the other hand, if you have no batteries or backup generation source, your home will be just as dark as your neighbours' home during the next electrical blackout. The irony of this happening might be too much to bear!

In theory, it seems simple, but just like everything else in life, the devil is in the details. Look at the overview shown in Figure 3-1 and let's follow the system through its operation.

All of the earth's energy comes from the sun. In the case of renewable electrical sources, the link is very clear: the sun's rays strike the photovoltaic panel, creating electricity. The sun's energy causes the winds to blow, moving the blades of a wind turbine and causing the generator shaft to spin and produce electricity. Heat from the sun causes water to evaporate, creating

Figure 3-4. An off-grid system requires a means of storing the electrical energy generated by the renewable sources for later consumption. A sizable battery bank provides the needed storage.

the rains that fall in the mountains. The rain becomes a stream that runs downhill into a micro-hydroelectric generator.

As well as being renewable, these energy sources are also variable or intermittent. In order to ensure that electricity is available when we need it, a series of wire cables, fuses, and disconnect switches delivers the energy to a battery storage bank.

Although there are many different types of storage batteries in use, the most common by far is the deep-cycle, lead-acid battery. You may be familiar with smaller ones used in golf carts or warehouse forklift trucks. Batteries allow you to store energy when there is a surplus and hand it back out when you're a bit short.

So why are we using a battery bank? What other means do we have of storing electricity? Great questions, simple answer: there is no other feasible method of storing electrical energy. Maybe down the road, but if you want an off-grid system now, batteries are the only way to go. Today's industrial deep-cycle batteries are a solid investment that should last 20 years with a minimum amount of care. At the end of their lives, the old batteries are recycled (giving you back a portion of their value) and new ones are installed.

Storing electrical energy is simple: just connect the renewable energy source to the battery and it immediately starts charging. Getting the energy back out is a bit more complex. First, electricity is stored in a battery at a low voltage or "pressure." You will probably know that most of your household appliances use 120 Volts, whereas off-grid batteries commonly store electricity at 12, 24, or 48 Volts. Second, the electricity stored in a battery is in direct current (D.C.) form. This means that electricity flows "directly" from one terminal of the battery to the other. Direct current loads and batteries are easily identified by a red "+" and a black "-" symbol marked near the electrical terminals.

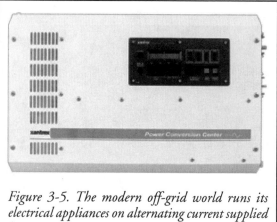

Figure 3-5. The modern off-grid world runs its electrical appliances on alternating current supplied by an inverter such as this. (Xantrex)

The electricity supplied by the utility to your home is "alternating current" (A.C.). This means the flow on the supply wires changes direction at a rate of 60 cycles per second or 60 Hertz. (Many of the terms used in electricity are named after the early inventors who discovered the physics, for example James Watt, Count Volta, and Heinrich Hertz.)

In order to increase the Voltage (pressure) of the electricity stored in the batteries and convert it from D.C. to A.C., a device known as an inverter is used. Without an inverter, your choices in electrical appliances and lighting would be reduced to whatever 12 Volt appliances you could find at the local RV store. Early off-gridders did in fact choose this path, but you shouldn't consider it except for the smallest of summer cabins or camps. A house full of middle-class dreams means a house full of 120 Volt, A.C. appliances.

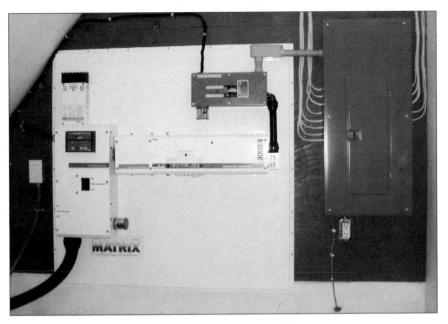

Figure 3-6. A renewable-energy system should be neat, simple, and well laid out, ensuring smooth sailing with electrical inspectors and your insurance salesman.

Standard electrical power also means standard wiring, standard electricians, and happy electrical inspectors who enforce safety standards.

That is the basic system. A supply of electrical energy from wind, water, or sunlight feeds low voltage electricity into a battery bank. The batteries store the electrical energy within the chemistry of the battery "cells." When an electrical load requires energy to operate, current flows from the battery and/or the renewable energy source at low voltage to power the inverter. The inverter transforms the direct current (D.C.) low voltage to higher 120 Volts alternating current (A.C.) to feed the house electrical panel and the waiting appliances.

The Changing Seasons

As a bad July sunburn will remind you, the amount of sunlight in summer is much greater than in winter. Simply put, the longer the sun's rays hit a PV panel, the more electricity the panel will push into the battery. The months of November and December tend to be dark and dreary by contrast. How does this affect the system and will there be enough energy in the winter?

Solar Insolation Map
Average Hours of Sun for the Worst Month Yearly

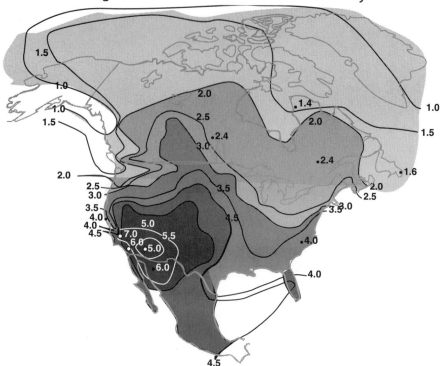

Figure 3-7. Sunlight hours per day are subject to extreme seasonal variability in most of North America, as a comparison of this map shows.

Seasonal variability is extreme in the northeastern section of North America. The two maps shown in appendices 5 and 6 show the average amount of sun hours in September and December across North America. These months are only separated by 91 days, yet the amount of sunlight in December is approximately half what it is in September and even less than half the amount we get in June. Obviously, the PV panels' output will reduce accordingly, and the amount of stored electricity will vary with it. This creates an odd paradox. There is too much electricity in summer and not enough in November and December. How do we design around this problem?

Hybrids (Winter Season)
Hybrid design simply means adding more than one source into our energy mix. In the example in Figure 3-1, we have PV, wind and micro hydro, plus a backup generator. This design is not typical, as most off-grid systems typi-

cally start with PV as the main renewable source, a backup generator second, and possibly a wind turbine third. For those of you lucky enough to have a year-round stream sufficient to operate a micro hydro system, that may be the only energy source you will require.

Grid-interconnected systems are typically PV based. Wind- and water-based sources are usually not connected owing to the rural nature of these energy sources.

Back to Watts and nuts and bolts for a second. Remember that we talked about consuming 3 to 6 kilowatt-hours of electricity per day. Now we have to look at what we produce, to see how well they match. Our PV panels' rating is 1200 Watts peak power output (28 Volts x 43 Amps D.C.). In reality, they tend to output approximately 950 Watts under ideal conditions, less if it is hazy, and nearly zero if the day is cloudy. In our home, the entire solar collector assembly is mounted on a sun tracker unit, which allows the panels to face the sun as it moves from early morning through late afternoon, winter and summer.

Referring to the worst month sun hours map in Appendix 5/Figure 3-7 for our location, we find the average sun hours to be 2.2 per day:

2.2 sun hours per day x 950 Watts output = 2,090 Watt-hours per day or approximately 2 kWh per day

With 2 kWh of production and an average consumption of 4 kWh per day, the system will lose 2,000 Watt-hours per day. If this were your bank account and you kept taking out more money than you put in, guess what would happen? Depending on how deep your pockets are, you'd run out of cash. The off-grid system batteries are no different. In fact, typical battery sizing assumes that you should be able to run your house "normally" for 3 to 5 days without having any input from your renewable energy sources. For our household, running average loads means the batteries need to supply:

5 days' supply x 4 kilowatt-hours per day = 20 kilowatt-hours' usable capacity

So, what happens at the end of four days? This is where the hybrid design comes in. You either have another renewable source pick up some of the load or rely on your backup genset. A wind turbine or micro hydro generator can operate in parallel, increasing total power generation.

If the secondary power source is a fossil-fuel generator, it will have to make up the deficit. Depending on the degree of automation in your system, you either manually start the backup generator (gas, diesel, or propane) or a generator control device starts the generator for you. In either case, the inverter now switches to battery-charging mode and fills the batteries back up. The house electrical loads automatically receive power from the generator during this charging time.

Once the batteries have reached full charge, the generator turns off automatically or you run out in your housecoat and slippers to shut it down. (I think the automatic feature is definitely worth the few extra bucks!)

If your system contains more than one renewable source, you will find that they tend to be complementary. A dull day in November often has brisk winds and conversely the air on a sunny summer day is hot, still, and stifling. However, do

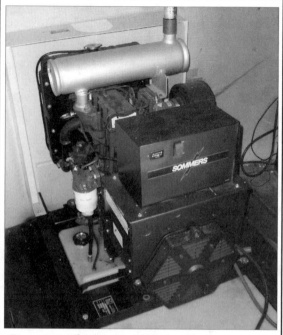

Figure 3-8. This large 7.5 kW diesel backup generator set (or genset) will last several lifetimes and be economical to operate as well.

not believe that having PV and wind will eliminate a backup generator; it will not. The combination will reduce the running time of the generator considerably, but it will not eliminate it. Our house still requires over 100 hours of generator time per year.

Hybrids (Summer Season)

During the summer months, the increase in sunlight hours, coupled with a lower need for lighting and less time spent indoors, creates a surplus of energy based on consumption levels of 3.5 kilowatt-hours per day.

Production:

6.0 sun hours per day x 950 Watts output = 5,700 Watt-hours per day or approximately 6 kWh per day

Surplus:

6 kWh/day produced – 3.5 kWh/day used = 2.5 kWh/day surplus

We may or may not need this surplus, depending on whether or not we require any air conditioning that day. As I mentioned earlier, an air conditioning unit uses an enormous amount of energy, approximately 1,100 Watts per hour operated. Based on a surplus of 2.5kWh/day, we should be able to operate the air conditioner for up to 2.5 hours per day without dipping into the energy bank or activating the genset.

On days when we do not need air conditioning, the surplus energy must go somewhere. You must consume this energy or the batteries will reach a fully charged state and eventually "overcharge." To prevent this from happening, a battery voltage regulator either "disconnects" the renewable source or connects it to a diversion load. A typical diversion load consists of an electric water heater supplied with low-voltage heating elements. The water heater is plumbed "before" the regular gas water heater.

During operation, the battery voltage regulator monitors the battery voltage or *state of charge*. When the batteries become full, it starts to divert surplus electricity to the electric water heater. The water starts to heat as it absorbs the extra energy produced by the renewable energy sources. Over the course of a day or two, the water can easily reach 140 degrees F. (60 C.) and then flows into the cold side of a regular gas water heater. As the incoming water is already hot, the gas heater remains on standby, thus conserving propane gas, energy dollars, and the environment. Energy is never wasted in this system!

Phantom Loads

As the name implies, phantom loads are any electrical loads that are not doing immediate work for you. This includes items such as doorbells (did you

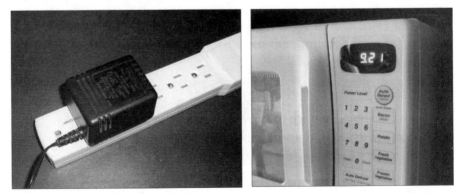

Figure 3-9. Phantom loads are unnoticed and often useless electrical loads that burn holes in your energy conservation plan. Examples are doorbells, "instant on" electronics, microwave ovens, or anything plugged into a "power adapter" like the one shown above.

know that your doorbell is always turned on, waiting for someone to push the button!), "instant on" televisions with remote controls, clock radios, and power adapters.

So what is the big deal? First, these devices are consuming energy without doing anything for you. That electric toothbrush was probably charged about 15 minutes after you put it back in the holder. During all the in-between hours when the device sits idle it is wasting energy. A television set that uses a remote control is actually "mostly on" all the time, just waiting to receive the "on" command from your remote.

While the total dollar cost for these luxuries is small on-grid (in the order of $5 to $10 per month for an average house), this consumption off-grid is unacceptable. By the way, this "small" bit of waste equals a few hundred million dollars a month in North America alone.

Another reason for the concern in off-grid installations is the inverter unit (described later). This device will go into a sleep mode when the last light is turned off at night, conserving a fair amount more energy. Any phantom load will keep the inverter "awake," consuming more energy than it otherwise would.

Phantom Load Management

Some phantom loads can simply be eliminated. Try a doorknocker instead of a doorbell and a battery-powered digital clock instead of the plug-in model. I have even heard that some people use manual toothbrushes.

Television sets and CD, DVD, and VHS players with instant on and remote control functions can be wired to outlets that can be switched off. This could mean having the electrician wire in the switch for you or using a power bar with an integral switch.

Many people cannot live without some phantom loads like fax machines or chargers for cordless phones, cell phones, PDAs and laptops. For these items, use

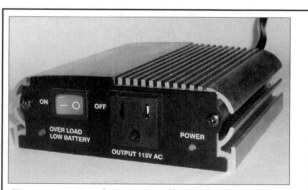

Figure 3-10. For those very small, "always on" loads, wire a small inverter such as this one to special outlets strategically placed throughout the house.

a separate wiring circuit which connects to specially marked outlets reserved for ESSENTIAL "always on" loads like the ones described above.

The power for these outlets comes from an inexpensive, 100-Watt inverter that is wired directly to the batteries. Such an arrangement ensures that the main inverter can go to sleep at night and still provide power for the small devices you want. If your home is already built, it may be possible to group all of these special loads at one central location and run the inverter to a power bar at that point.

Take note about this "small" inverter powering phantom loads. If you load it up with all of your toys, it is possible to burn up nearly 100 Watts of power. Over the course of 24 hours, this is a lot of juice:

100 Watts x 24 hours per day = 2,400 Watt-hours or 2.4kWh

A load of 2.4 kWh is almost half of the daily total energy production of most renewable energy systems. Tread lightly and limit the use of this power to truly essential items!

Metering and Such

At this point you are probably wondering if I run around the house with a note pad and calculator, chastising Lorraine for using her hair dryer too long or making the toast too dark, while recording every volt and who knows watt. Actually, we hardly look at the system at all. Once you install your equipment and load the house with all the required electrical goodies, within your average production limits, the system will almost take care of itself. If we use more energy than we produce, the generator may run for a while. If we make more energy than we use, the next shower is free because of the savings in hot water. The system is almost invisible.

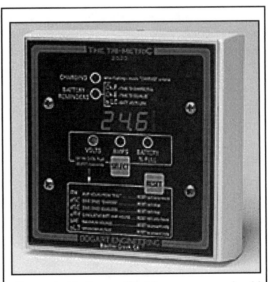

Figure 3-11. Every off-grid power system should have metering capabilities. A proper meter will lead to increased load control knowledge and system reliability. (Courtesy Bogart Engineering)

For grid-interconnected designs, the system *is* invisible. The only notification you get that all is well is a statement from the utility advising you that you have a credit of $500 on your electrical bill. (Of course some people just have to keep running outside to watch the electrical meter running backwards.)

As with any piece of complex machinery, a bit of care and management is required. I would include a multi-function meter in the mix. This unit will monitor the energy produced and consumed and calculate all the nasty mathematics necessary to tell you how much juice is really in the battery. Monitoring the meter and other system parameters will provide a comprehensive snapshot of the health of your own power station.

Is It Economical?

How far the installation is from the electrical grid determines the break-even point for most off-grid homeowners. If your hydro utility is more than ½ mile (0.8 km) from your house, it may pay for itself from the second you turn on the first light. And you can add to this the benefits of no hydro bill, minimal environmental pollution, and the feeling of self-sufficiency you get the next time your neighbor's lights go out during that dark and stormy night

If you are planning on constructing a new home and the building lot is not yet purchased you may have another savings angle. Property values are determined by location, location, and location. If that location happens to be in a beautiful lake or mountain setting with the grid two miles away, you can be sure the land will not be as desirable as it might otherwise be. Use this to your advantage. Many people have never heard of an off-grid system or are unwilling to use one. Construction costs of a lengthy grid extension drive property prices down, perhaps below what the renewable energy system will cost to install.

Chapter 4 will showcase a number of renewable energy installations and outline the costs for each system. By analyzing each of these lifestyles and comparing it to your own, you will be able to determine a suitable budget that matches your personal requirements.

Figures 3-12 and 3-13. Environmental stores offer a wide array of ecologically responsible products including energy-efficiency devices. Arbour Environmental Shoppe in Ottawa, Canada, has a perfect mix of useful products to help lessen your environmental impact.

Figure 3-14. Solar-power radios, energy-auditing software, photovoltaic-powered toys, and eco-friendly household products are but a few of the resources and educational products carried by many environmental outlets. Some of these products are kits which make great gifts and can even distract kids from the television set for a few hours—an interesting way of reducing energy consumption.

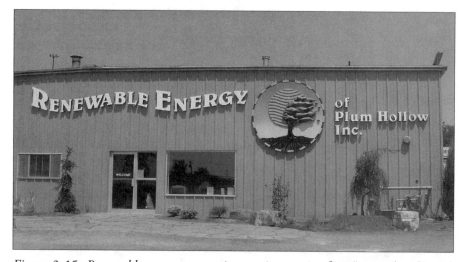

Figure 3-15. Renewable-energy generation equipment is often "co-marketed" with complementary products such as wood stoves, swimming pools, or heating and air-conditioning equipment.

Selecting a Dealer

Education and ambition are the engines that get the renewable energy wagon rolling; but you still need a dealership to sell the wagon in the first place. There are thousands of technologies designed to save or generate energy that "guarantee" they are excellent investments. Sadly, the guarantee isn't always valid. Although most dealers are honest, reputable people who are an integral part of the community and recognize that fleecing their clients is not in their long-term best interests, others aren't quite as principled. And even if the dealer is doing a respectable job, competing technologies can make the purchasing process a complex one requiring expert and unbiased advice.

When faced with the prospect of spending money, whether a few dollars or a few thousand dollars, seek the advice of a reputable dealer. Local retailers who have been in business for a number of years and provide a clean, neat storefront are likely to provide honest and unbiased service. As you would with any sizable purchase, consider obtaining references from past clients as well as visiting installations to see exactly how well the product operates in the real world.

Energy-efficiency products (such as compact fluorescent lamps and low-flow shower heads) tend to be relatively inexpensive and may be pur-

Figure 3-16. When shopping for a renewable-energy products dealer, look for a store that is well maintained and stocked. Get references from current and past customers and try to arrange a site visit with a previous client. A dealer that is hesitant to offer this service may not be one you wish to deal with.

chased through local environmental or hardware stores. Renewable-energy generation equipment is often "co-marketed" with complementary products such as wood stoves, swimming pool and spa products, or heating and air-conditioning equipment.

Just like every other retail product in North America, renewable energy systems may be purchased via the Internet from reputable dealers and used-equipment traders or through auctions on Web sites such as eBay. Unless you are an experienced home handyman (or woman) it may be wise to stick with local dealers you can trust. While it may be possible to have an inverter serviced over the phone by someone at a call center a thousand miles away, having a friendly supplier just down the street is very comforting. This is especially true if a critical piece of equipment goes "poof" and you are left in the dark.

Many dealers not only sell the equipment but offer *qualified* installation and repair service. Warming hot water with the sun sounds like child's play until you examine the technology required to harvest the sun's energy and transfer it safely and reliably into your water tank. Differential temperature controls, expansion tanks, vacuum solar collectors, and a myriad of other bits and pieces must be carefully orchestrated and installed, requiring quality workmanship to ensure a long and trouble-free operating life. Many jurisdictions require licensed contractors to perform the installation work, demanding building permits and insurance certificates prior to completion. Many dealers will not have such skilled staff in-house, but will generally have a co-operative working arrangement with appropriate contractors.

Dealers are more than warehouses for the products they sell. They are experienced, professional, and educated people who want to be sure that you are satisfied with their products. While you may be able to purchase goods at a lower price, quality dealers are worth their weight in gold for the installation, operation, and lifelong servicing of their products.

4
A Showcase of Selected Off-Grid Homes

Introduction

Before delving into the complexities and components of renewable energy electrical and heat generating systems, a review of lifestyle and system costs would be prudent. One of the first questions people ask regarding off-grid electrical systems is, "How much does it cost?". I have heard people say that they live a completely "standard" lifestyle having spent only $5,000 on their off-grid system. But I have also seen and heard media reports that a "normal" lifestyle can be supported only by spending upwards of a quarter million dollars if you're disconnected from the ubiquitous electrical grid As with most things in life, the truth lies somewhere in the middle.

In order to break through some of the hype, this chapter will demonstrate a number of off-grid systems, ranging all the way from humble portable power units that can be taken camping to a luxuriously equipped home suitable for even the most discerning lifestyle.

In order to find the truth it is necessary to understand that the issue of system cost is directly related to needs and wants: people living in mansions pay higher energy bills than retirees living in a condo. It stands to reason that a large home with numerous occupants (especially teenagers) and high energy demands will translate directly into a more expensive system. Conversely, reduce any or all of these factors and you will save money.

Figure 4-1. The size and cost of your renewable energy system is directly related to your needs and wants. Whether you want to add some TV time to those rainy days out camping or outfit a year-round home, energy demand will dictate final costs. Bonnie Klein and her partner Dave McDonald use this PV array to charge 6 Volt golf cart batteries. An 1,800 Watt inverter provides power to a stereo, laptop computer, microwave oven, cordless tool battery charger, and television for about the same cost as a small gas generator, and with none of the noise or pollution.

There is no correct system price, but rather a range of prices according to your needs and wants. Each of the selected dwellings in this chapter has been chosen to illustrate the type of lifestyle you can expect for a given capital expenditure. At the same time, the reader is cautioned that these are 2005 prices and will vary depending on dealer markup, the "do-it-yourself" factor, or market trends in your area. In all cases, new equipment was installed and third-party installation labor costs have been factored in.

The Portable System

Figure 4-2. Using the caloric renewable energy contained in one raisin will provide enough muscle power to shake this LED flashlight, generating electricity and providing approximately 20 minutes of light.

Portable RE systems can fit into any budget or system size. Starting with the ubiquitous solar-powered calculator at one dollar, renewable energy technologies are now available for just about any item. For fewer than ten dollars, you can now purchase flashlights (Figure 4-2) and radios that require nothing more than a bit of sunlight or human muscle power to wind or shake the device into operation.

Figure 4-3 (right). This 150 Watt power box can be used to power small home appliances during a power outage, charge cell phones or laptop computers, and keep modern grid-powered devices running during a camping trip.

Figure 4-4 (left). This 1200 Watt power eliminator is literally an off-grid system on wheels. A unit of this size can provide power at the cottage or hunting camp.

With the recent development of low-cost, high-power electronics, emergency backup power sources can be purchased at most hardware and department stores. Small units such as the 150 Watt portable power box shown in Figure 4-3 are perfectly sized to power a small radio or desk lamp, recharge cell phone or laptop batteries, and even jump- start cars in a pinch. This model is currently available for $75 or less.

The 1200 Watt power eliminator shown in Figure 4-4 is available for under $200 and is literally an off-grid system on wheels. Using its large internal battery and alternating current inverter, the unit is capable of powering small tools, circular saws, or an energy-efficient fridge during power outages. People with small cottages or hunting camps often use these devices to power a couple of compact fluorescent lamps, a small water pump, and a fridge without having to pay to update wiring and install a more expensive, permanently installed system.

Figure 4-5. Recharging the battery eliminator units shown above is easy if they are coupled to an appropriately sized photovoltaic (PV) panel. After connecting the panel to the battery inputs, place the panel in direct sun for several hours to recharge the power unit's battery bank. A voltage regulator (as discussed in Chapter 10) may be required. Consult the battery eliminator manual for correct PV panel and regulator requirements.

Photovoltaic panels such as the model shown in Figure 4-5 can be coupled to many models of battery eliminator boxes in order to recharge the internal battery. The manufacturer will recommend a specific size or power rating of PV panel to match the charging requirements of the unit. When large PV panels are connected to a battery bank, a device known as a charge controller may be required to prevent overcharging. Make sure you consult the power eliminator data manual for correct operation and connection information. Overdischarging or overloading these devices will greatly reduce the life expectancy of the unit.

Figure 4-6. If you just can't leave your cell phone or Blackberry at home when you go camping, perhaps you need to take a stress management course. Alternatively, these lightweight photovoltaic panels can be used to charge the batteries of a GPS, an MP3 player, and other small electronic devices. (Courtesy Spheral Solar)

Figure 4-7. Even if your home or cottage isn't off-grid, sometimes it just doesn't make sense to extend electrical wires under paved parking lots or over extended distances. A completely self-contained lighting system incorporates PV panels, a battery bank, and a charge controller, often for less cost than a line extension. (Courtesy Phantom Electron Corporation, www.phantomelectron.com)

4.1 The Beevor Family Cottage
Seasonal and weekend solar systems
System Cost: Electrical   US $2,400.00
               Solar Thermal n/a

The Beevor family enjoyed "camping" in their 600 square foot (56 square meter) off-grid cottage but found the tedium and inconvenience of oil lights and lack of TV for the rainy days a bit tiresome. Paul decided to do something about it and installed a small 12 Volt direct current system in the spring of 1999.

"Even with such a small system, everything about the cottage became a pleasure," enthuses Margaret. "We have gone from prehistoric times to living the life of luxury!" With children underfoot, having a TV and VCR and doing away with the outhouse can certainly change how life at the cottage feels.

"Our current system comprises two Siemens SP65 photovoltaic cells installed with a roof mounting system," continues Paul. "The power output of the two panels is 130 Watts, which is sufficient for our small and occasional loads. We use the cottage for three seasons, so the lack of winter sun is not a concern."

The output from the PV panels is regulated using a ProStar 30 amp voltage regulator to prevent battery overcharging conditions. The energy is stored in gold cart-style, deep cycle batteries. Electrical power is routed to a 12 Volt

Figure 4.1-1. The Beevor family enjoyed "camping" in their 600 square foot (56 square meter) off-grid cottage but found the tedium and inconvenience of oil lights and lack of TV for the rainy days a bit tiresome.

fuse box which supplies the cottage with direct current power suitable for the small loads. 12 Volt recreational vehicle lights and appliances are used.

"We also wired the cottage for standard 120 Volt AC supply in case we decide to add an inverter at some future time," says Paul. "In the meantime, I plan to add a Shureflow 12 Volt direct current water pressure pump and accumulator tank to provide pressurized water for the toilet and two sinks." As with most small cottages, the Beevors use a propane cook stove and a wood stove for heating.

The Beevors' low-cost energy system proves that good things can come in small packages.

Figure 4.1-2. Paul Beevor decided to do something about the lack of electricity and installed a small 12 Volt direct current system in the spring of 1999. The family's off-grid system comprises two Siemens SP65 photovoltaic cells installed with a roof mounting system. The cottage is used for three seasons per year, so the lack of winter sun and resulting energy is not a concern.

Figure 4.1-3. "Even with such a small off-grid system, everything about the cottage became a pleasure," enthuses Margaret. "We have gone from prehistoric times to living the life of luxury!"

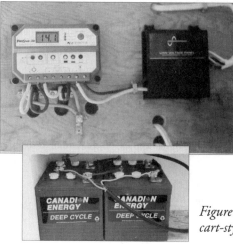

Figure 4.1-4. The output from the PV panels is regulated using a ProStar 30 amp regulator to prevent battery overcharging conditions. Voltage is fed to the cottage wiring as 12 volts direct current, necessitating the use of low-voltage RV or camping-style appliances and lights.

Figure 4.1-5 Electrical energy is stored in golf cart-style, deep cycle batteries.

4.2 Fairmont Hotels' "Kenauk Cabins"

A luxury resort company goes off-grid
System Cost: Electrical   US $7,500.00
              Solar Thermal   n/a

Anyone who is familiar with the Fairmont hotel chain (www.fairmont.com) will know that it is the largest luxury resort company in North America. One of their properties, the famous Chateau Montebello is a stunning red cedar castle which is the largest log building on the planet. Located in the town of Montebello, Quebec, Canada the resort attracts a variety of people, many of whom come to enjoy the natural beauty along the Ottawa River.

For those people who still want topnotch service but prefer to "rough it," the people at Fairmont have started to include eco-destinations in their array of product offerings. One of these is a series of off-grid eco-cabins (Figure 4.2-2) located north of Montebello that are accessible only by traveling down a 20 mile (32 km) gravel road. These cabins are located on a private lake offering the ultimate in peace and quiet that the countryside has to offer (Figure 4.1-1).

System installer Mike McGahern of Ottawa Solar explains: "The cost to run the electrical grid to these cabins would be enormous, well beyond the value of the property. Running a generator 24/7 to run a few lights and power a stereo was out of the question from an environmental standpoint.

Figure 4.2-1. Fairmont has started including eco-destinations and cabins in its array of product offerings. The cabins located on this private lake offer the ultimate in peace and quiet that the countryside has to offer.

The only realistic option was solar."

Each cabin was equipped with a non-tracking PV array (Figure 4.2-4) with a nominal rating of 660 Watts. Maximum power point trackers (MPPTs) were installed to increase the energy available to charge a small 24 Volt, 530 amp-hour battery bank. A 2500 Watt inverter provides the cabin with standard 120 Volt electricity to power the various appliances.

Propane gas does the high-energy work of space and water heating (Figure 4.2-9) as well as powering the cook stove and refrigerator, while the off-grid system takes care of the rest. "The electrical appliances in the cottage are limited to necessities, taking the cabins to just one step above camping," Mike continues. "There are approximately six compact fluorescent lamps inside and on the exterior of the building. In addition, a deep-well 120 Volt water pump supplies pressurized drinking and washing water. There is also a two-way radio that is powered by the off-grid system to call for room service should the need arise."

This is, after all, a Fairmont property.

Figure 4.2-2. Located north of Montebello, Quebec, Canada these off-grid eco-cabins are accessible only by traveling down a 20 mile (32 km) gravel road.

Figure 4.2-3. Each cabin is equipped with a non-tracking PV array mounted in a traditional pole-mount configuration.

Figure 4.2-4. The array is constructed using six 110 Watt PV panels interconnected to operate at 24 Volts DC.

Figure 4.2-5. The completed PV array provides a rated output of 660 Watts which in turn charges a 530 amp-hour battery bank. Low-voltage direct current is stepped up to 120 volts alternating current using a Magnum Energy inverter/ battery charger.

Figure 4.2-6. The inverter, MPPT voltage regulator, circuit breaker, and battery bank are located in a crawlspace area under the cabin.

Figure 4.2-7. All of the electrical components are mounted on a single "power board," which makes installation and wiring interconnection a snap for the installing electrician.

Figure 4.2-8. There is insufficient electrical energy produced by this system to operate an air conditioning unit, although in the backwoods a ceiling fan can be equally appreciated.

Figure 4.2-9. (above) During the heating season, a propane-powered direct-vent heater provides the cabin with warmth. The off-grid system powers a self-contained distribution fan housed within the heater.

Figure 4.2-10 (right). Energy-efficient compact fluorescent lamps provide all of the interior and exterior lighting for the cabins.

4.3 The Houston/Lefebvre Earthship 3

An affordable entry into solar-powered energy independence

System Cost: Electrical   US $13,250.00
                Solar Thermal   n/a

You wouldn't be inclined to think of a pile of worn-out old car tires as part of a sustainable lifestyle or particularly energy-efficient building material, but this is exactly what Hillary Houston and her partner Raymond Lefebvre used to build their earth-bermed home.

Raymond explains that old tires can be carefully filled with an earthen mixture and rammed or compressed to form a solid building wall. Stacks of tire upon tire form the walls of the house and provide a large thermal mass, which is a prerequisite for solar thermal heating.

Most tire-based, earth-bermed homes are constructed in a bungalow-like design (Figure 4.3-2) and feature large amounts of south-facing glazing. Because of the radiused shape of the tires, wall construction does not have to follow standard 90° corners. This allows interior rooms to feature a more organic, curvy look.

Raymond explains: "It took approximately 800 tires to build our 1200 square-foot home, and you can't use just any tire. We learned fairly quickly that you can't just have the junkyard drop off a load of tires and expect to use them, as uniformity is extremely important."

"Once we got the hang of selecting the proper tires for our project, we made absolutely sure that we didn't bring in any tires that were not going to end up inside a wall," interjects Hillary. "Early in the process we received 150 tires we couldn't use. It was a steep learning curve."

In northern cold-weather climates, the exterior walls are clad with a suitable insulating material and then parged with natural earth or concrete

Figure 4.3-1. You wouldn't think of a pile of worn-out old car tires as part of a sustainable lifestyle or particularly energy-efficient building material, but this is exactly what Hillary Houston and her partner Raymond Lefebvre used to build their earth-bermed home.

stucco materials (Figure 4.3-4). The tremendous mass of the interior walls (Raymond estimates them at one quarter million pounds) provides a combination heatsink and energy radiating system. During the daylight hours sunlight is admitted into the home through the south-facing glazing, brightly illuminating the interior (Figure 4.3-9) while the wall structure absorbs the heat energy, preventing the interior space from overheating. During the dark hours the opposite condition occurs and heat energy is reradiated from the walls into the living space.

As a result, the heat demand of this home is relatively low, with the majority of the energy being provided by the building's passive solar envelope and an airtight woodstove (visible at the rear in Figure 4.3-8). Strategically placed propane wall heaters (Figure 4.3-17) provide backup heat and are used to take away any chill on overcast days.

Electrical power for the house is provided by a 600 Watt photovoltaic array (Figures 4.3-2 and 4.3-13) mounted next to the solar glazing on the metal-clad roof. The 24 Volt direct current output is wired directly into a 1300 amp-hour battery bank (Figure 4.3-14) which in turn supplies a Xantrex 2500 Watt modified square wave inverter (Figure 4.3-15). The entire home is wired for 120 Volt operation, including the one-half horsepower deep-well pump. A two-stroke gasoline-powered 4000 Watt generator provides battery-charging power during the dark months of November and December.

Figure 4.3-2. The south-facing wall of Hillary and Raymond's earthship home collects both solar thermal energy through the vast glazing area and electricity from 600 Watts of photovoltaic panels.

A tour through the photo gallery makes it very clear that Hillary and Raymond are satisfied with their lifestyle and have shown that full-time off-grid living does not have to break the bank.

Figure 4.3-3 (right). The exterior walls of the earthship are parged with a colored concrete that gives the home its adobe-finished look.

Figure 4.3-4 (above). In this view, the "lumpy" surface of the tires can be seen beneath the final parged coating of the exterior wall. Large roof overhangs are used to protect the tires and finished wall surface from the ravages of excessive moisture in this location. A protective barrier for the thermal insulation can be seen at the ground wall barrier. Also note the satellite television dish, proof of a no-compromise lifestyle.

Figure 4.3-5. The north-facing wall is completely bermed by the surrounding earth, providing mechanical support for the building as well as additional thermal buffering from the elements. The roof is punctured by skylights which can be opened during the summer season (or on sunny winter days) to provide additional cooling for the house. Screen doors and windows in the east and west walls are the air intake vents.

Figure 4.3-6. Hillary and Raymond welcome you into their home on a blustery and cold late winter day. The outside temperature is approximately 26°F (-3°C), while a toasty 73°F (23°C) greeted us on the inside without the benefit of any heating source except the sun.

Figure 4.3-7. Living inside a greenhouse might seem a bit odd, but nothing beats picking fresh salad greens in the middle of winter.

Figure 4.3-8. The open-concept design and high sloping roofline make the home appear much larger than its 1200 square-foot area suggests.

Figure 4.3-9. The south-facing view forms a magnificent vista overlooking the forest and fields full of deer.

Figure 4.3-10. Raymond shows off their high-efficiency refrigerator and freezer unit. Because the couple wanted to stay within a very tight budget for their off-grid system, choosing very energy-efficient appliances became even more critical. Note the pop cans embedded in the scratch-coat parging. They take up space and reduce the quantity of finish-parging materials required.

Figure 4.3-11. An electric dishwasher was not included in the financial or electrical plans for this home. Hillary demonstrates the most efficient way of getting dishes clean

Figure 4.3-12. The peaked ceilings and large expanse of glass contained in this home provide a light and majestic feel that belies the fact that the entire south-facing wall is below grade level. Perhaps Tolkien considered this design for his subterranean hobbits.

Figure 4.3-13. A 600 Watt photovoltaic array connected in a 24 Volt series circuit provides approximately 90% of the home's yearly energy consumption. A 4000 Watt gasoline generator provides the balance of energy required during the dark days of November and December.

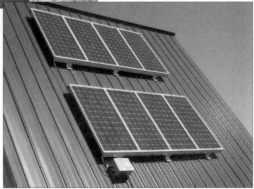

Figure 4.3-14. Electrical energy is stored in this 1300 amp-hour deep cycle battery bank. A well-sealed, ventilated walk-in closet provides a suitable storage area.

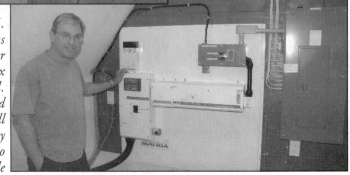

Figure 4.3-15. Raymond checks the electrical power status on his Xantrex inverter board. This prewired system contains all of the necessary components to operate a renewable energy-based home. The installing electrician simply mounts the panel to the wall and connects it to the battery bank, photovoltaic panel, backup generator, and standard house electrical circuit breaker panel shown at the right of the photograph.

Figure 4.3-16. Electrical wiring within an earthship must be run inside conduit for protection. In this view, the electrical outlet and supply are mounted to the rough-coat masonry covering the tire wall. The finish parging will be applied directly over the electrical wiring conduit.

Figure 4.3-17. Two propane heating units are installed, one at either end of the house. They are used primarily for backup heating when the couple travels or to take the chill off on those cool fall nights.

4.4   The Kemps' Lake View Stable

A mid-sized sustainable home and horse stable
System Cost: Electrical   **US $31,000.00**
                     Thermal   **US $ 4,600.00**

There is no point in arguing with a woman who has made up her mind. When the time came for Lorraine and me to talk about selling our old Victorian house we had pretty much made up our minds that the new home would be sustainably built and energy efficient and would look as if it had been built 100 years ago. That much we could agree upon. Where the difficulty lay was in choosing the building lot.

I was aiming for a nice quiet cottage-style lot on a river. Lorraine had another plan, and she dragged me kicking and screaming to a lot she already owned. Now, a normal person would have jumped at the opportunity to avoid purchasing land. My holdup was the fact that Lorraine is the "baby" of 11 in her family and the lot is on the family farm.

I have one sister. As Lorraine and I discussed matters with the building inspector I realized that this location would have me within a stone's throw of 50+ nieces and nephews and assorted in-laws who would want to visit the horses, borrow a cup of sugar, and stay for the afternoon.

A dozen years later, I have come to enjoy this sort of community living

Figure 4.4-1. Designing a traditional country-style farmhouse while integrating passive solar thermal technology can be a challenge. The large roof overhang (also visible in Figure 4.4-2) forms the front porch and provides a means of shading the house from direct sunlight in the summertime while allowing the low-angle winter sun to penetrate into the building envelope.

arrangement. This was reinforced during the ice storm of 1993 (grid power was down for 13 days) and the blackout of 2003 (we were eating ice cream at the time and didn't find out until someone phoned). On both occasions, it was nice to be able to offer a hot meal or coffee while acting as a hub in the community.

But I digress. Our 3,300 square foot home was built to look like a traditional country farmhouse (Figures 4.4-1 and 4.4-2) from the early 1900s. We had a pretty good idea of the general layout and, working with our architect, we were able to develop a home that required the least amount of building material while giving us the greatest square footage. We decided on a two-story design with a completely finished basement. This made it possible for the house to be built around a catalytic wood stove (Figure 4.4-18) which permits heat to circulate naturally throughout the home without the need for energy-robbing circulating fans.

Although many of the building techniques we used in 1993 were cutting edge at the time, several of these materials and designs have worked their way into the mainstream. For example, ultrastrong flooring trusses were used to span the 28 ft (8.5 m) width of the living room without the need for support posts or beams. Although the trusses were unusual at the time they're quite common now.

Designing a traditional country farmhouse style of home while integrating passive solar thermal technology was a challenge. The large roof overhangs visible in Figures 4.4-1 and 4.4-2 formed the front porch and provided a means of shading the house from direct sunlight in the summertime while allowing the low-angle winter sun to penetrate into the building envelope.

In keeping with our sustainability requirements, we opted for a siding material made from wood scraps which were formed into an artificial clapboard siding. The house is insulated primarily with blown-in cellulose which is manufactured from recycled waste paper products. Areas that were difficult to insulate were tackled with spray foam (icynene) insulation which also forms its own vapor barrier.

Interior wood trim and flooring were harvested from locally grown forests and processed at local sawmills. In addition, local tradesmen and subcontractors were used in an effort to keep our money in the local economy.

The domestic water system was based on a standard drilled well with a deep-well submersible pump and a large water pressure accumulator tank to minimize pump cycling. Low-flow and ultra-low flow appliances were selected to keep our water consumption well below half the national average of 91 gallons (345 liters) per person, per day. Using natural and phosphate-free cleaners and a septic tank with an effluent filter and leaching bed allows waste water to percolate through the earth and right back into the water table to become usable again.

Rain water is collected and automatically fed through drip irrigation to perennial flower beds. When grass trimming is called for, a solar-charged, battery-powered 24 Volt electric mower is used. This mower works symbiotically with the seasons: more sun in the summer permits charging without a concern for energy consumption. When we have heavy cutting work or fertilizing to do, we put a call out to the horses and they're more than happy to oblige.

Although energy efficiency and being off-grid were front and center in our list of wishes, we did not want to sacrifice our lifestyle. We carefully assessed the electrical load of all the appliances required in our day-to-day life. We then examined the equipment necessary to generate the electrical power required and prepared a financial budget.

The electrical generating equipment comprises a photovoltaic array (Figure 4.4-20) with peak electrical rating of 1200 Watts made from an array of 16 individual panels rated 75 Watts each. The manufacturer's rating

Figure 4.4-2. Lorraine and Bill Kemp along with Shadow and Cedar enjoy a pretty standard lifestyle while powering their home with renewable energy from the sun and wind.

is based on an optimum amount of light falling on the panels, but in reality the sun's energy and thus electrical output always fall below this ideal rating. On a clear, bright sunlight day our 1200 Watt array will output between 950 and 1000 Watts.

The array is mounted on a solar tracking device which follows the sun on its daily path from east to west. The tracker is also equipped with an automatic elevation device which causes the panels to change angle as a function of the seasons. During the winter months, the photovoltaic array is oriented almost perpendicular to the ground (vertical), while during the summer months the panel will lie almost flat as the sun approaches solar noon.

In addition to the photovoltaic array, we have a Bergey 1500 W wind turbine (Figure 4.4-24) mounted on a 100 ft (30 m) tall guyed, lattice tower. Completing the electrical generating mix is a 10 kW diesel generator (in the building shown in Figure 4.4-23) which is fueled with a mixture of 30% commercial-grade biodiesel. We would have preferred to use a 100% biodiesel blend, but in this area temperatures can reach as low as -40°F (-40°C), making such a choice impossible.

On a yearly basis our photovoltaic panels provide approximately 80% of our total electricity requirements, while the wind turbine provides 15% and the backup generator provides the remaining 5%. Because of local weather conditions, neither the wind turbine nor the photovoltaic panels provide significant output during the month of November and part of December. It is during this six-week period that the backup generator becomes a necessity.

The relatively low energy output of the wind turbine is due to the area having a fairly low annual wind energy rating. We knew that the wind turbine would not supply a significant percentage of our yearly energy requirements, but it does reduce the generator run time considerably. We estimate that generator operating time has been reduced by approximately 300 hours per year due to the installation of the wind turbine, which reduces fuel consumption as well as maintenance costs and wear.

Electrical energy from all of the electrical sources is fed into a battery bank (Figure 4.4-22) with a gross capacity of approximately 1,000 amp-hours, providing approximately three days of power without any input from the renewable resources. The control system is designed to automatically start the generator and recharge the batteries should they reach a predetermined low-power level. Low-voltage power from the batteries is fed to an inverter bank with a total output capacity of 4 kW, which in turn supplies household electrical needs.

Electrical or plug load energy and space heating are only part of the mix of energy that is required in a home. Most off-grid homes rely on propane gas for appliances with large heat demands such as a hot water heater, clothes dryer, cooking stove, and backup heat source.

Approximately one-third of a home's total energy requirement is used to provide domestic hot water. In our case, the requirement for hot water is increased because Lorraine and I love nothing more than to soak in the hot tub while listening to the loons calling on the lake. The most common fuel of choice (or necessity) for the off-grid home in North America is expensive propane, but not wanting the resulting greenhouse gas emissions to kill off these lovely birds, we chose to install a vacuum tube-based solar thermal collection system (Figure 4.4-25). This unit reduces our yearly hot water heating fuel consumption by approximately 60% and provides the thermal energy required to keep our well-insulated hot tub (Figure 4.4-15) nice and toasty during its three seasons of use. In most regions of North America it is also possible to reduce domestic hot water heating fuel requirements by at least 60% through the installation of a solar system.

Cooking is carried out with an old-fashioned style of propane stove (Figure 4.4-6) as well as a microwave oven and slow-cooker crock. On the advice of a friend, we have just begun to add solar cooking into the mix (Figure 4.4-26). Later this summer I plan to add a rack to the PV tracker to hold the solar cooker. The PV tracker will ensure that the solar cooker remains perfectly aligned with the sun. Although Lorraine's desire is to actually do serious cooking with the unit, I am excited about the prospect of not overheating the house and running our air conditioning any longer than necessary.

Over the years we have had many visitors, some of whom come over for the first time at night, when the wind turbine and PV panels are hidden in

Figure 4.4-3. Bill and Cedar offer a tour of their home. Anyone knowing Bill realizes that the first stop in any tour is in the kitchen for a ubiquitous cappuccino.

the darkness. The comments are always the same. "I thought you didn't have any power out here," people say, as they wonder how the automatic yard lights and electric garage door openers can possibly function. Of course I enjoy this little game, showing off a pretty standard array of appliances crammed into our typical country home. Living life off-grid is definitely not the same as living life without power.

Figure 4.4-4. Although energy efficiency is doubly important in off-grid home designs, living a Spartan lifestyle isn't necessary. Balancing energy production and demand is important in developing a successful off-grid system. And yes, there is even room for a cappuccino machine.

Figure 4.4-5 (above). The house plan calls for plenty of open space, allowing both air and light to circulate freely. An efficient floor plan reduces construction materials and cost.

Figure 4.4-6 (left). Major heat-producing appliances, such as this cook stove, are normally powered by propane gas. The small microwave oven (on top of the cook stove) is used occasionally for reheating and defrosting. A larger model could have been chosen had there been a desire to use one more regularly.

Figure 4.4-7. Baking bread in the old cook stove may be reminiscent of grandma, but with free electricity and the relative efficiency of automatic bread makers, along with the requirement for less elbow grease, Lorraine prefers the modern approach. A food processor, blender, and standard Sears refrigerator complete the mix of kitchen appliances.

Figure 4.4-8. African violets love diffuse, indirect light. Thanks to the layout of the roof overhangs, late spring, summer, and early fall sunlight never comes directly into the house. Conversely, the sun's low winter angle shines deeply into the house interior, providing welcome warmth.

Figure 4.4-10. A satellite dish is one way to stay connected in the woods, providing TV as well as high-speed Internet links to the urban world and beyond.

Figure 4.4-9. The living room is the center of the audio and visual system. A home theater system with large screen TV, CD, DVD, tape, surround sound amplifier, powered subwoofer, and satellite system works perfectly with today's sine wave inverter power.

Figure 4.4-11. Household chores need not require muscle power. A standard central vacuum system with powered "beater bar" forms part of the household mix of appliances.

Figure 4.4-12. The Staber washing machine is the most efficient model in North America, although many front-loading machines are quickly catching up.

Figure 4.4-13. Commuting 20 minutes each way to the "local" gym is no way to live sustainably. It's far better to have one in the house and invite friends to stop by. A mix of equipment such as a treadmill, elliptical trainer, stereo, and TV are all supported by the off-grid system.

Figure 4.4-14. The horses require pressurized water, supplied by the house pressure system, as well as lights and even their own stereo.

Figure 4.4-15. With Lorraine doing all that work in the last few slides, she has earned the right to relax in the hot tub, in water which is heated and circulated using the power of the sun.

Figure 4.4-16 (above). Bill's office has all of the devices one would expect. A computer, printer, fax machine, and photocopier are fairly standard fare. In addition, some not-so-normal equipment such as electronic test and measurement equipment and large-scale drafting machines all operate off-grid.

Figure 4.4-17 (left). We expected the library sitting area to be a bit dark, so a Sun Pipe (see Chapter 2.4) was installed to brighten things up a bit. This unit provides the equivalent light output of a 400 Watt lamp without the resulting heat loss of a typical skylight or need of electrical power.

Figure 4.4-18. This view of the living room and staircase shows how the heat from the catalytic wood stove can circulate through the upper and lower floors. Heat rising up the wide staircase displaces cooler air, which flows down the stairs, forming a continuous loop. Air will continue to loop until thermal equilibrium is reached between the two floors.

Figure 4.4-19. This propane-powered, freestanding stove provides a cozy setting for dinner parties. It is also the backup heat source (furnace) when the wood stove is not used or during extended traveling. Many people cannot believe that this little unit is capable of heating the entire house, but it is!

Figure 4.4-20. Eighty percent of the home's electrical energy is produced by this photovoltaic (PV) array and sun tracking device. The panel generates approximately 1,000 Watts of power for as long as the sun shines.

Figure 4.4-21. With Bill leaning on the PV array, you can get a pretty good idea of the size of these units. The array is approximately 8 ft high by 16 ft long (2.4 m x 4.8 m).

Figure 4.4-22 (below). The electricity provided by the deep cycle battery bank is what keeps the house running when the sun sets and the wind stops. There is sufficient energy storage in this battery to power the house for approximately three full days without any input from the PV array or wind turbine.

Figure 4.4-23 (right). When the battery bank shown in Figure 4.4-22 is discharged to a certain level, the diesel generator housed in this outbuilding will start, quickly charging it back up. This generator is operated on a mixture containing biodiesel, which is a renewable, low-emission fuel.

Figure 4.4-24 (left). This wind turbine is rated 1,500 Watts and will generate power in non-linear relationship with wind speed. Simply stated, wind turbines are not for every location, and considerable study of Chapter 7 is recommended before you purchase and install one. This unit provides 15% of the total energy requirements of the house.

Figure 4.4-25 (right). Solar thermal hot water heating systems such as this collector system are discussed in Chapter 5. The vacuum tube array shown here reduces domestic hot water fuel requirements by over 60% and provides thermal energy for the hot tub.

Figure 4.4-26 (left). A solar cooking unit such as this model completely eliminates the need for fossil fuel energy for cooking, replacing it with free sun energy. Think of these models as the ecologically friendly version of the barbeque.

Figure 4.4-27 (right). Using the sun to dry clothes reduces cost and airborne emissions while giving clothes that great "smell of the outdoors" without the use of artificial fragrance sheets.

Figure 4.4-28 (left). PV-powered electric fencing protects approximately 25 acres of horse paddock area. The combination of PV- and rechargeable battery-powered electric fencers makes great environmental and economic sense, eliminating the cost and waste of disposable batteries.

4.5 Hubicki Home "Unter den Birken"

5,200 square feet of sustainable luxury

System Cost: Electrical   US $42,000
                Thermal   US $24,000

It was a splendid fall day when Lorraine and I arrived at Tina and Mike Hubicki's luxurious off-grid home (Figures 4.5-1, 2, 3 and 4). Located on 65 acres of mixed hardwood and cedar forests, it sits atop a moraine and surveys a vista of pasture and farmlands as far as the eye can see.

"Our family had a dream of moving to the country and living in harmony with the land and our local ecosystems. We decided to explore healthy and sustainable options for housing, heating, and power generation," explains Mike. "It was our vision to build this sustainable, efficient home and open its doors to others to gain needed information and reassurance when developing their own projects."

Tina and Mike began researching their dream home through by scanning magazines, searching the Internet, and visiting other homes, developing a wish list of the principles of sustainable living. "Our list of sustainable principles includes a healthy home environment and energy-efficient structure, using local and recycled materials and integrating the home into the surrounding ecosystem," says Mike. "Once an efficient structure was developed, we would then use renewable energy and careful resource allocation to power the home."

Figure 4.5-1. Tina and Mike Hubicki's luxurious off-grid home is located on 65 acres of mixed hardwood and cedar forests. It sits atop a moraine and surveys a vista of pasture and farmlands as far as the eye can see.

WINNER
Canada's
Energy Efficiency
AWARDS
2005

"We committed to altering as little of the site as possible and integrated our home, driveway, and landscaping into the existing site," continues Tina. "Our gardens include mostly native perennials and vegetables (Figure 4.5-6) and the newly built eco-pool (Figure 4.5-14) uses rainwater and plant filtration systems (Figure 4.5-15) rather than environmentally harsh and unnecessary chemicals."

Because Mike uses the house as home and office, the family opted for a total living area of 5,200 square feet. Heating such a large house without the use of energy-robbing fans and duct work was a challenge, especially when the goal was to use renewable sources for this energy requirement. The Hubickis chose an in-floor hydronic heating system which is coupled to a Danish wood gasification/oil boiler (Figures 4.5-22 and 23) and 60 vacuum solar thermal heating tubes (Figure 4.5-21). The energy output from these sources feeds directly into an insulated 620 gallon (2347 liter) hot water storage tank (Figure 4.5-24), providing thermal buffering which prevents the boiler from rapid cycling and inefficient operation.

The addition of the 60-tube solar thermal system provides the home with sufficient energy for domestic hot water between mid-May and mid-October, at which time the wood boiler provides the extra muscle needed for winter heating. The oil backup system (Figure 4.5-23) is required when the family travels.

The Hubickis' electrical needs are met with a combination of a one kilowatt African Power wind turbine and 12 Solerex PV panels wired into a 24 Volt array, providing 900 Watts of rated power (Figure 4.5-17). The PV array is mounted on a tracking controller which ensures that the panels remain in constant alignment with the sun.

Figure 4.5-2. Tina, Mike, their daughter and pets enjoy their dream home in a rural setting north of Lake Ontario, near Cobourg. The property is nestled in a spot with breathtaking views and clear access to lake-effect wind and sun power.

Energy is stored in a deep cycle battery bank (Figure 4.5-19) installed in an insulated cabinet located in the garage/workshop area. Two 4,000 Watt Xantrex sine wave inverters (Figure 4.5-20) provide utility-standard 120/240 Volt power to the home. A 15 kW Coleman propane-powered backup generator (Figure 4.5-18) completes the mix and supplies supplementary energy during poor weather periods in the late fall and early winter.

The decision to develop a fully off-grid home was reinforced by the fact that a utility line extension and transformer were going to cost approximately US $16,000. When the savings from solar and high-efficiency heating systems were taken into account, Mike estimated a payback of 9.8 years (10.2% return on investment) with no need to worry about rising energy prices.

"We moved into our home on a cold January day in 2003. It was hard to imagine that we were living in our dream home using all of the appliances North Americans take for granted, but doing it off-grid," says Tina. "Winter is perhaps our favorite time of year now." With such a beautiful home and grounds, the Hubickis have found their paradise, "unter den Birken."

Figure 4.5-3. "We decided to explore healthy and sustainable options for housing, heating, and power generation," explains Mike. "It was our vision to build this sustainable, efficient home and open its doors to others so that they could gain needed information and reassurance when developing their own projects."

Figure 4.5-4. "Our list of sustainable principles included building a healthy home environment and energy-efficient structure using local and recycled materials and integrating the home into the surrounding ecosystem," says Mike. "Once an efficient structure was developed, we would then use renewable energy and careful resource allocation to power the home."

Figure 4.5-5. This northeast view shows the south-facing, passive solar glazing of the home and garage as well as renewable energy electrical and thermal components.

Figure 4.5-6. "We committed to altering as little of the site as possible and integrated our home, driveway, and landscaping into the existing site," continues Tina. "Our gardens include mostly native perennials and vegetables."

Figure 4.5-7. The spacious home was designed as a universally accessible bungalow with all normal living spaces on the main floor. The design incorporates regional materials such as locally milled pine boards and recycled stone.

Figure 4.5-8. The home combines the best energy efficiency and healthy home characteristics that the Hubickis could afford. Room layout, walls, doors, and windows were all considered before developing the final plan.

Figure 4.5-9. Tina is very pleased with the kitchen area, which features all of the standard appliances one would expect in a home.

Figure 4.5-10. A microwave oven is built into the kitchen island and wet bar. Units such as these are relatively energy efficient, as they are operated for short periods of time, reducing total energy demand. One problem with most microwave ovens is that the time- of-day clock forms a "phantom load" which consumes an excessive amount of electrical energy without providing any value. Ensure that devices like

this are wired to wall sockets that can be switched off after use.

Figure 4.5-11. As with most off-grid homes, propane gas is used for the large heat-demand appliances such as the cook stove and oven.

Figure 4.5-12. An EnergyStar, high-efficiency front loading clothes washer was chosen to reduce electricity, soap, and water consumption. A propane-powered clothes dryer supplements the outdoor clothes line.

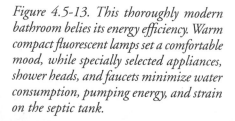

Figure 4.5-13. This thoroughly modern bathroom belies its energy efficiency. Warm compact fluorescent lamps set a comfortable mood, while specially selected appliances, shower heads, and faucets minimize water consumption, pumping energy, and strain on the septic tank.

Figure 4.5-14. The "eco-pool" and clarifying pond use recycled tires for earth retention and steps along the inside of the pool. Water is circulated through the clarifying pond by an energy-efficient pump, providing natural filtration. A copper-silver water ionizer provides final sterilization before recirculating water back to the pool.

Figure 4.5-15. The clarifying pond filters particulate matter from the pool water as well as acting as a catchment basin for summer rain, which provides makeup water.

Figure 4.5-16. To reinforce the agricultural theme, the detached garage was built in the Ontario barn style complete with silver steel roof and red barnboard siding. The garage contains the complete off-grid electrical system and batteries.

Figure 4.5-17. The Hubickis' electrical needs are met with a combination of a one kilowatt African Power wind turbine and 12 Solerex PV panels wired into a 24 Volt array, providing 900 Watts of rated power. The PV array is mounted on a tracking controller, ensuring that the panels remain in constant alignment with the sun.

Figure 4.5-18. A Coleman 15 kW propane-powered generator supplies battery charging and backup power requirements in the event of a system failure.

Figure 4.5-19 (above). The 24 Volt battery pack comprises two series strings of 12 two Volt batteries, providing a total gross energy storage of 3,000 amp-hours or 64 kilowatt hours of energy. Deep cycle batteries must never be discharged completely, as this severely reduces their operating life. Normal day-to-day operation should be limited to 40% energy withdrawal or less.

Figure 4.5-20 (left). Two Xantrex 4,000 Watt sine wave inverters provide the house with utility standard 120/240 Volt power. An Outback MX60 maximum power point tracker (MPPT) increases solar energy output by approximately 30%.

Figure 4.5-21 (left). This close-up view of the garage details the two banks of 30 vacuum tube solar thermal collectors, which supply the house with domestic hot water energy between mid-May and mid-October.

Figure 4.5-22 (below). Mike is shown loading the Tarm wood gasification boiler, which provides the extra muscle needed for winter heating. This unit is one of the most energy efficient and environmentally friendly wood-burning units on the market.

Figure 4.5-23 (above). During extended travels or time away from home, the wood fuel will gradually burn down. A control system in the Tarm monitors the boiler temperature and signals a backup or secondary heating source to take over. In this view, Mike is identifying the oil burner unit.

Figure 4.5-25 (below). A Lifebreath heat recovery ventilator (HRV) provides clean air for the home by capturing waste air from strategic locations in the home and exhausting it to the outdoors. Fresh replacement air is then drawn into the house, while an exchange membrane transfers heat from the stale outward-bound air to the cooler incoming air.

Figure 4.5-24 (above). Heated water/glycol solution from the Tarm boiler is fed to this 620 gallon (2347 liter) insulated thermal storage tank. The purpose of the storage tank is to "buffer" thermal energy from the solar thermal vacuum tubes and boiler. Heat is then slowly drawn away to provide space and hot water heating energy.

5
Heating and Cooling with Renewable Energy

Price signals capture everyone's attention. When oil was a couple of dollars a barrel very few people worried about how much of it they consumed to operate cars and heat homes. In the early 1970s automobiles weren't much smaller than the tanker ships used to flood North America with Saudi Arabian crude, and home design and construction gave little attention to energy efficiency and related fuel consumption.

In 1973 an organization known as OPEC unilaterally raised oil prices, providing the wake-up call that shook Americans out of their fuel consumption complacency. Prices soared and those who heated their homes with oil went from being energy pacifists to being scared silly almost overnight. As if that wasn't bad enough, all Arab oil producers stopped shipping oil to the United States as a way of protesting American policy regarding Israel. Price increases of over 500% coupled with uncertain supplies exposed the vulnerability of the United States economy, which depended then as it does now on foreign oil supplies.

Following considerable political posturing by the Nixon administration, the embargo was lifted and life returned to an uneasy state of normal. In the years following the energy crisis considerable work was done to improve the efficiency standards of homes and automobiles. Unfortunately politicians have very short memories, and the crisis of 1973 was quickly forgotten as fuel prices and supplies stabilized, causing most North Americans to return to their gluttonous ways.

The beginning of the 21st century has brought about a new energy crisis fostered not by embargoes but rather by concerns about the environment, supply, and terrorism. Although the economy is still completely addicted to oil and demand continues to rise unabated, OPEC remains relatively friendly to the West. It should; after all, American consumers have drained the astounding sum of $7 trillion from the economy simply to burn Arab oil and watch it go up in smoke. According to a report in the October 25, 2003 issue of *The Economist*, this estimate does not include subsidies of staggering cost, such as the ongoing military presence in the Middle East as well as development subsidies and tax breaks to the large oil companies.

Environmental degradation continues unabated, as neither the government nor oil companies consider it a cost to be added to the balance sheet. Canada exports nearly 8% more oil to the United States than does Saudi Arabia. The province of Alberta, through its phenomenally dirty tar sands oil extraction program, reaps the short-term economic windfall even though the process makes it one of the largest greenhouse gas emitters in the world. With governments addicted to increasing gross domestic product and job creation, energy efficiency and the environment are of no more concern than they are to the Saudi Arabian oil sheiks with whom they compete for market share.

The long-term view of energy is to move the world economy away from nonrenewable resources to clean-burning, carbon-neutral renewable fuels such as ethanol, biodiesel, and perhaps hydrogen for use in transportation infrastructure. Wind turbines, photovoltaic panels, and clean-burning biomass provide complementary energy sources for the production of electricity. The transition has already started. In 2003 Germany announced that it had already decommissioned one nuclear power plant as a result of increased use of wind turbine and photovoltaic electricity generation technologies. On the residential home front many renewable energy technologies such as solar domestic hot water systems are currently available and economically viable.

Renewable Fuels

Burning any fuel has some negative impact on the environment. Fossil fuels are nonrenewable and contribute smog-producing chemicals and greenhouse gases—mainly carbon dioxide—into the atmosphere. Some nonrenewable fuels are better than others. Natural gas and propane are the best, while low-grade coal is the worst. Consider that a natural gas or propane cook stove allows the burning fuel to vent *into* the house. Try *that* with coal or oil!

Wood and wood pellets, dried corn, biodiesel, and ethanol are different from fossil fuels in that they are renewable, easily replaced energy storage units. A growing plant absorbs nutrients and water from the ground as well as carbon dioxide from the atmosphere. Photosynthetic processes within the plant convert carbon dioxide gas into carbon, which is stored in the structure of the wood or oil seed. When dry wood or biofuels are burned, carbon is consumed, releasing heat and carbon dioxide gas. The good news is that burning the plant products properly exhausts no more greenhouse gases into the atmosphere than simply letting the tree or plant rot on the forest floor or farmer's field. However, allowing even the best wood stove to burn smoldering, smoky fires releases ash and other pollutants into the atmosphere. (See Chapter 16, "*Biodiesel Home Brewing and Co-operatives*", for more information on this subject.)

Solar Energy – Something for Everyone

If you are designing a new home or retrofitting an old one, an even better energy source to consider is the sun. A properly designed and oriented home absorbs free energy on sunny days, helping reduce your reliance on burning anything. Both passive and active solar heating systems can be used. Passive solar heating is just that: passive use of properly oriented walls, windows, and architectural house features. Active solar heating involves a series of solar collection and storage units designed to increase the "density" of the captured energy. Within the broad range of renewable fuels and "solar thermal" technologies are a number of options for the urban homeowner to consider, each with its own pros and cons. While the number of configurations is virtually infinite, let's assess those technologies that follow the *eco-nomic* path:

- 5.1   Passive solar space heating
- 5.2   Active solar air heating
- 5.3   Active solar water heating
- 5.4   Solar pool heating
- 5.5   Active solar space heating
- 5.6   Heating with renewable fuels
- 5.7   Space cooling systems
- 5.8   Systems that cannot be used off-grid

5.1
Passive Solar Heating

Solar energy is free, non-polluting, and renewable. Why don't more people use it? My guess is that most people think that houses using solar energy have to look like something George Jetson would live in. Well, don't worry. Geodesic dome designs aside, you won't have to live in a house that upsets your local town hall planning committee or causes you to be barred from block parties for life in order to live with a passive solar system. Another possible reason for being afraid of solar energy is its variability and the fact that your building contractor isn't able to quantify it. If you require heat, just install a big furnace and then forget about it, right? Everyone knows it's not quite that easy with solar energy.

Not so fast. Living with solar energy does not require a house full of complicated electronics and miles of glass. What passive solar energy does require is following some fairly simple guidelines to allow your home to take full advantage of the sun's energy.

Step 1 – House Orientation

Orienting the house to collect solar energy may seem obvious, but it's not. Most people give no thought to ensuring that the house is oriented in such a way as to capture as much sun as possible in the winter and provide proper shading in the summer. This is how it's done:

- Using a compass at the proposed building site, locate magnetic south. Consult the magnetic declination chart in Appendix 4 and determine the compass correction for true or solar north and south. Using this heading, place the long axis of the home at a right angle to it.
- Place deciduous trees in a path between the summer sun and the house. This provides shading for the house and glazing. When the autumn winds blow, the leaves fall from the trees, allowing the welcome winter sun indoors.
- Plant pine, cedar, or other evergreen trees to the north and east to provide a windbreak. In treeless areas, wind and storm blocks can be made using rock outcroppings, hills, or even your neighbor's house.

These orientation strategies apply whether you are in the country or the city. If you are designing a home in a city, try to evaluate the sun's path over your lot and take into account any shading effects caused by future buildings. The Solar Pathfinder shown in Figure 6-10 works wonders if you are concerned about your home's view of the sun.

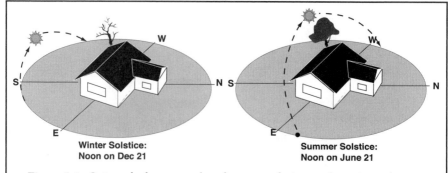

W

S   N S   N

E   E

Winter Solstice: **Summer Solstice:**
Noon on Dec 21   **Noon on June 21**

Figure 5-1. Orient the house to take advantage of winter solar gain and summer shading.

Step 2 – Insulate, Insulate, Insulate

If you haven't taken the time to review Chapter 2, "Energy Efficiency", go back and read it. A house that is under-insulated and prone to air leakage will quickly put an end to your quest for energy efficiency. **Remember that it is far less expensive to conserve energy than to acquire more energy to heat a substandard house, even if the energy is free!**

If you live in an area where summer air conditioning loads are high, consider adding radiant barrier insulation in the attic, as discussed in Chapter 2.

Step 3 – Window Design

Windows are the passive solar collectors of the house. Vertically oriented glass will catch the low winter sun and miss the high summer sun angle. This means you can create a "normal" looking home and still get maximum solar efficiency.

- Place the majority of the desired glazing on the south side (long axis) of the house.
- Choose windows that are of the highest quality you can afford. Refer to Chapter 2 for information on window construction. Remember that triple-glazed glass with Krypton gas fill is the most efficient available, reducing nighttime energy loss.
- Do not use low-E glass on the south side of the house or on any window that you wish to use as a solar collector.
- Ensure that window units that can be opened are installed on the side of the house facing the prevailing winds as well as on the opposite side of the house. You can open these windows to provide nighttime cross-flow cooling of the home.

Figure 5-2. This home, located in Eastern Ontario, is designed to capture the maximum amount of winter sun energy, while avoiding the high-angle summer rays.

Figure 5-3. This lovely veranda also acts as the south window summer shading system. The roof overhang was designed to allow sunlight to enter the house structure from mid-September through mid-March.

Figure 5-4. This newly built "earthship" design is very heavily glazed by northeastern standards. It will require plenty of shading and cross-flow ventilation to keep it from overheating in summer. (Note the photovoltaic panels mounted to the right of the windows.)

- Limit the amount of glazing on the north, northeast, and northwest sides of the house.
- Do not over-glaze the south side. Excess window area equates to excess heat.

Step 4 – The Finishing Details

These finishing touches will assist in making the job work "just right":

- Provide proper home ventilation using an air-to-air heat exchanger unit as outlined in Chapter 2.
- Provide sun shading over the east-, west-, and south-facing windows as required, to prevent the summer sun from heating the structure.
- Use summer sunscreen blinds similar to the units shown in Figure 2-10.
- Install a thermal absorber and buffer system in the living area. The sun will quickly overheat a properly insulated house, making it unbearable. A solar mass such as cement/tile flooring or walls, finished in dark colors, will absorb heat, helping to moderate temperatures. During the evening hours, this stored energy is radiated back into the room, helping to keep temperatures constant.

The key to ensuring that your passive solar design does not cause extreme cycling of interior temperatures is to balance the solar collector area (windows) with the thermal storage and buffering mass of the floor and wall systems. When these components are not in balance the results are very similar to a greenhouse: hot during the day and cold at night.

An educated architect or heating contractor will be able to assist you in developing a heat-loss calculation for your home. Alternatively, a home inspection auditor who is capable of performing an air leakage test may be able to assist with this calculation. Armed with this information, you will be able to calculate what percentage of the yearly heating load can be expected from solar energy. With the sun providing approximately 1000 W per square meter of window area, it's a shame not to take advantage of this free energy source.

For readers who would like more detailed design information on the subject, an excellent book is available that teaches construction techniques, theory, and mathematical modeling of heat gain and loss. *The Passive Solar House* by James Kachadorian (ISBN 0-930031-97-0, Chelsea Green Publishing Company) is labeled as "a building book for the next century" and is highly recommended for anyone who wishes to push the building envelope.

5.2
Active Solar Air Heating

Figure 5-5. The Solarwall® system capitalizes on the fact that a building's exterior cladding will be exposed to the energy of the sun and might as well capture that energy for fresh air heating and ventilation. (Courtesy Conserval Engineering Ltd.)

A simple, unique solar heating concept is produced under the trade name Solarwall® by the firm Conserval Engineering Ltd. This cladding material is mounted on the south, east, or west walls. Tiny perforations allow outdoor air to travel through the exterior surface. During the day, as outside air passes through the panel and along its inside surface, it absorbs the sun's energy, warming the air which in turn rises. The preheated, warm air is then drawn into the building's ventilation system, reducing the load on the heating system.

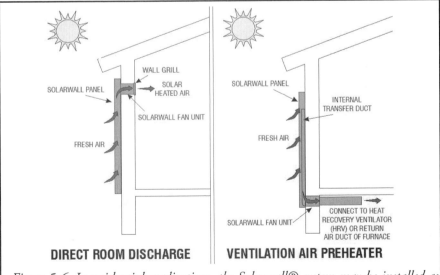

Figure 5-6. In residential applications, the Solarwall® system may be installed as a direct room discharge unit or as the preheater for the intake of a furnace or heat recovery ventilator system.

Figure 5-7. This view of a partially completed home shows the Solarwall® prior to the installation of the fieldstone siding desired by the homeowner. The design will work well with the stone facing, providing a very aesthetic finish.

During sunny days the Solarwall® preheats intake air by between 30°F and 50°F (17 and 28° C), depending on the desired flow rate. Snow can reflect up to 70% of solar radiation, into the Solarwall®, increasing the system's performance further. The system also works on cloudy days, as diffused light can provide up to 25% of the total radiation available on a sunny day.

The Solarwall® can even help cool your building in the summer. The cladding structure shades the interior wall and any hot air that accumulates at the top of the unit escapes through the perforations, keeping the wall surface cooler than it would be if it were exposed directly to sunlight. Electrically (or manually) controlled dampers allow unheated air into the building, maintaining air quality. The dampers can also switch to the wintertime position, providing preheated air intake.

Figure 5-8. The Cansolair solar heating unit may be installed as a retrofit device on the south-facing wall or roof. Using a thermostatically controlled fan unit (shown above right), the system can provide preheated intake air to the building ventilation system or directly to a room The unit produces approximately 2,900 kWh of energy per year based on 2,200 hours of sunlight. At a hydro cost of 10 cents per kWh, this equates to almost $300 per year of free energy.

Figure 5-9. The Solarwall® system has been adapted to act as a metal roof structure as well as an air solar thermal collector. (Courtesy Solar Unlimited Inc., Utah)

A slightly modified version of the vertically installed Solarwall® is the roof-mounted system shown in Figure 5-9. If you are building a new home or planning to re-roof an existing house, the incremental capital cost may be negligible.

An Alternative to the Solarwall® system is a retrofit product known as the Cansolair unit. The Cansolair is a cross between the Solarwall® system and a greenhouse, contained within a package of the same size as a sheet of plywood. The heated air may be delivered to the home's heat recovery ventilator or via a thermostatically-controlled room ventilation fan shown in Figure 5-8.

The economic case for solar thermal heating systems depends on many factors including the age of your home, its heating demand, and its thermal efficiency. There is little point in installing a system that may only provide a small percent of total heat demand because of a thermally leaky house. On the other hand, an energy-efficient home that has a correspondingly lower heat demand may receive a significant portion of its total yearly thermal energy from such a system. It will be necessary to compare the heat loss of your home with the heat production of the solar thermal unit to determine how much purchased energy, in the form of oil or gas for example, the system will offset. This is a complex calculation and is best left to professional heating and cooling contractors and their computers. If you are willing to give the calculations a shot, then consider purchasing the book *The Passive Solar House* discussed above.

5.3

Solar Water Heating

Figure 5-10. Solar thermal water heating systems are used throughout the world and with good reason. Approximately one-third of the energy bill for a home is related to hot water heating costs. Almost everyone is aware that Arizona and Florida can achieve huge savings with solar systems, but what about the majority of people who live in more northerly climates? It may come as a surprise, but if you live in Anchorage, Alaska or Inuvik in Canada's arctic you can still achieve a 35% reduction in hot water heating costs by using a solar thermal water heating system. (Courtesy EnerWorks Inc.)

When you ask most people about solar systems almost everyone will talk about heating for domestic hot water supply, and with good reason. Approximately-one third of the energy bill for a home is related to hot water heating costs. According to a Florida electrical utility, for every $1000 you pay in energy costs, $300 goes towards water heating. This figure is based on 2002 data as shown in the pie chart in Figure 5-12.

While everyone expects solar energy to make economic sense in the tropical or desert areas of the southern United States, what about the populated areas to the north, including Canada and its arctic regions? The maps shown in Figure 5-12 indicate the percentage of hot water heating energy that can be achieved in different geographic locations. Throughout the densely populated northeastern section of the United States and Canada, homeowners can achieve an average of 55% of their water heating demand from solar. If you happen to live in Anchorage, Alaska or even Inuvik in Canada's arctic region, you can still achieve an impressive 35% average savings in annual water heating costs.

Figure 5-11. These maps of Canada and the United States show the average annual solar energy that can be captured to reduce home hot water heating costs. Solar thermal hot water systems are one of the most economic means of putting the sun's energy to work. A typical family of four can expect to receive a return on investment of approximately 10% to 14% if they are located in an area with a minimum of 50% solar heating penetration. For those lucky enough to live in the Arizona desert or sunny Florida, the return on investment can exceed 20% according to industry figures. (Courtesy EnerWorks Inc.)

You may be wondering if your house will have hot water only for the percentage of time indicated in the maps in Figure 5-11. All solar thermal systems work in conjunction with your existing hot water heating system, providing "pre-heating" of the water supply. For example, if the desired hot water temperature is 140°F (60°C), the solar water heating system preheats the incoming cold water supply to 110°F (43°C), reducing the fuel demand of your hot water heater, while maintaining adequate supplies of hot water.

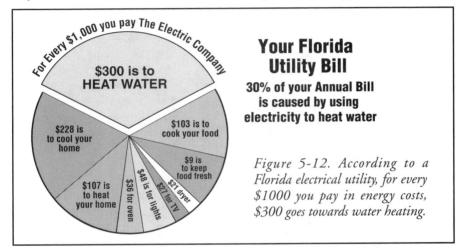

Your Florida Utility Bill

30% of your Annual Bill is caused by using electricity to heat water

Figure 5-12. According to a Florida electrical utility, for every $1000 you pay in energy costs, $300 goes towards water heating.

Before Installing a Solar Thermal System...

Most people practice their Luciano Pavarotti while taking a shower; I expect that the engineering minds at Renewability Energy Inc. focus on just how much hot water is actually going down the drain rather than singing. In fact the hot water used for household tasks such as laundry, dishes, showers, and baths requires enormous amounts of energy, but much of the heat contained in the water is not extracted before the job is completed. If the drain water were cold, the energy efficiency of the task that uses the hot water would be 100%. However, this is never the case, and a considerable amount of energy actually goes down the drain in the form of warm waste water.

The Power-Pipe™ heat recovery system is an amazingly obvious device which transfers energy from warm waste water to incoming potable cold water, which in turn supplies the home's hot water heater. As shown in Figure 5-13, a copper drainpipe is mounted in a vertical orientation, allowing waste water to pass through on its way to the sewer. The vertical orientation is important in order to ensure that drain water "skins" along the inside diameter of the pipe as it makes its way to the sewer, increasing the surface temperature of the pipe. A second smaller pipe is wrapped in a tight spiral around the vertical drainpipe. The bottom or inlet side of the spiral pipe is connected to the incoming cold water supply. As the warm drain water pours down the larger vertical pipe, heat is transferred to the spiral water supply pipe. Potable water exiting from the top of the spiral assembly is thus pre-warmed and may be sent to the hot water heater or plumbing fixtures, reducing their energy requirements (see Figure 5-14).

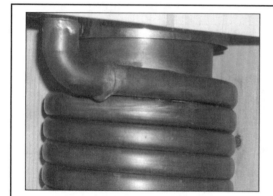

Figure 5-13. A close-up view of the top section of the drain and spiral wound supply pipes of the Power-Pipe system. After installation the entire assembly is wrapped in insulation, preventing heat loss and increasing thermal efficiency. (Courtesy Renewability Energy Inc.)

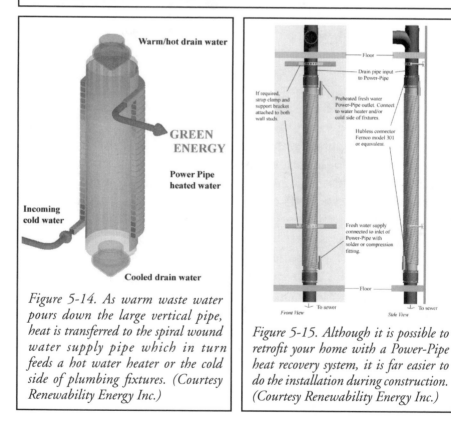

Figure 5-14. As warm waste water pours down the large vertical pipe, heat is transferred to the spiral wound water supply pipe which in turn feeds a hot water heater or the cold side of plumbing fixtures. (Courtesy Renewability Energy Inc.)

Figure 5-15. Although it is possible to retrofit your home with a Power-Pipe heat recovery system, it is far easier to do the installation during construction. (Courtesy Renewability Energy Inc.)

Understanding the Basics of Solar Domestic Hot Water Systems

Solar hot water heaters work in conjunction with your existing fossil-fuel or electrically operated water heating tank, supplying a percentage of your total hot water heating requirements. They will reduce your hot water heating costs, greenhouse gas emissions, and smog and atmospheric pollutants, contributing to a cleaner and more sustainable environment.

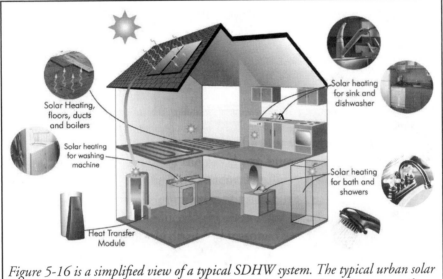

Solar Heating, floors, ducts and boilers

Solar heating for washing machine

Heat Transfer Module

Solar heating for sink and dishwasher

Solar heating for bath and showers

Figure 5-16 is a simplified view of a typical SDHW system. The typical urban solar water heater consists of roof-mounted panels which capture the sun's energy as heat. A circulator pump causes a transfer fluid to absorb this energy and pass it to a heat exchanger which in turn heats water in a storage tank. Solar-heated water is most often used in the kitchen, laundry, or bathroom, although it can also provide heat for space, pool, and spa applications as discussed later in this chapter. (Courtesy EnerWorks Inc.)

Energy Collection

Solar collectors are best mounted on the roof, facing solar south and mounted at an angle approximately equal to your geographic latitude in much the same manner as electricity-producing photovoltaic panels (see Chapter 6). Solar collectors are available in a number of different configurations, although flat plate designs (see Figures 5-10 and 5-17) are the most common. Evacuated tube collectors (see Figures 5-18 and 5-19) make up the balance, although newer collector configurations show up on the market from time to time (see Figure 5-20).

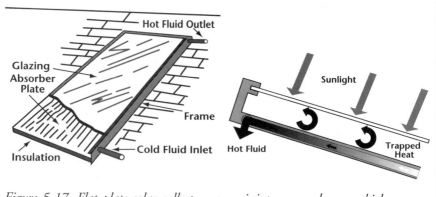

Figure 5-17. Flat plate solar collectors are miniature greenhouses which capture sunlight, trapping heat between a glazing material and an insulated rectangular box. A series of thin pipes attached to a dark-colored heat absorbing plate is placed inside the collector box. A circulating fluid such as water or antifreeze passes through the pipe assembly and heat collector plate, absorbing the trapped heat. If the heated fluid is potable water, it is supplied directly to the water heating unit. In severely cold climates an antifreeze solution is used instead of water, requiring the use of a heat exchanger to transfer heat energy to the hot water system.

Figure 5-18. A single evacuated tube solar thermal collector. A heat absorber plate is bonded to an apparatus known as a heat pipe. The assembly is placed inside a glass tube which is sealed and evacuated, creating a vacuum. The collector plate absorbs the sun's energy, causing a fluid inside the heat pipe to boil. Heat is then transferred to a condenser bulb mounted at the top of the unit. If you have ever made coffee and placed it inside a thermos bottle you know how long the drink stays hot. In a similar manner, heat cannot escape from the evacuated tube collector, providing a system that is not affected by extremely cold weather or high water temperatures. (Courtesy Carearth Inc.)

Figure 5-19. Evacuated tube solar collectors are most commonly mounted to a "header" in groups of ten. The transfer fluid flows through the header, absorbing heat from the condenser bulb located at the top of the evacuated tube. The header assembly is well insulated, preventing heat loss.

Figure 5-20. Solar thermal collection systems can also go high-tech. This Power-Spar™ from Menova Engineering Inc. uses a spherical mirror to concentrate the sun's energy directly onto a fluid transfer header, which has the effect of amplifying the sun's energy. Additionally, the Power-Spar™ contains a unique solar tracking unit which ensures that the mirror is correctly aimed at all times. The Power-Spar™ may also be fitted with photovoltaic panels to produce electricity at the same time as heat energy.

Flat plate solar collectors tend to be less expensive than their evacuated tube cousins, explaining their relative popularity. However, for the same energy output flat plate collectors tend to require more surface area for mounting than evacuated tubes. In addition, evacuated tube configurations operate at higher thermal efficiency and are therefore better suited for operation in extremely cold environments. If neither of these considerations affects you, the selection of which collector to use can be based strictly on price.

Regardless of which solar thermal collector you choose, it must have an unobstructed view of the sun between the hours of 9 a.m. and 3 p.m. throughout the year. Before purchasing a system, ensure that your site provides the necessary solar exposure. In addition, the collectors must be mounted so that they are facing as close to solar south as possible, bearing in mind that a phenomenon known as magnetic declination must be considered when using a compass for alignment. Magnetic declination maps provide a correction to compass readings in order to locate true north and solar south. Consult Appendix 4 to determine the magnetic error in your location.

Remainder of System

There are numerous types of solar water heating systems available, including the batch storage units shown in Figure 5-21. Although common throughout the developing world, these units are not considered aesthetically pleasing

Figure 5-21. Batch solar and passive open-loop systems use natural convection to circulate water through the collectors and water storage tank. Although not common in North America, these units are ideally suited for three-season cottages or where freezing temperatures and visual aesthetics are not a concern. (Courtesy Carearth Inc.)

to the North American eye, and they are susceptible to freezing. They are, however, extremely simple and relatively inexpensive and may be ideally suited for three-season cottages in northern environments or full-time residences in the south.

The most common configurations in use today are active closed-loop systems (Figure 5-23) that use a pump, temperature sensors, and electronic controls to regulate fluid circulation through the solar collector(s) and heat exchange mechanism.

The drain back system shown in Figure 5-22 features a standard hot water tank with a non-pressurized secondary water storage tank mounted on top. Thermal sensors mounted at the solar collector panel and the hot water tank are connected to an electronic device known as a temperature differential controller. This unit measures the difference in temperature between the solar collector and the hot water tank. When the collector temperature is higher than the hot water tank *and* below the desired or maximum water temperature of the system, the circulator pump is activated, transferring heat to the tank. When the solar collector cools, for example at night or on a cloudy day, the differential controller will turn off the circulation pump, causing water to flow back into the secondary storage tank. As the main water heater tank is equipped with a heat exchanger, water in the secondary storage tank does not come into contact with the potable hot water supply.

Drain back systems are relatively simple and theoretically immune to freezing. However, under unusual condi-

Figure 5-22. The drain back system is relatively simple and theoretically immune to freezing, making it a popular choice for domestic solar thermal hot water systems in moderate climates. Water contained in a storage tank is circulated through the solar collector and the hot water tank, which is equipped with a heat exchanger unit. When an electronic controller stops the circulation pump, water will drain back into the storage tank, preventing freezing. (Courtesy Alternate Energy Technologies, LLC)

tions it is possible that water could freeze within the solar collectors or the plumbing before draining back into the secondary storage tank. For extremely cold climates, the closed-loop glycol system is preferred.

The closed-loop glycol system is the preferred configuration in areas where freezing is a concern. A mixture of food-grade propylene glycol antifreeze (it *must* be safe; soft ice cream contains 30% of the stuff) and water continuously remains in the solar collectors, plumbing lines, and heat exchanger. A differential electronic controller and temperature sensing devices are connected to an alternating current circulation pump in the same manner as in the drain back system described above. It is also possible to replace the

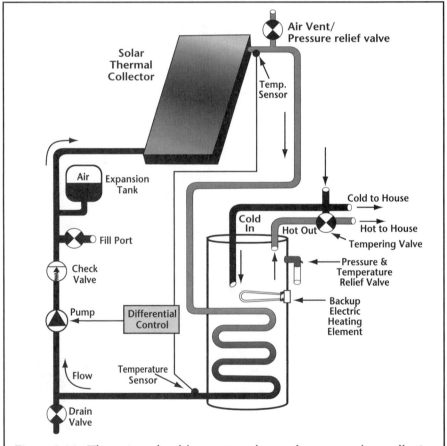

Figure 5-23. The active, closed-loop system shown above uses a heat-collecting antifreeze fluid which is circulated through the solar collectors and heat exchanger. Energy is stored in either a single-storage water heater tank (indicated) or in a two-tank configuration, according to the manufacturer's design.

1. Self-Limiting Solar Collectors

1a Freeze Protection

Household Hot Water

Washing

Showers/Bathing

Dishwashing

Laundry

Solar Storage Tank

Auxiliary Storage Tank (existing or new, electric, gas, propane or oil)

2 Heat Exchange Module

2a Passive Anti-Fouling Protection

Figure 5-24. A mixture of food-grade propylene glycol antifreeze which remains in the solar collectors and plumbing lines at all times makes the closed-loop glycol system completely freeze-resistant in any climate. A differential electronic controller and temperature sensing devices activate a circulation pump, transferring energy to an existing hot water tank via a heat exchange module. (Courtesy EnerWorks Inc.)

alternating current circulation pump with a direct current model connected to a small photovoltaic panel mounted in the vicinity of the solar thermal collectors. This configuration eliminates the need for temperature sensors and the differential electronic controller by activating the circulation pump whenever the sun shines.

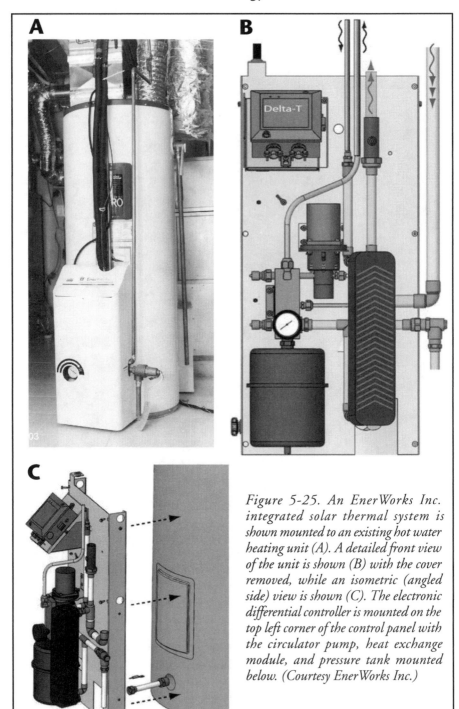

Figure 5-25. An EnerWorks Inc. integrated solar thermal system is shown mounted to an existing hot water heating unit (A). A detailed front view of the unit is shown (B) with the cover removed, while an isometric (angled side) view is shown (C). The electronic differential controller is mounted on the top left corner of the control panel with the circulator pump, heat exchange module, and pressure tank mounted below. (Courtesy EnerWorks Inc.)

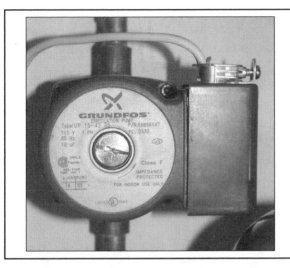

Figure 5-26. The Grundfos brand of centrifugal-type hydronic heating pump is commonly used in solar hot water systems because of its low power consumption, high reliability, and low maintenance. Other manufacturers such as Taco and Hartell provide similar models.

Circulation Pump

Centrifugal-type hydronic heating pumps are used in solar hot water heating systems to circulate a heat collecting fluid. They are chosen because of their low power consumption, high reliability, and low maintenance. For long life it is recommended that bronze or stainless steel pumps be used in this application even though they are slightly more expensive than cast-iron models.

These pumps are readily available from any plumbing supply store that specializes in hydronic heating systems.

Plumbing

Solar thermal systems are generally plumbed using standard domestic copper plumbing pipe which is wrapped in an insulation sleeve material such as Rubatex brand Insultube. An alternative material is PEX tubing manufactured for the hydronic (in-floor) heating industry.

To minimize heat loss it is recommended that plumbing lines be routed inside the building envelope as soon as possible after leaving the solar thermal collectors. If you are planning on building a house, it is a good idea to pre-plumb lines from the location of prospective solar collectors to your mechanical room. The cost is insignificant and will save a considerable amount of grief during a future installation process.

Check Valve

A one-way check valve is installed on the outlet port of the circulation pump to prevent the convection flow of warm fluid from the hot water storage tank to the cold solar collectors at night. Check valves may be spring or gravity operated. Gravity-operated check valves must be installed in the orientation shown in the installation manual. In applications using an AC circulator pump, the spring-operated check valve is preferred, as it is less susceptible to internal leakage caused by minute particles circulating in the heat collecting fluid.

Fill and Drain Ports

One fill and one drain valve must be provided to allow the filling of the circulation system with water or glycol solution. In order to fill and pressurize the circulation system, a positive-displacement high-pressure pump is required to overcome the vertical "head pressure" related to the force of gravity and the expansion tank air pressure.

Figure 5-27. An expansion tank allows the water or glycol solution in the closed-loop system to expand and contract as a function of its temperature.

Expansion Tank

An expansion tank allows the water or glycol solution in the closed-loop system to expand and contract as a function of its temperature. Without an expansion tank the buildup of pressure would cause the plumbing lines to burst when the circulation fluid is heated by the sun.

Hydronic expansion tanks such as the model shown in Figure 5-27 contain a rubber bladder which is in turn connected to the plumbing circulation

Figure 5-28. A pressure relief valve will prevent an explosion of the plumbing lines in the event of an excessively high pressure buildup.

Figure 5-29. An air vent is installed at high points in the circulation system and is used to remove trapped air (left). A pressure gauge (right) indicates that the circulation system is operating within acceptable pressure limits.

lines receiving the circulation fluid. With the tank empty the air chamber is pre-charged to approximately 15 psi (103 kPa) using a bicycle pump or compressor. As the circulation fluid expands and contracts as a result of the changing temperature, the pressurized air bladder maintains a relatively constant pressure in the plumbing system.

Your hydronic heating component supplier will be able to determine the optimum size of expansion tank based on the quantity of circulation fluid in the system.

Pressure Relief Valve

Pressure relief valves are designed to prevent the catastrophic explosion of plumbing lines in the event of excessively high pressure buildup. A setting of 50 psi (345 kPa) is common for most hydronic heating systems. All pressure relief valves including the model shown in Figure 5-28 are equipped with a drain port which must be connected to a capture bucket, floor, or sanitary drain.

Pressure Gauge

A pressure gauge indicates that the circulation system is operating within acceptable pressure limits. The normal operating pressure is approximately 15 psi (103 kPa) greater than static (pump off) pressure of the system. The static pressure is determined by the vertical height of circulation fluid weight and is based on the relationship of pressure rising 1 psi (6.9 kPa) for every 29 inches (0.74 meters) of plumbing rise above the gage height. For example, the static pressure for a plumbing line that rises 14 feet (4.3 meters) is 5.8 psi (40 kPa).

Static Pressure = (14 feet rise x 12 inches/foot) / 29 inches per psi
= (168 inches) / 29 inches per psi
= 5.8 psi

The operating pressure will vary somewhat depending on the temperature of the circulation fluid: the warmer the temperature the higher the pressure reading; the cooler the fluid the lower the pressure.

The pressure gauge may also be used to detect leaks in the circulation system. A slow continuous drop in static pressure over time indicates a fluid leak.

Air Relief Valve

Air trapped in the fluid circulation lines will lower the heat exchange efficiency or stop the circulation of fluid altogether. To remove trapped air, a relief valve (Figure 5-29, left) is installed at high points in the fluid circulation lines and at the solar collector panels. During commis-

Figure 5-30. The heat exchanger transfers energy from the solar collectors to domestic potable water in the storage tank.

sioning, these valves are opened, allowing the trapped air to escape to the atmosphere. Automatic air relief valves may be installed as an option.

Atmospheric gases are often dissolved in the circulation fluid and make their presence known by a gurgling sound in the plumbing lines. If the circulation system is completely free of air there should be no sound at all. During the yearly maintenance review of the system it may be necessary to activate the relief valves to eliminate any trapped air.

Heat Exchanger

The heat exchanger transfers energy from the solar collectors to domestic water in the storage tank. Heat exchangers may be external to the storage tank, such as those shown in Figure 5-25, or internal, as shown in Figure 5-30.

Heat exchangers are available as single- or double-wall designs, indicating the number of mechanical barriers between the heat-absorbing circulation fluid and the potable water in the storage tank. Most plumbing codes require the use of double-wall heat exchangers to prevent the contamination of potable water in the event of a leak.

Figure 5-31. Temperature sensors are located on the hot fluid outlet line of the solar collector and the bottom or cold line of the water storage tank. They feed temperature information to the electronic controller module.

Heat exchangers work when the circulation fluid is at a temperature higher than that of the water contained in the storage tank. Heat transfer is also affected by the surface area (size), thermal conductivity, and flow rating of the heat exchanger. If you are designing your own solar thermal system, contact a hydronic heating contractor to determine the appropriate size for your installation.

Antifreeze

To prevent the collector circulation fluid from freezing, an antifreeze solution of 50% propylene glycol mixed with 50% demineralized or distilled water is used. Common automotive antifreeze containing ethylene glycol must not be used because of its toxic nature.

The collector loop is filled with the antifreeze mixture using a positive displacement pressure pump. Sufficient fluid is introduced into the collector circulation loop to develop approximately 15 psi (103 kPa) of working pressure (static or pump off pressure plus 15 psi = working pressure) as indicated on the pressure gage.

Temperature Sensors

Temperature sensors are located on the hot fluid outlet of the solar collector and the bottom or cold heat transfer pipe of the water storage tank. The sensors, known as "thermistors," change their electrical resistance as a function of temperature, thereby feeding information to the electronic controller module.

The temperature sensor shown in Figure 5-31 is applied to the cold side of the thermal storage tank plumbing as indicated in Figure 5-23. Prior to attaching the temperature sensor to the plumbing lines, thermally conductive grease supplied with the controller module is added to ensure a proper temperature reading. The sensors are secured in place with a screw-type hose clamp as indicated in the picture.

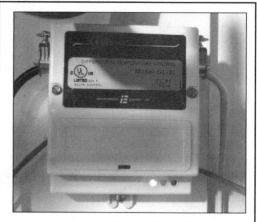

Figure 5-32. The brain of a solar thermal system is a device known as a differential temperature controller.

Electronic Control Module

The brain of the solar thermal system is a device known as a differential temperature controller. This unit monitors the circulating fluid temperature at the outlet of the solar thermal collector and on the cold side of the water storage tank. When the temperature of the solar thermal collector outlet exceeds the cold-side temperature of the storage tank by a predetermined number of degrees, the circulation pump is started and the heat transfer process begins.

A differential temperature in the order of 10°F to 25°F (5°C to 14°C) is required in order for adequate heat transfer to begin. Electronic controls such as those shown in Figure 5-32 also contain a high-temperature safety limit or cutoff circuit which will shut the circulation system down once the water storage tank reaches a predetermined setting. The upper limit is adjustable and is normally set to 170°F (77°C).

Setting the turn-on differential too

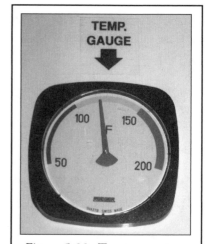

Figure 5-33. Two temperature gauges are recommended in order to measure the circulation fluid temperature at the inlet and the outlet of the heat exchanger, with a third gauge installed in the hot water storage tank as shown.

low will cause the pump to "short cycle," turning on and off the circulation pump. Setting the differential point too high will waste thermal energy.

The correct setting will cause the controller to short cycle only in the early morning, late afternoon, or during a cloudy day. During the remainder of a sunny day, the pump should operate continuously during the peak solar window.

Temperature Display

Two temperature gauges are recommended in order to measure the circulation fluid temperature at the inlet and the outlet of the heat exchanger, with a third gauge installed in the hot water storage tank as shown in Figure 5-33.

A temperature differential of approximately 20°F (11°C) across the heat exchanger indicates effective operation of the system. Monitoring the maximum temperature of in the water storage tank provides an indication of the effectiveness of the high-limit cut-out circuit.

The House Side of the Plumbing System

So far we've concerned ourselves with the heat-absorbing side of the circulation system. The output of the solar thermal storage tank may be fed to a fossil-fuel or electrically powered water heater or directly to the house hot water supply.

Regardless of which installation is chosen, a temperature-limiting tempering valve should be added to the hot water outlet prior to feeding the household hot water circuits. The reason for this is that the solar thermal system will produce hot water of varying temperatures depending on the amount of sunlight. If the solar thermal system is incapable of meeting the desired temperature set point, a backup heating element installed in the solar tank or a regular secondary hot water heater will heat the water to the desired temperature. On the other hand, if there has been a significant amount of sunny weather the water temperature in the storage tank may exceed the desired set point temperature, scalding anyone who is not paying attention. A tempering valve will "mix in" an amount of cold water to ensure that the hot water supplied to the house is at the desired temperature, improving system safety.

Choosing and Installing a System

A qualified, experienced dealer will be able to help you select the water heating system that best meets your specific needs and can assist you in determining what state, municipal, or local utility tax incentives and rebates may apply.

(You can also review the Database of State Incentives for Renewable Energy—DSIRE—at www.dsireusa.org.) If you have unpleasant recollections of some fly-by-night organizations selling solar thermal systems during the peak of the oil crisis, today's licensed professionals should be able to alleviate your concern.

It is wise to select dealers who are able to provide warranty service for both parts and labor and who have a satisfactory track record installing systems. Whenever possible, ask to see a previously installed system and try to discover if the owner is satisfied with both the installation and after-sales service.

Installing a solar thermal system is not the same as going to the hardware store to purchase a regular water heater. The dealer will have to perform a site evaluation, taking measurements and determining pipe routing and other details required to complete the installation. You will become an active part of this process.

Because you are making modifications to the structural and plumbing systems of your home, it will be necessary to install components that are certified by Underwriters Laboratories in the United States or the Canadian Standards Association to ensure fire and electrical safety. Purchasing approved products also lets you sleep at night knowing that product designs have met construction-quality standards. In addition, you must also obtain a building permit from your local inspection authority.

For local installation contractors and equipment suppliers in the U.S. contact the American Solar Energy Society at www.ases.org. In Canada contact the Canadian Solar Industries Association (CanSIA) at www.cansia.ca.

Operating a Solar System

Once the system is up and running you will be able to monitor its operation by observing the water temperature in the storage tank as well as the differential temperature at the inlet and outlet ports of the heat exchanger as discussed above.

You can also monitor the system for slow, minute leaks by watching the pressure gauge over a period of time. System pressure will rise and fall with a change in circulation fluid temperature; however the average "resting" pressure should remain constant over time.

In extremely sunny weather it is very likely that the system will produce more hot water that you can possibly use. When this situation occurs the electronic control module will stop the circulation pump, preventing the storage water tank temperature from rising above preset "over-temperature" limits. Circulation fluid trapped in the solar thermal collectors of antifreeze-based

Figure 5-34. Manufacturers have developed numerous methods for preventing stagnation of the antifreeze circulation fluid. EnerWorks has developed a thermally activated vent which opens when stagnation conditions develop, admitting fresh ambient air. (Courtesy EnerWorks Inc.)

systems will continue to absorb energy, increasing their operating temperature. Drain back systems are not subject to this condition.

This brings about a phenomenon known as "collector stagnation." If this condition persists the propylene glycol solution may prematurely break down, leaving a sticky, sludgy deposit in the plumbing components. To prevent this from happening manufacturers have developed a number of novel solutions. For example, EnerWorks has developed a "smart metal thermal actuator" which is installed on the underside of the company's flat plate solar collectors. During normal operation heat is drawn away from the collector by the circulating fluid and the actuator remains closed. Prior to stagnation occurring, excess heat in the collector assembly causes actuators located at the bottom and top of the rear side of the panel to open, allowing fresh ambient air to ventilate and cool the collector assembly. This operation is shown in the photograph in Figure 5-34.

An alternative to having this heat energy "wasted" when the water storage tank is filled is to include a bypass valve that will direct energy to other areas of the home that may be able to put it to work. For example, the hot fluid could be plumbed to a baseboard hot water radiator unit to provide auxiliary space heat in the home. Alternatively, a second heat exchanger could be plumbed into a hot tub or swimming pool, providing an excellent use for this energy. In either instance, your heating contractor will install the necessary valve and control equipment to ensure that domestic hot water is heated first with the auxiliary loads operated secondly.

An examination of the antifreeze circulation fluid is recommended at least every two years. There are two ways of testing this fluid. Samples may be sent for analysis to the Dow Chemical Company under its free annual fluid analysis program. The company will provide a complete summary report with a results table and a cover letter explaining the condition of the sample and outlining necessary maintenance procedures for their products. (Contact Dow at: http://www.dow.com/heattrans/index.htm).

ACUSTRIP® Company, Inc. offers a simplified fluid test kit for propylene glycol antifreeze solutions. A sample of the antifreeze solution is placed in a clean glass jar. A test strip (Figure 5-35) is immersed in the fluid and then compared to a color chart, indicating percentage propylene glycol concentration and pH (acidity) level. Ideal concentration levels are 50% propylene glycol to 50% demineralized or distilled water. The pH of the solution should be in the range of 8 to 10. If the glycol concentration is too low or the pH level too acidic, the system must be partially drained and new glycol added.

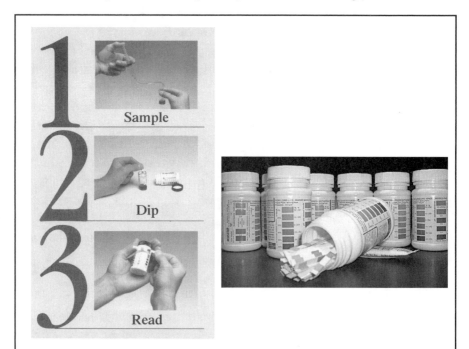

Figure 3-35. ACUSTRIP® Company, Inc. offers a simplified fluid testing program for propylene glycol antifreeze solutions. This test kit should be used at least once every two years.

5.4
Solar Pool Heating

Heating a swimming pool using the heat of the sun is probably one of the most obvious and cost-effective applications for solar energy. According to the Canadian Solar Industries Association there are more than 250,000 swimming pools in Canada, of which approximately 40% are heated. It is further estimated that homeowners spend more money to heat their pools than their homes. Switching from fossil-fuel or electric heating to a solar thermal system would recoup the initial capital in less than 2.6 years, yielding a return on investment of approximately 38%. According to the GO Solar Company (www.solarexpert.com), using natural gas for pool heating costs in excess of $2000 for a typical swimming season even in sunny southern California.

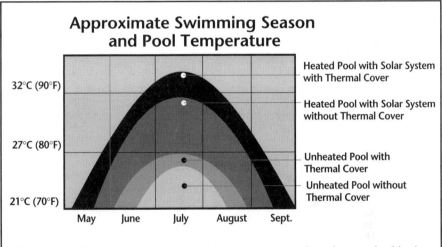

Figure 5-36. *Using a solar thermal pool heating system and insulating solar blanket not only increases the pool temperature but can extend the swimming season by up to two months in the U.S. northeast and Canada.*

The graph shown in Figure 5-36 indicates the increase in swimming pool water temperature as well as the extension of the swimming season when a solar thermal heater is used in conjunction with a solar blanket. A swimming pool that uses both of these items will be more comfortable and will save an enormous amount on heating bills. You will also eliminate tons of greenhouse gas emissions, literally! Figure 5-37 shows typical greenhouse gas emissions (in short tons: 1 t = 0.91 metric tonnes) during a single swimming season.

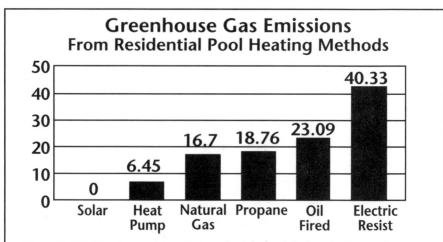

Figure 5-37. *Heating a swimming pool with fossil fuel is similar to throwing money into a lake: very expensive and not a really bright thing to do. Heating a swimming pool with electricity in Southern California creates 40 tons (36.3 tonnes) of greenhouse gas emissions in a single swimming season while costing the homeowner in excess of $2000. A solar thermal pool heating system can reduce both of these figures to near zero.*

Figure 5-38. *Solar thermal collectors used for swimming pools may be as simple as these flexible rubber pad units manufactured by Enersol or the high-efficiency flat plate collector design discussed above. A typical collector surface area is approximately 60% to 70% of the swimming pool surface area.*

The Pool Heating System

To get the maximum benefit from your solar pool heating system, the collectors must be mounted on a south-facing roof or structure which is ideally angled to match your geographic latitude. It is also possible to place the collectors on the ground, although there will be a resulting drop in heating efficiency, requiring either the installation of additional collectors or an acceptance of reduced system performance.

The required collector area will depend on the surface area of the swimming pool, the desired water temperature, and the amount of shading from trees and neighboring buildings. As a general rule of thumb the collector surface area should be approximately 60% to 70% of the surface area of the pool. For example, a swimming pool that measures 10' x 20' (3 m x 6.1 m) will require a minimum solar collector area of:

$$\textit{Minimum solar collector area } = \textit{10 x 20 feet x 0.6}$$
$$= \textit{120 sq. ft. (11 m}^2\textit{)}$$

With few exceptions, the existing filter pump will be capable of circulating pool water through the solar panels. In areas where freezing is a concern, ensure that the system is equipped with drain spigots and properly sloped plumbing lines to ensure complete water drainage for winter storage.

Controlling the temperature of the swimming pool maybe accomplished in one of two ways. The first method incorporates nothing more complex

Figure 5-39. This three-season drain back solar pool heating unit extends the swimming season by two months and increases the water temperature when used in conjunction with a solar pool blanket.

than a timer to turn the circulation pump on and off in conjunction with sun up and sun down times. The advantage of this configuration is simplicity and low cost. The disadvantage is reduced pool circulation time and lack of control over pool water temperature.

An automatic solar pool heating control consists of an automatic diverter valve which channels water to the solar collectors based on differential temperature control and desired pool temperature. When the pool reaches the desired operating temperature, the diverter valve redirects the water flow from the solar collectors to the filtration system.

In either design an auxiliary fossil-fuel or electric pool heater may be added to the system. In this configuration the solar collectors will preheat the water before it reaches the backup heating unit. Fossil fuel or electricity will be used only to make up the difference between the solar thermal output temperature and the desired pool water temperature set point.

Figure 5-40. With few exceptions, the swimming pool pump should provide sufficient circulation of water through the solar collector panels. Temperature control may be as simple as a day/night timer turning off the pool water circulation at night.

5.5
Active Solar Space Heating

Before looking at how an active solar thermal system figures into the design, let's review the basic operation of a typical hydronic or hot water heating system.

Hydronic in-floor heating systems are becoming increasingly popular, in part because of the continuous, even warmth it provides throughout the home. As the heat is developed in the flooring material, even hard-to-heat products like ceramic tile stay nice and toasty for cold toes. Many people who suffer from airborne allergens find that their symptoms are greatly reduced with hydronic heating, as there is no dusty air being blown about the house through the centralized fan-forced heating system. Lacking a fan, hydronic systems are also very quiet and energy efficient.

The concept behind hydronic heating is quite simple. Heated water (or an antifreeze mixture) is pumped through a series of flexible plastic pipes located under the flooring or embedded in the insulated concrete slab of your

Figure 5-41. With hydronic heating, hot water or antifreeze solution is pumped through plastic tubing located in or under your home flooring.

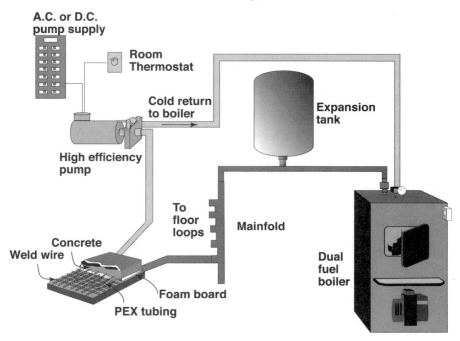

Figure 5-42. A typical hydronic heating system is shown in schematic view.

house. A zone thermostat turns a very tiny, energy-efficient pump on and off as heat is required in each zone. For areas where freezing is a problem, environmentally friendly (propylene glycol) antifreeze is added to the water. This is also required for outdoor boiler heating systems described later in this chapter.

If you prefer to have a forced-air system or would like to add central air conditioning, a hot-water-to-air-heat exchanger can be used. This unit installs in the blower/ductwork assembly and transfers heat from the boiler to the circulating air in the same manner as a conventional forced-air furnace. A second air exchanger can be installed to add air conditioning to the mix.

Another advantage of hydronic heating concerns the boiler unit. Most boilers provide two "heating loops," one for hydronic space heating and the second for domestic hot water heating. These boilers can provide 100% of your space and water heating requirements using renewable resources such as biodiesel, wood, wood pellets, and even dried corn. Dual-fuel boilers are also available, typically combining an efficient wood-burning stove with an oil or electric backup energy supply.

Figure 5-42 illustrates a dual-fuel fired boiler which provides the energy for both space heating and, optionally, a second plumbing circuit for domestic hot water. The dual-fuel fired boiler has a wood-burning section on the top and an oil-fired burner on the bottom. Provided the wood fire is maintained, the boiler is programmed not to start the oil burner unit. This allows you to supply up to 100% of your heating needs from a renewable source.

Hot water or a water/antifreeze mix leaves the boiler and enters a manifold line which contains an expansion tank that is placed in the system to allow the water to expand or contract in volume, depending on temperature. The manifold provides a number of outlets to feed the various zones or area heating loops, each equipped with its own circulation pump and thermostat.

Figure 5-43. Active solar heating panels such as these evacuated tube thermal collectors can supplement your renewable energy hydronic heating system. The amount of solar energy developed during the winter and the heat loss of your home will determine the performance and cost effectiveness of the installation.

The heating loops consist of, for example, a flexible plastic pipe known as PEX, manufactured by companies such as REHAU Inc. (www.rehau-na.com). PEX pipe may be encased in insulated cement flooring over grade or affixed directly to the bottom of interior subfloors. Heating an insulated concrete slab provides a high level of thermal mass and helps to regulate the room temperature, thus providing much higher levels of comfort than central hot air furnace designs, where air temperature fluctuates as a result of the furnace cycling on and off.

PEX lines should be no longer than approximately 300 feet (91 meters), as the heat dissipates with distance. Cool fluid exiting the heating loop is drawn by a high-efficiency circulation pump and returned to the boiler water inlet. Hydronic heating contractors use circulation pumps from manufacturers such as Grundfos (www.grundfos.com) which are rated between approximately 40 and 80 watts at 120 volts.

For on-grid systems, Grundfos manufactures one of the finest circulation pumps that has been proven to operate reliably over many years of service. Although a rating of 40 to 80 watts may appear acceptable, each pump can be expected to operate at a 50% duty cycle requiring 0.5 to 1 kWh of energy per day. It must also be remembered that each zone (area controlled by one thermostat) has at least one pump. For a system with six pumps, the energy requirement may exceed 3 kWh per day; a sizeable portion of off-grid wintertime energy production.

An alternative, low-energy selection is the El-Sid (Electronic Static Impeller Drive) pump from Ivan Labs. This pump does not have a conventional motor and spinning shaft but uses an electronic drive circuit to cause a magnet embedded in an impeller to spin. Being completely sealed, it will last "forever." The El-Sid pump is also very efficient. A 10 Watt, 12 Volt D.C. pump can circulate water through two 300 foot (90 meter) heating loops. To put this in perspective, most people use 15 Watt CF lamps for each table lamp. A direct current-driven pump allows for either direct connection to the battery bank or connection through a converter module (see Chapter 11). Connection directly to the battery bank saves electrical energy by reducing inverter losses. It also ensures operation of the heating system in the event of inverter failure.

The controls for a hydronic heating system couldn't be easier. The pump is connected directly to a zone thermostat which in turn powers the pump on and off based on heat demand.

Heat loss calculations, PEX tubing layout, and other plumbing factors necessitate a discussion with your heating contractor to determine design and installation details.

Factoring in Solar Energy

As you learned above, a solar thermal hot water heating system such as the model outlined in Figure 5-23 will provide between 35% and 90% of your hot water requirements, depending on geographical location. What these figures do not tell you is when this energy is available and how it is dispersed over the year.

Appendix 5 and Appendix 6 illustrate the average sun hours per day for North America for the worst month and yearly average respectively. In the sun belt of the United States the average number of sun hours per day remains relatively constant throughout the year, meaning that solar energy is available throughout the summer and winter months. However, despite the availability of solar energy, home heating requirements in Arizona and

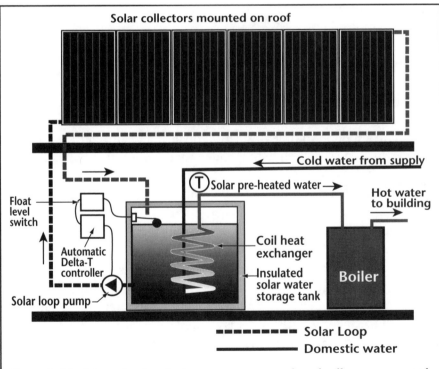

Figure 5-44. A large insulated solar water storage tank and collector array supply preheated water to a hydronic heating system. The practicality of such a system will depend to a large extent on the heat loss (energy requirement) of the building. It is not practical to store heat energy from one season to the next.

Florida are minimal. In the northeastern United States and most of Canada solar energy is concentrated over the long summer hours. Once fall arrives and winter storms set in, sunlight hours per day drop dramatically, coincident with increasing heat demand.

A solar thermal system energy output is balanced to provide an *annual average reduction* in purchased hot water heating fuel as shown in Figure 5-12. Therefore, a solar thermal system will not meet 100% of the wintertime hot water requirement and will have little additional energy for space heating.

Many people believe this is a problem that can be corrected by size: simply add more solar thermal panels and away you go. That's far too easy an answer. Firstly, a larger system may provide ample hot water for space and other domestic requirements at the expense of having an enormous amount of excess energy that could not be used in the summer. Secondly, the increase in capital cost for such a large configuration would almost certainly not be economical.

Perhaps storing some of this excess heat during the summer might solve the problem? Read on.

Thermal Storage and Heat Loss

It stands to reason that the amount of the energy required for space heating will be dependent upon how much heat the home loses. Heat loss is a function of how well the home is constructed and includes such variables as building envelope volume, insulation quality, and air tightness. In addition, the difference between indoor and outdoor temperatures adds a further complication.

There is also a difference between temperature and heat that must be understood prior to tackling the challenge of thermal energy storage. If you heat 1 cup (237 ml) of water over a flame, its temperature will rise more rapidly than 1 quart (1 L) heated by the same flame. Therefore, as the mass of water increases, so does the amount of heat energy required to achieve the same temperature.

In order to store heat, a mass (or weight if you prefer) of a substance will be required. Fortunately, water is an ideal substance as it has an unusual ability to store large amounts of heat energy, a phenomenon known as having a *high specific heat*. For example, the specific heat of lead is nearly 30 times lower than that of water, even though its density is higher. The measure of the specific heat of water is 1 BTU per pound per Fahrenheit degree (1 calorie per gram per Celsius degree). Expressed in English, this means that for every pound of water that is heated by 1°F, one BTU of energy will be

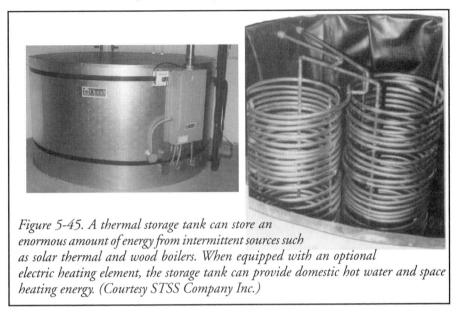

Figure 5-45. A thermal storage tank can store an enormous amount of energy from intermittent sources such as solar thermal and wood boilers. When equipped with an optional electric heating element, the storage tank can provide domestic hot water and space heating energy. (Courtesy STSS Company Inc.)

added. Conversely, for every pound of water that is cooled by 1°F, one BTU of energy will be removed. (As the majority of heating contractors continue to work with British Thermal Units or BTUs, let's stick with this system. For discussion purposes, 1 BTU = approximately 1 kilojoule or 239 calories).

For example, a tank of water weighing 1,000 pounds (120 gallons/ 455 L) which is heated to 50°F (28°C) above room or ambient temperature will have 50,000 BTUs of added energy. Assume for a moment that the tank is installed in a room which has a heat loss of 10,000 BTUs per hour. If we transfer 10,000 BTUs per hour from the storage tank to the room, it is possible to maintain a constant temperature in the room for approximately five hours. Once all of the heat energy is removed from the tank, the room temperature begins to fall.

Stored Energy (BTUs) ÷ Heat Loss (BTU/Hr) = Available Heating Time (Hours)

and:

Heat Loss (BTU/Hr) x Desired Heating Time (Hours) = Heat Storage Required (BTUs)

The simplest way to determine your home's heat loss is to have a heating contractor or home energy auditor runs the calculations for you. It is not possible to generalize the heat loss for all homes and locations other than to say the more extreme the winter temperature and the larger and more poorly built (older) the house, the greater the heat loss. A northern location provides fewer sun hours per day resulting in lower heat production and a colder climate requiring higher levels of heat energy.

A well-built home in wintry Montana may, for example, lose 400,000 BTUs of heat energy per day. In order to store this amount of energy for a one-month period the storage tank must be large enough to carry 12 million BTUs of energy:

400,000 BTU/day Heat Loss x 30 days' Desired Storage = 12,000,000 BTU

A storage tank containing water heated to 200°F (93°C) or 130°F (68°C) above ambient or room temperature would have to weigh approximately 92,000 pounds (42,000 kg) in order to store this much energy. To hold this mass of water the tank would have a capacity of 11,500 gallons (43,500 L). This is simply impractical.

On the other hand, if the heat loss is considerably lower, perhaps due to a more temperate climactic zone or better-built house, a smaller thermal storage unit may be practical. Furthermore, multiple energy sources can supply heat to one central storage tank. For example, the large storage tank shown in Figure 5-45 can be sized to hold up to 1,600 gallons (6000 L) of water with a weight of 13,000 pounds (5897 kg). Assuming a water temperature of 100°F above ambient, this equates to approximately 1.3 million BTUs of energy, enough to supply a family of four with a couple of weeks' worth of domestic hot water.

The manufacturer STSS Company Inc. of Mechanicsburg, Pennsylvania has developed a variety of sizes of these unique storage tanks, each with the ability to be shipped flat and then expanded at the final destination. Water/glycol solution can be circulated in a closed loop through the spiral heat exchanger, supplying domestic hot water and space heating.

An integral electric heating element and thermostat provide backup energy to maintain tank water temperature. Heat exchanger coils may also be routed to a wood- or oil-burning boiler, heat pump, and solar thermal collectors for additional energy input and storage.

The complexity of thermal storage units dictates that a comprehensive heat loss analysis of your home be completed prior to sizing the energy input sources and possible thermal storage system. This applies to solar thermal heating as well as to traditional heating sources such as fossil-fuel and electric heating supplies.

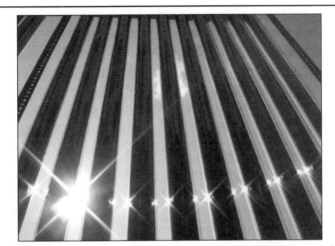

Figure 5-46. Solar thermal systems are the most economically feasible means of harnessing the sun's energy provided they are installed professionally and for the correct application. These evacuated tube solar thermal collectors provide the homeowner with free domestic hot water and a return on investment of 15% or more. (Courtesy Carearth Inc.)

Given the difficulty in cross-season energy storage (from summer to winter), solar thermal systems do not generally make *eco-nomic* sense in areas where winter heat demands are high.

5.6
Heating with Renewable Fuels

If you ask people to name all the renewable home-heating fuels, chances are they will stall immediately after they mention "firewood." In many parts of the developing world wood is as scarce as the proverbial hen's teeth, leaving people to search for alternatives such as dried peat or cow dung. Fortunately we don't have to chase Bossy with a shovel in order to lengthen our list of renewable heating options.

Chapter 16 examines biodiesel as one renewable and clean-burning fuel for home heating and transportation requirements. For those heating with regular fossil furnace oil it may come as a surprise to learn that No.2 automotive diesel fuel, heating oil and its renewable partner, biodiesel are virtually interchangeable.

Biodiesel is a vegetable oil or animal fat derivative that is clean burning as well as carbon neutral, which means that it contributes no net carbon dioxide greenhouse gas emissions into the atmosphere. With North American production of biodiesel running in excess of 30 million gallons per year (79 million liters per year) and numerous states offering it as a clean alternative to fossil-fuel heating oil, demand is starting to heat up.

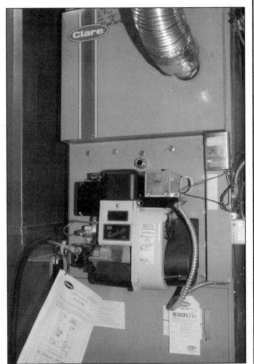

Figure 5-47. Biodiesel is a vegetable oil or animal fat derivative that is clean burning and carbon neutral, which means that it contributes no net carbon dioxide greenhouse gas emissions to the atmosphere.

Wood, Wood Pellets, and Corn

Heating with wood goes back thousands of years. Europeans who settled in the New World along the east coast and in Pennsylvania experienced winters in North

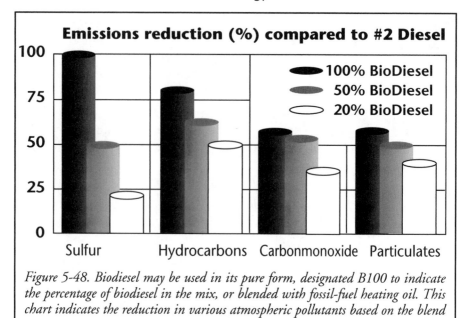

Figure 5-48. Biodiesel may be used in its pure form, designated B100 to indicate the percentage of biodiesel in the mix, or blended with fossil-fuel heating oil. This chart indicates the reduction in various atmospheric pollutants based on the blend concentration.

America that had a ferociousness never seen in Europe. Rapid population growth coupled with poor energy efficiency of both wood-burning appliances and homes caused heavy deforestation and left towns with a grimy, sooty pall. It's no wonder that the majority of the population eventually jumped on the oil, natural gas, and electric heat bandwagon. Write a check, set the thermostat, and you have instant comfortable central heating—what could be better? Why would anyone take a step "backwards" into the wood-heating scene?

Since cavemen started using fire, there have been several advances in wood-heating technology. These advances have included new wood-burning stove designs that increase the amount of heat recovered and at the same time lower environmental pollution. Besides saving the environment, these technologies can also save your back; when you couple the energy-conservation techniques discussed in Chapter 2 with clean, high-efficiency wood-burning appliances, you reduce the amount of wood you have to cut, split, pile, and burn.

When you have decided on wood (which for the sake of brevity "wood" includes wood pellets and dried corn fuel sources) as your main or secondary source of home heating, check with your stove supplier or local building inspector for installation details. Since the installation of wood and pellet

Figure 5-49. Firewood stacked neatly in your garage or yard is like money in the bank. When properly burned in an EPA-certified wood heater it is a clean-burning energy-efficient heat source (Canada utilizes the EPA certification system). Wood pellets and dried corn also provide a clean-burning, renewable heat source with the added benefit of automatic stoking in many models of heating unit.

stove systems is complex and varies from region to region, no specific installation details are provided in *The Renewable Energy Handbook.* It is highly recommended that a contactor or professional installer do the installation work. Use the details provided here to increase your knowledge and understanding of wood-burning appliances and to help you purchase the best unit to meet your needs.

Wood-Burning Options

Burning wood is not just about tossing a log into a fireplace or refurbished wood stove from the antique dealer and expecting to heat your house. It takes a high-tech wood-burning stove to heat cleanly and efficiently for a long period of time. There are several types of stove that fall into this category:

- Advanced combustion stoves
- Catalytic stoves
- Wood pellet and corn stoves
- Russian or masonry heaters
- Wood furnaces and boilers

Designers and manufacturers of these units have as their primary aim the safe and clean extraction of every BTU of energy available from the wood you load into the stove. The reduction in frequency of wood loading lowers

Figure 5-50. This Vermont Castings (www.vermontcastings.com) catalytic wood stove easily heats this 3,300 square-foot, ultra-energy efficient home. Located in the very cold reaches of eastern Ontario, the home requires only two cords of firewood per heating season.

costs by reducing the amount of wood purchased or cut and processed.

The increase in efficiency comes first from creating an airtight burning chamber, causing controlled combustion, and then from developing a secondary combustion process, burning the smoke emitted from the fire before it reaches your chimney. (The masonry stove differs from this concept and will be discussed later in this chapter.) Although the EPA wood heater certification program was created to reduce air pollution, it resulted in added benefits like higher efficiency and increased safety. On average, EPA-certified stoves, fireplace inserts, and fireplaces are one-third more efficient than older conventional models. That means one-third less cost if you buy your wood and a lot less work if you process it yourself.

An airtight burning chamber is required to limit the amount of oxygen reaching the fire, lengthening the burning time and evening out the heat flow from the wood stove. This eliminates traditional fireplaces as efficient heat sources, as the energy they produce is negated by the massive amounts of cold makeup air, drawn from the outdoors, that is required to keep the fire burning. A slow, smoky fire also increases airborne pollutants and the risk of chimney fires as a result of unburned fuel coating the chimney as creosote.

The idea of burning the smoke before it reaches the chimney may sound

Figure 5-51. As this picture illustrates, there is an amazing array of wood-burning stoves and fireplaces available for purchase. A reputable dealer such as Embers located in Ontario, Canada is a good place to start your quest for home-heating energy self-sufficiency.

like snake oil, but it's true. Modern wood-burning appliance technology is aimed at developing ways of capturing the unburned fuel present in the emission of smoke. Smoke is the result of burning wood decomposing into "clouds" of combustible gases and tar. Applying additional oxygen and heat or special catalyzing materials causes a secondary "burn." This results in increased heat output and lower atmospheric emissions as well as a decrease in fuel dollars and backache.

Advanced Combustion Units

Advanced combustion wood stoves expand on the concept of a simple box stove that has been in use for over a hundred years. Simple stoves allow the heat and smoke of the fire to travel in a direct path up the chimney. The advanced combustion stove places an "air injection" tube into the smoke path. Secondary inlet air is drawn in through the tube, increasing the oxygen content of the smoke and causing it to burn. The smoke then travels through a labyrinth path, radiating heat before it exits via the chimney. This approach is like adding a turbocharger to a car's engine: free horsepower from otherwise wasted exhaust.

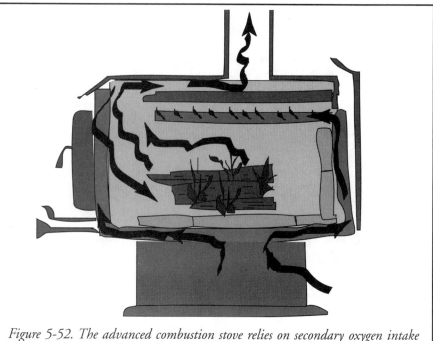

Figure 5-52. The advanced combustion stove relies on secondary oxygen intake and smoke burning to increase efficiency.

Catalytic Units

Once you understand how a catalytic wood stove operates, you can impress your friends by telling them how an automotive catalytic converter works. The process is really the same except that in the case of a car the unburned fuel is in the exhaust. Unburned wood smoke is routed through a catalytic device, a ceramic disk made with a myriad of honeycomb holes running through it. The disk is coated with a special blend of rare earth metals that have the unique feature of lowering the combustion temperature of the exhaust gases when mixed with a secondary supply of atmospheric oxygen.

As the smoke (or exhaust) passes through the catalytic device, it is mixed with oxygen from an air inlet and it ignites. If you watch the wood-burning stove's catalytic converter operating, you will see it glowing red hot and engulfed in flickering flames as the smoke is being consumed. A quick glance at the chimney on a crisp, cold January afternoon verifies the operation, with nothing but wisps of steam being emitted.

The downside of catalytic technology is the requirement to move a control handle from the bypass to operating position once the stove has reached operating temperature. Although this may sound like a minor issue, laziness

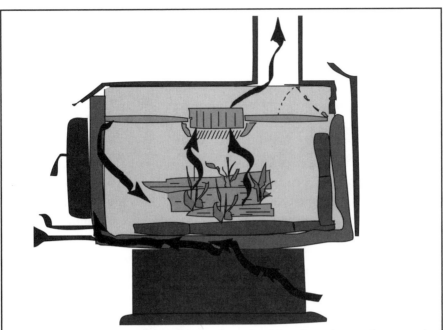

Figure 5-53. The catalytic wood stove uses a device made with special rare earth metals which allows wood smoke to burn in the presence of a secondary air supply.

Figure 5-54. The wood pellet or corn stove uses a motor-driven auger to feed precise amounts of waste wood pellets into a fire pot.

or neglect in operating this control will negate the energy-efficiency and emission-reduction capabilities of the stove. (I will personally admit to leaving the converter turned off in the early fall and spring, when home heating demand is much lower.)

Another consideration is the cost of replacing the catalytic converter after burning approximately 10 full cords (4' deep x 4' high x 8' long. / 1.2 m x 1.2 m x 2.4 m) of firewood.

Wood Pellet and Corn Stoves

Pellet stoves use dried corn and waste wood by-products from manufacturers of furniture, lumber, and other wood products. The waste material is ground and pressed together using naturally occurring resins and binders to hold the rabbit-food like pellets (Figure 3-49 inset) together. As a waste product, biomass fuel pellets offer excellent synergy, heating your home while reducing landfill waste and needless greenhouse gas emissions at the same time. Pellets are convenient, as they are supplied in neat and compact dog food-size bags which can be stacked in your garage ready for use. Simply scoop a bunch into the hopper of the stove about once per day and the controlled feeding unit will automatically deliver the right amount of fuel to the burner. The only downside is: no electricity, no fire. A biomass stove draws electrical power 24 hours per day and may require too much energy for the off-grid system to supply during the dark, dreary days of winter. Use caution when evaluating the energy demands of these devices.

Most biomass stove manufacturers provide an automatic battery backup system to operate their units in the event of a power failure. This device operates in the same manner as an uninterruptible power supply (UPS) for a home computer. During normal operation the pellet stove operates from utility power and the UPS recharges simultaneously. When a power failure occurs, the UPS draws on a battery and inverter system to generate electricity, powering the stove. The UPS must be sized in accordance with the electrical load of the stove and the estimated number of hours before utility power returns.

Russian or Masonry Units

When we think of Russia, we think of winters in Siberia. In that neck of the world, serious heat is needed, and there is no fooling around with wimpy stoves. The masonry heater is a serious unit. It is designed with the firebox and chimney lined with refractory brick and the flue routed through a labyrinth path designed to slow the smoke on its way to the chimney. An external facing of brick, stone, or adobe completes the design and increases the mass of the unit.

When these units are operated, a fast, furious fire is ignited in the firebox. Smoke and its accompanying heat zigzag through the flue passages, giving up energy to the surrounding masonry work. Depending on the locale, one fast fire per day is all that is required to heat the masonry, with the stonework slowly radiating its stored heat into the house.

The masonry unit achieves its environmental passing grade by creating a hot, fast fire. Fast firing of a stove will generate the same amount of heat energy as a slow fire, but in a shorter amount of time. The resulting hot fire consumes more oxygen than a slow, smoldering fire, burning more completely and emitting fewer pollutants.

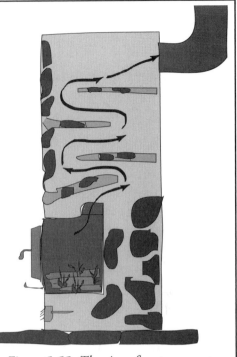

Figure 5-55. The size of a masonry stove makes it the centerpiece of the house. A popular design from the Old World, it provides heat long after the fire has burned down. (Courtesy Temp-Cast Enviroheat)

The downside of masonry heaters is their cost, the need to keep re-firing, and their size and weight. Assuming you have the space and the floor joists to support one of these big guys, they offer a warm and pleasing welcome to any home.

Wood Furnaces and Boilers

Wood furnaces and boilers come in two varieties: indoor and outdoor. Outdoor wood boilers are super-large wood stoves surrounded by a tank of water or antifreeze solution. Heat from the burning wood is transferred to the fluid, which is then pumped into the house for space and water heating. A control system located in the house regulates the temperature of the boiler by dampening the fire. Most units also provide an alarm signal to indicate when it's time to re-stock the unit. Furnaces, which provide warm air, are available as indoor units only.

Indoor wood boilers and furnaces are often dual-fuel fired, like the unit shown in Figure 5-57 from Benjamin Heating Products (www.benjaminheating.com). These units are very similar to outdoor models except for their size, emission of pollutants, and appetite for firewood. A major advantage of indoor, combination-fired boilers is the ability to supply heat even after the fire has burned down. The Benjamin unit contains both a high-efficiency wood-burning stove and an oil-fired backup burner which is only used when the wood-fired section cannot supply sufficient heat to maintain the desired boiler temperature. Units are also available with fan-forced heating which supplies a standard central air plenum or as hot-water boilers for use with hydronic heating.

The Specifics of Wood Heating

The most popular style of heating system is the central warm air distribution design. A thermostat commands a centralized furnace to heat an air plenum or chamber. An electric fan circulates room air through the hot plenum to a series of ducts throughout the house. When the thermostat senses the room temperature is warm enough, the furnace turns off, causing the house to cool,

Figure 5-56. Outdoor wood-fired boilers such as this model keep the wood and chips outside. Although these units are popular they are also inefficient, gobbling up vast quantities of wood while spewing out considerable amounts of atmospheric pollutants because of their inefficient burning chambers, which cause smoky, smoldering fires.

Figure 5-57. The combination boiler provides space and water heating in one compact unit. This model burns wood and oil efficiently and provides maximum flexibility of fueling options. (Courtesy Benjamin Heating Products)

Figure 5-58. The Multi-Heat automatic stoker boiler is especially suitable for effective and environmentally friendly firing with biofuels such as wood pellets and corn. The unit's efficiency of approximately 90% rivals that of high-quality oil and gas furnaces. The large hopper must be filled approximately once or twice per week depending on heat requirements. (Courtesy Tarm USA Inc.)

whereupon the cycle repeats.

Hot water or steam boiler systems replace the hot air with water. These systems use either radiator units to transfer heat to the room or hydronic, in-floor hot water pipes as shown in Figure 5-41.

A wood stove or masonry unit uses direct radiation and convection to distribute its heat.

When determining your choice of heating system, consider the following:

- Will the renewable heating system be your primary heat source or will it be supplementary? If it is to be your primary heat source, you must consider backup heating methods, particularly in colder areas subject to freezing.

- If you live in a more temperate area and travel little, and renewable heating is to be your primary heat source, consider pellet or masonry heaters. These models can provide sufficient heat for up to one-and-a-half days. The Tarm USA Inc. Multi-Heat stoker boiler shown in Figure 5-58 can operate for up to one week without refueling. Heating for longer periods of time will require a friendly neighbor to stock the stove or an alternate heat source with automatic controls that activate when the fire dies.

Figure 5-59. A wood stove placed in a central location near open areas of the home facilitates airflow, helping to ensure an even flow of heat. Remember to plan access for wood supplies coming in and ashes going out.

- Whichever method you choose, consideration must be given to the amount of electrical energy required to operate the heating system. Convection wood stoves require no electrical energy but must be tended frequently. On the other hand, the Tarm stoker boiler will operate independently for up to one week, albeit at a cost of electrical power that may be in short supply during the dark winter months.

Anyone who has ever been to a cottage that is heated with an old Franklin or cook stove cannot imagine using wood as a primary source of heat. These old stoves gobble up armfuls of wood and never seem to get the place really warm a lot of the time. At other times it's the opposite problem: there is too much heat and belching smoke, making you wonder if your cottage has been transported straight into the depths of hell.

Many people use wood or wood pellet stoves as their primary heating source. Combining one of these stoves with proper energy-conserving techniques and passive or active solar heating will give you the best heating synergies. The answer to avoiding the cottage nightmare is good planning and the use of quality equipment.

Figure 5-60. A wood pellet stove offers the look of a traditional wood-burning stove with the convenience of only having to add a few scoops of pellets into a hopper to provide equivalent heating capabilities. (Courtesy Harman Ltd.)

There are several systems for using wood as a primary heating source:
- Convection space heaters
- Wood-fired furnaces (single- or dual-fuel combinations)
- Hydronic or hot water in-floor heating

Convection Space Heating

Convection space heating is the most common type of wood heating installation. No electricity is required to operate this type of unit, which is doubly good when the grid fails. Locate the wood stove in a central part of the house, with open passages to the upstairs and other rooms. An example of a well-placed unit is shown in Figure 5-59, where heat from the stove can easily move around the home and up the open stairwell.

A common problem with space heating units occurs when people purchase a model too large for their needs. Make sure you discuss size with your dealer, and explain your requirements regarding primary or secondary heating as well as your floor plan. Most first-time wood stove users cannot believe that a small unit is capable of heating an entire house, but it is! On the other hand, if you are the sort who likes to wear shorts in January, maybe this isn't such a bad thing.

If your heating area is enclosed, or broken into many rooms, it is possible to use small "computer fan" units to blow air down hallways or between rooms. These units are available from wood stove dealers. Many building codes allow for grates to be cut between the ceiling of the heated room and the floor above, allowing natural air circulation.

Another option is to consider a pellet stove in tight locations. Because of their precise electronic controls, it is possible to lower burning rates below that of a typical wood space heater without degrading environmental operation and still provide a glowing hearth. An example of a Harman Ltd. space-heating pellet stove is shown in Figure 5-60.

A wood pellet stove hopper holds up to one-and-a-half days' worth of fuel. This offers the added advantage of not having to run back home to re-stock the stove or bugging your neighbor should you decide to stay out late at night.

Homes with an existing central furnace provide an automatic means of supplying backup heat. When the fire goes out the thermostat turns on and the furnace kicks in. This also provides an alternate means of moving the heat from around the wood stove. Figure 5-61 illustrates such a design. Heat from the wood stove pools at the ceiling level. An existing or additional cold air return or "suction duct" is located near the ceiling. The furnace fan

Figure 5-61. The air distribution provided by the central furnace fan provides the best method of moving heat away from the wood space heater. A ceiling fan is a good choice for cathedral ceilings or off-grid homes.

is left in "manual" or low-speed mode, drawing excess heat from the room and circulating it throughout the house.

An effective alternative for homes without centralized air distribution or constructed with cathedral ceilings is to place ceiling fans in the heated area, blowing warm ceiling air downward and mixing it with cooler room air.

Wood-Fired Hot Air Furnaces

Wood-fired furnaces operate in the same manner as fossil-fuel units. The main advantage of wood furnaces is the fan-forced circulation system that allows even heat distribution. All models contain automatic dampers, thermostatic fan control, and safety shutdown dampers. Due to the air-heating plenum design, central air conditioning may be added without difficulty.

Most models of wood-fired hot air furnaces are designed as dual-fuel units. These systems switch from wood to a fossil/biodiesel fuel (OK) or electric resistance heat source (not allowed off-grid) automatically as the fire dies down.

Before purchasing one of these models ensure that it carries the EPA certification label. Older designs of wood/electric furnaces that were sold during the height of the 1970s oil crisis are woefully inefficient and will eat you out of house and home.

Fan-forced hot-air furnaces are generally not recommended for off-grid applications. This recommendation stems from the energy requirements of the blower motor during the winter months when renewable energy electrical production is reduced. Possibly the most efficient fan-forced model is the Benjamin model similar to the one shown in Figure 5-57. Its energy requirements are:

250 watt motor x 7 hours per day - 1.75 kWh per day consumption

Depending on your location and heating requirements, you may need more or less fan running time per day. This simplified example shows that even a unit having a very small, energy-efficient fan requires a large percentage of your winter electrical power generating capacity.

A Word About Outdoor Wood Boilers

The outdoor wood boiler shown in Figure 5-56 is becoming a common sight in rural homes and farms in the northeastern United States and Canada. Although these units are popular they are also very inefficient, gobbling up vast quantities of wood while spewing out considerable amounts of atmospheric pollutants because of their poorly designed burning chambers, which cause smoky, smoldering fires. Many people say that the wood is free, so who cares about efficiency? I personally know several people who own these monsters and I am amazed by the number of hours they require to cut, split haul, and load wood for the cavernous maw of these units. If you add up all of the labor, fuel, and chain saw expenses required to keep these units fed, you will find them anything but cheap.

Many towns and municipalities are outright banning these devices, and rightly so. If you care about the environment, forget about purchasing one of these environmental disasters and consider a solid fuel combustion unit with an EPA certification providing low combustion emissions.

Wood Fuel – The Renewable Choice

Nature provides us with an endless supply of wood. Dead trees rotting on the forest floor produce the same amount of greenhouse gases as if you had burned the wood in your stove. Sustainable woodlot management requires the thinning and cutting of damaged and dying trees. Even in a small acreage,

Figure 5-62. Propane fireplaces, radiant heaters and freestanding units such as this model require no electricity and can be sized to meet the heating requirements of any house. They also look better than the $5,000 furnace in the basement.

this "waste wood" will provide sufficient fuel for the largest house.

The species and quality of the wood you burn will have a major impact on the ease of use of your heating system. Freshly cut, wet wood can contain up to 50% moisture by weight. Attempting to burn this wet wood will result in difficult ignition, with reduced heat output and greatly increased pollution. On the other hand, properly dried and seasoned wood will ignite rapidly and provide nearly twice the warmth with less work.

Firewood should be cut and split (at least in half) early in the spring and properly stacked. Piles of wood should be neatly arranged in covered rows, allowing space between each successive row for air to blow through. The summer warmth and breezes will quickly reduce the weight of this wood by half, saving your back when you bring it inside next winter and at the same time ensuring that the stored energy is not being used to boil off water and sap in unseasoned wood. During the seasoning process, wood will start to crack and split on the ends and turn a grayish color. Picking up two pieces of well-seasoned wood and banging them together will create a clear ringing tone. Doing this with freshly cut wood will result in a dull "thud."

The type of wood you purchase or cut will also make a difference. Softwoods such as white birch and poplar are fine for fall and spring when you are just a bit chilled. However, winter heating requires serious heating woods such as maple, oak, elm, and ironwood. These woods have a higher density than softwoods, resulting in higher heat output. Pick up an armful of white birch and an armful of maple. The maple feels twice as heavy as the birch, and guess what? It puts out about twice the energy for the same volume. As wood is sold and trucked by volume, purchasing the hardest woods will save you money and reduce the number of trips to the woodshed. Just remember to get at least a little mix of the softer woods for those fall and spring days when the sun is providing some of the energy heating mix.

Purchasing Firewood

Your local dealer will offer you a confusing blend of softwood, mixed wood, mixed hardwood, fireplace cords, face cords and full cords, green or seasoned. It is important to understand what all this means; otherwise your experience with heating in winter may make you think you are back in colonial Pennsylvania.

Firewood is sold in "face cords" or "full cords." The wood is typically delivered pre-cut in 16" (41 cm) logs. When stacked in three rows, a full cord of wood should measure 4' deep x 4' high x 8' long. (1.2m x 1.2m x 2.4 m). A face cord is, as the name implies, one row or "face" of a full cord, which is equal to one-third of a full cord (16" deep x 4' high x 8' long / 0.4m x 1.2m x 2.4 m). But be careful. If the logs of the face cord are cut into 12" (30 cm) lengths, it will yield only a quarter of a full cord.

Figure 5-63. Firewood should be properly stacked and covered and allowed to season for at least six months to drive off moisture. Dry firewood should be a grayish color and have cracks on the ends.

If firewood is sold "green" it means that the wood is not seasoned. If you can get a discount for buying green firewood and have the time to season it, go right ahead. Wood can also be purchased in 4' or 8' lengths. Provided you have the tools, time, and stamina to cut and split your own wood, the savings may be well worth the effort. To quote Thoreau, "Wood heats you twice, once when you cut it and once again when you burn it."

How Much Wood Do You Need?

The best way to figure this out is to try. A well-insulated house, such as our 3,300 square foot, ultra-insulated home, uses only two full cords (4' x 4' x 16'/ 1.2m x 1.2m x 4.8m) of mainly hardwood (and a bit of softwood) per year. A smaller, turn-of-the-century stone home just down the street, which has no insulation to speak of, uses six full cords. If you have an inefficient outdoor boiler, you may use 3 to 4 times the amount of wood used by a high-efficiency indoor stove. It just depends. The best thing to do is to buy more than you think you may require. Covered wood will not rot and should last up to three years, after which it will start to decay. So purchase a bit more and let it sit like money in the bank. You can always use it next year.

Ecology and Wood Heating

Many people argue that heating with wood moves the source of pollution from the "clean-burning stove" to the chain saw and splitter. While there is no doubt that forest management and wood harvesting takes energy (plenty of it according to Thoreau) the environmental impact can be managed with the proper attitude and correct tools and supplies.

The best way to reduce wood harvesting energy consumption, whether it's your energy or chain saw fuel, is to reduce wood consumption. A home that is properly constructed or retrofitted for energy efficiency, along with the most efficient wood heater you can afford, can reduce wood consumption by a factor of 2 or 3 times.

Proper wood lot management is an important area to consider and consultations with your local forestry management association will ensure sustainable harvesting practices.

As for the gasoline-powered tools, consider the following options:
- Purchase chain saws or other gas-powered tools that are equipped with catalytic converters to reduce emissions.
- Small two-cycle engines are notorious for spewing oil and other air pollutants. Manufacturers such as Jonsered (www.jonsered.com) offer several environmentally friendly products which have their own Environmental

Product Declaration, meaning that each product will be documented for manufacturing, packaging, recycling, and waste disposal.

- Purchase biodegradable chain saw oil and grease. These products won't damage the ecosystem or your lungs. Chain saw oil is vaporized during the sawing process, creating an oil-based aerosol. Biodegradable oil products lessen the health impact of breathing this potential hazard.
- Do not leave engines idling unnecessarily.
- Use ethanol-based gasoline to reduce CO_2 emissions.

You cannot eliminate the energy requirement and the pollution created in harvesting wood, but with a little care and attention to detail these can be dramatically reduced.

Figure 5-64 To reduce the environmental impact of harvesting wood, use sustainable wood lot management practices, ethanol-based gasoline in catalytic converter-equipped chain saws, and biodegradable vegetable-based chain oils and greases.

5.7
<u>Space Cooling Systems</u>

Central air conditioning is by far the largest and least efficient load in the home. It does not belong in an off-grid home.

The best way to keep your home cool is to stop the heat from getting inside in the first place. This might sound simplistic, but a well-insulated home with shading that blocks the summer sun is well protected against overheating. Open windows at night to create a cross-breeze and cool the house. In short, follow all of the guidelines discussed in Chapter 2, "Energy Efficiency", paying particular attention to:

- light-colored roof and exterior walls
- upgraded insulation in walls and roof
- upgraded radiant barrier insulation in attic rafters
- low-emissivity (low-E) windows that are not required to aid in winter heating
- window shading
- nighttime cooling using fans and natural cross-flow ventilation

Evaporative Coolers

For those lucky enough to have relative humidity levels below 30% in the summer, an evaporative cooling unit may be the ticket. These devices are very common in the southwestern region of the U.S. and use very little electrical energy.

Working in the same manner as perspiration cooling our bodies on a hot day, air blown through a wet pad is cooled as the water evaporates. An electric fan draws outside air to this wet pad, humidifying and at the same time cooling it. The conditioned air is blown into the home causing hot, stale air to be expelled outside. Because cooling only works with dry outside air, these systems do not work well where summertime relative humidity exceeds 70%.

Most evaporative coolers are rooftop installations that use a bottom discharge blower to channel conditioned air into the home. Rooftop-mounted units are less expensive than ground-mounted configurations. Evaporative coolers can also be installed as add-ons to conventional refrigeration-type central air conditioning or window-mounted units.

Evaporative coolers require large amounts of moving air to cool a home. For desert climates a unit capable of supplying three to four cubic feet per minute of airflow is required for every square foot of floor area. For most other climates this airflow can be reduced to two to three cubic feet per minute.

Refrigerant-Based Air Conditioning

Most A/C systems are sized far larger than needed. This is done in an effort to make sure that "you're getting what you paid for": a very fast, obvious cooling of the indoor air. It also ensures that you don't call the installer back because the A/C isn't working well. On the other side of the coin, a large unit cools the air but does not have sufficient time to reduce indoor humidity levels. A smaller unit running for a longer period will ensure lower indoor humidity and temperature.

All air conditioners sold in North America must have an EnergyGuide label similar to the one shown in Figure 3-49. Use this label as a guide to compare the energy efficiency of different models.

An alternative labeling program is known as the Electrical Efficiency Rating (E.E.R.) and is calculated by dividing the BTUs of "cooling power" by the watts of electrical energy used. An air conditioner with a higher E.E.R. rating is more energy efficient and less expensive to operate.

Don't get caught up in a quest for the lowest-cost air conditioner at the expense of energy efficiency. This is false economy, as any first-cost savings will be absorbed by higher operating charges time and time again. These rules apply to window air conditioners as well as central cooling systems.

Room air conditioners are less expensive to operate than central units if you only need to condition one or two rooms. When selecting a room air

Area to Be Cooled (Ft²) / (M²)	Capacity (BTU per hour)
100 to 150 / 9.3 to 14	5,000
150 to 250 / 14 to 23	6,000
250 to 300 / 23 to 28	7,000
300 to 350 / 28 to 33	8,000
350 to 400 / 33 to 37	9,000
400 to 450 / 37 to 42	10,000
450 to 550 / 42 to 51	12,000
550 to 700 / 51 to 65	14,000
700 to 1000 / 65 to 93	18,000

conditioner, match the area to be cooled to the rated capacity of the unit as shown in the accompanying table.

Although these units are very large loads on an off-grid system, they may be just the ticket to bring down the humidity and temperature to a reasonable level. One plus for these small units is that they tend to be used only when there is a surplus of energy. This normally occurs on those long, hot summer days that make the photovoltaic panels so happy and energy productive.

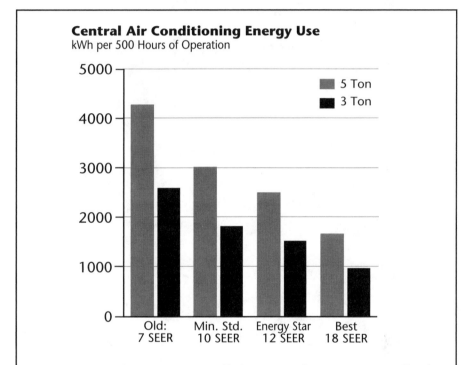

Figure 5-65. Purchasing an energy-efficient air conditioning system will reduce operating costs by more than half. Purchase models carrying the ENERGY STAR label or with a higher efficiency level based on EnergyGuide or EnerGuide ratings. (Source: Washington State University Extension Energy Program)

Figure 5-66. Now you see it, now you don't. This one-ton (12,000 BTU) air conditioner is only required occasionally. Installing it permanently in a wall-mount design as shown here reduces the amount of encroachment on your living space and allows for easy hiding.

5.8
Systems that cannot be used Off-Grid

Geoexchange technology is more commonly known by its many nicknames such as geothermal, heat pump, and ground-source heating. No matter what you call it, geoexchange technology offers a virtually endless supply of renewable solar energy that is stored just below the earth's surface since approximately 50% of the sun's energy is absorbed into the ground.

Figure 5-67 The air-to-air heat pump is a cross between a central air conditioning unit and a geoexchange heat pump. Heating and cooling energy is transferred between indoor and outdoor air rather than between indoor air and the earth or well water. The outdoor component contains the compressor and heat exchange coil while the indoor unit contains the air handler, blower, and second heat exchange coil. Insulated refrigerant lines run between the two components.

The basic concept is easy to understand. In winter, heat energy is drawn from the earth through a series of pipes called a loop. An antifreeze solution circulates through the piping loop, carrying the earth's natural warmth to a compressor and heat exchanger inside the home, hereinafter called a "heat pump."

The heat pump concentrates the earth's energy and transfers it via duct work or a hydronic heating system for space heating requirements. If domestic hot water is also desired, the heat pump can supply energy directly to any thermal water storage tank.

According to the U.S. Environmental Protection Agency and Natural Resources Canada, "**They are the most energy-efficient, environmentally clean, and cost-effective space conditioning systems available.**" Because geoexchange systems burn no fossil fuels onsite, there are no locally produced atmospheric emissions. If the electricity to run the geoexchange unit comes from hydroelectric sources, greenhouse gas and other emissions are reduced to zero.

There are approximately 500,000 installations in the United States and 35,000 in Canada resulting in annual energy savings of approximately 4 billion kWh of electricity, which eliminates 20 trillion fossil-fuel BTUs and slashing greenhouse gas emissions by approximately 3 million tons (2.7 million tonnes).

Air-to-Air Heat Pumps

An air-to-air heat pump operates in the same manner as a geoexchange heat pump with the exception that heating and cooling energy is transferred between indoor and outdoor air rather than between indoor air and the earth or well water. The common central air conditioning unit is an example of a unidirectional heat pump that can only provide space cooling.

That Was the Good News

Heat pumps use electrical power to operate a compressor which extracts approximately two-thirds of the home's required heating energy from the ground. Because this earth-based energy is free and the heat pump puts more energy into the home than it consumes from the electrical grid, heat pumps have an energy efficiency rating of greater than 100%. For every kWh of electricity supplied to the heat pump, between 2.8 and 6.7 kWh of energy are supplied to the home. This efficiency factor is known as the *coefficient of performance* (COP) and is calculated by dividing the heat output (in watts) by the energy input.

Unfortunately, even though these units are very efficient, the compressor and various electrical components require an enormous amount of electrical energy at a time when the winter sun will not be cooperating. Although you might be tempted to run a backup generator all winter long to offset the energy crunch, it simply will not work.

The banned list of heating and cooling appliances includes:

- Geothermal heating/cooling systems
- Central air conditioning units
- Electric heat of any type (including plug-in room heaters). Do not be fooled by advertisements that indicate that a given model is 100% efficient or better than the rest. They are all banned from off-grid houses.
- Most central furnaces, unless they are operated as a backup or secondary source of heat
- Hydronic (in-floor) heating, unless an electrical load calculation for the pump and control system has been carefully determined and allowed for in the electrical energy generation plan.

6
Photovoltaic Electricity Generation

The process of capturing and using solar energy is as old as time. Like most things in the modern world, the simplicity of capturing the sun's energy has been elevated to new technological heights. Unlike solar heating collectors that are used to warm fluids running within the collector, photovoltaic cells *magically* convert light from the sun directly into electricity. The photovoltaic cell used in renewable-energy systems is definitely the product of rocket scientists, powering communications satellites to ensure everyone on the planet can watch reruns of *Friends*.

Figure 6-1. Photovoltaic cells are the backbone of renewable-energy systems for homeowners. Since prices and supporting technologies have improved, hundreds of thousands of homeowners are now living lightly on the planet. (Courtesy Sharp Solar)

The term photovoltaic is derived from the Greek language "photo," meaning light, and "voltaic," voltage which assists the flow of electricity. Friends simply call them "PV" cells for short. Bell Laboratories discovered the PV cell effect in the 1950s. It didn't take the folks at NASA very long to figure out that PV cells would be an ideal means of producing electricity in space. Many missions later, PV cells have improved in performance and have come back down to earth in price. Nowadays, the technology is used in watches, calculators, street signs, and renewable-energy systems for the urban homeowner.

Figure 6-2. PV technology is everywhere you look. Watches, calculators, and even street signs use this ultrareliable technology to provide electricity wherever the economics makes sense.

What is Watt?

For the home renewable-energy system, PV products are relatively standardized, allowing even a novice to make accurate comparisons between product lines. There are currently four major product technologies that should be seriously considered for home use: single crystalline, polycrystalline, laminate (roof shingles) and string ribbon cell. Other cell technologies such as "thin film" are an option provided product warranty and manufacturer financial strength to honor the warranty period are acceptable.

Figure 6-3. PV cells are similar to transistors, except they're larger. Individual cells are polished, interconnected, and mounted in a frame, creating a PV module.

PV Cell Construction

PV cells are transistors or integrated circuits on steroids. Most people have seen the latest microcomputer chip used in PCs. It's a silicon wafer about the size of your thumbnail that holds several million transistors and other electronic parts. PV cells start out the same way as chip circuits, but they are kept in the oven until they're much larger, approximately 4" (10 cm) in diameter. The baked silicon rods are sliced into thin wafers which are polished and assembled with interconnecting electrical wires.

Figure 6-4. If too much whiskey is added to the batter, PV cells get a bit wavy. Flexible modules are lightweight, easily transported, and rugged, making them ideal for boats, RVs, and trips to the cottage. They can even be fabricated as roofing shingles if you are concerned about "curb appeal." (Courtesy Spheral Solar Power Inc.)

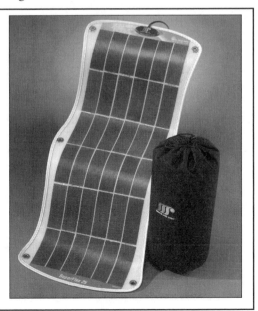

If we were to take one wafer, expose it to bright sunlight, and connect it to an electrical meter, we would measure 0.6 Volts DC (Vdc) of electrical pressure. A voltage higher than the battery rating is required to "push" electrons and charge the battery. For example, to charge a 12 Vdc battery, at least 15 Vdc is required, plus some additional voltage for electrical losses in the system. For this reason, PV cell manufacturers typically connect 36 cells in series to create an additive voltage. (Maybe you should reconsider reading Chapter 1.3 now). A grouping of PV cells thus arranged and mounted in a frame is known as a module.

36 cells in series x 0.6 Vdc per cell = 21.6 Volts Open Circuit

This voltage appears to be a bit higher than our target voltage of "just over 15 V." An interesting phenomenon of PV cells or other electrical generators is a reduction of voltage when the cell is under load, for example when charging a battery. In addition, heat also causes PV cell voltage to drop. When the voltage of a PV cell or series of cells is measured without a connected load, we call this the "open circuit" voltage. Manufacturers often refer to this as "Voc."

When the module is at its maximum-rated power output, the voltage is less than the open circuit rating. This is known as the "voltage at maximum load" and is typically 17 Vdc for a 12 V unit, double for a 24 V model.

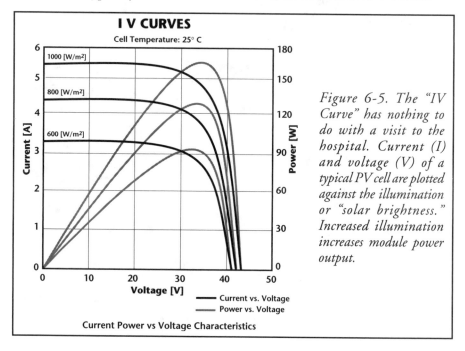

Figure 6-5. The "IV Curve" has nothing to do with a visit to the hospital. Current (I) and voltage (V) of a typical PV cell are plotted against the illumination or "solar brightness." Increased illumination increases module power output.

In order to complete an electric circuit, we must have a source of voltage and current flow. A PV module will cause current to flow within the cells and out of the supply wires to the connected load. As an example, a Sharp (www.sharpusa.com) model NE165 has a rated current of 4.77 Amps (A) at a rated voltage of 34.6 Vdc under ideal conditions and rated illumination.

34.6 Vdc x 4.77 A current flow = 165.04 Watts (W) of output power

This module is therefore rated at 165 W. However, there's one small exception: if it's dark or very cloudy outside, the power output is zero. Obviously light intensity has to play into our equation. The standard light intensity should naturally be that of the sun, as this is the source of energy the panels will use. But where should we measure our sunlight? In Alaska during winter, or perhaps the Bahamas in July? PV manufacturers have taken the liberty of helping us out here. PV lighting intensity has been standardized by industry agreement so that all manufacturers are using the same "solar energy" levels to prepare their data tables. Without this standardization, it would be impossible to determine which PV panel would be the best for your application.

Our sun has very generously decided to output a nice round 1,000 W of energy per square meter at noon on a clear day at sea level. In order to test the intensity of PV panels, light sources have been developed that create this same level of light intensity or *flux*. The light is beamed onto the test PV panel, and its electrical ratings are recorded in the product data sheets. As a consumer, this greatly helps determine the value of one PV panel relative to another.

Figure 6-6. PV systems are upgradeable. This homeowner installed new single-crystalline modules to increase the capacity of the original polycrystalline units located in the center of this array.

PV Output Rating Caution

Use caution when comparing module power ratings. Manufacturer power ratings are based on ideal sunshine conditions, which rarely occur in the real world. It is wise to derate module power ratings by 20-40% based on local atmospheric conditions. Check with local system owners or a reputable dealer to determine the "Real Power Rating" of your proposed installation.

Manufacturer Module Rating x 0.6 = Real Power Rating
...OR...
Real Power Rating ÷ 0.6 = Manufacturer Module Rating

PV module output can also be increased using a technology known as maximum power point tracking. Chapter 10 covers this technology and will discuss how to ensure that your expensive PV panels always operate at their peak efficiency.

PV Module Lifespan

Almost all PV module manufacturers provide a written guarantee for 20 to 25 years or more. The manufacturers are obviously quite sure that their products will stand the test of time. The reason for this certainty is the same one that explains why old transistor radios last so long. The semiconductor technology of the cell wafer results in very little wear and degradation.

The standard warranty term from Siemens states that any module that loses more than 10% power output within 10 years or 20% within 25 years will be repaired or replaced. (This is known as a *Limited Liability Warranty*, more commonly known as the fine print; be sure to read this detail carefully to ensure you understand the warranty terms.) Cell technology and quality of workmanship are very high in the industry, so be sure to purchase cells with the best possible warranty for your money.

The cells themselves are quite fragile. To protect them from damage and weather, the cells are bonded to a special tempered-glass surface and sealed using a strong plastic backing material. (Laminate or flexible "roof shingle" systems such as those offered by Uni-Solar (www.uni-solar.com) replace the glass surface with a tough, flexible polymer.) The entire module is inserted into an aluminum, non-corroding housing to form the finished assembly. Once a grouping of modules, called an array, is mounted to a roof or to a fixed or tracking rack, it should stay put forever.

PV Module Maintenance

Let it rain. That pretty well summarizes what you need to do to maintain your PV array. In the wintertime, ice and snow may build up on the glass. Don't smash the ice to remove it, or you run the risk of smashing the glass too. A quick brushing with a squeegee will take off the loose, highly reflective white snow. Once this coating is removed, the sun will quickly warm the panels, melting any ice even at –4 °F (-20 °C) or lower.

PV Module Installation Checklist

Place your PV module in the sun and collect power—that's it, that's all? Well, not quite. Although PV modules are well designed and last a long time there are several issues that you should considered before choosing where and how to install them:

1. Calculate your electrical-generation requirements.
2. Ensure the site has clear access to the sun.
3. Decide whether to rack or track.
4. Eliminate tree shading.
5. Consider snow and ice buildup in winter.
6. Ensure that modules do not overheat in hot weather.
7. Decide on the system voltage.
8. Locate the array as close to the batteries as possible.

Step 1 - Calculate Your Energy Requirements

This is square one in your quest for a supply of renewable energy. Let's start the ball rolling with a question: How much energy do you need? If you skipped over the chapter on energy efficiency, it's quite likely that you are consuming significantly more electricity than necessary, resulting in equally high generation system requirements and costs. Remember the rule of thumb: for every energy dollar you put into energy efficiency, you reduce electrical generation capital costs by between three and five dollars. You cannot afford to sidestep the energy efficiency process. Sorry!

Start by checking your electrical bills over the last few months or call your utility to find your average monthly electrical consumption. A figure of 20 to 40 kilowatt-hours per day (kWh) is average and may be considerably higher if you have electric heat, electric hot water, or central air-conditioning.

If your current electrical consumption is a bit on the high side you will quickly find out that you need pockets as deep as Bill Gates in order to generate enough electricity to power your house.

To quickly recap, let's consider the case for compact fluorescent light bulbs one more time. As discussed earlier, a compact fluorescent light bulb uses approximately 4.5 times less energy than a standard incandescent bulb. Assume for a moment that your home requires 600 W of lighting in the main

Figure 6-7. This grid-interconnected, ultra-high-efficiency home is equipped with a 2.4 kW roof-mounted photovoltaic array. It is located in Finland at 63.5 degrees North latitude, proving that photovoltaic panels work in all climates and not just in the Nevada desert. This home uses one-tenth the heat energy and one-quarter the electrical energy of the typical Finnish home. (Courtesy International Energy Agency)

floor area. If this light is generated using incandescent bulbs, we need 600 W of PV panels to power them. If PV panels are selling for $4.00 per watt, the cost is $2400. With compact fluorescent lamps, however, we need only 130 W of bulbs to create the equivalent brightness, for a PV module cost of $520. Efficiency is **ALWAYS** cheaper than additional generation.

Another factor in determining the size of your PV generation system is any grants or low-interest loans that may be available in your jurisdiction. The California renewables buy-down program provides the homeowner with a grant of up to 50% with a maximum dollar limit. Political realities being what they are, it is always wise to get onto the gravy train while you have the chance.

Lastly, discuss this decision with your dealer. Get advice about system sizes for your roof area, equipment package prices, and other market variables that are impossible to define in this book.

As there is no standard or typical system size, we will assume an electrical energy-generation objective of 4,000 Watt-hours (4 kWh) per day.

To give you a bit of guidance on this value, Lorraine and I run our household on between 3 and 5 kWh per day, less if we are traveling, more if we have friends over for a late-night party. If you have teenagers who stay up all night, maybe you should stay grid-interconnected. To define "typical" is pretty hard, but here are some values that I can share from other installations:

- **Weekend cottage running a few lights, small T.V., boom-box stereo, and no fridge**: a 12 Volt D.C. system is fine with two to four PV modules and a small battery bank. This system will generate 200 Watts peak power and 1,000 Watt-hours per day energy.

- **Full-time, seasonal cottage with four people, requiring refrigeration and limited water pumping**: a 12 Volt system is fine, possibly using a mix of direct current and alternating current outlets for various loads. If the cottage will "grow" over time, wire for 120 Volts A.C now. Use a small inverter for 1,500 Watts peak power and up to 2,000 Watt-hours per day of energy. The system will require a PV array of approximately 500-Watts peak output.

- **Full-time residence with four people, large energy-conserver philosophy, no dishwasher or clothes dryer, a small television and stereo, high-efficiency clothes washer, and 120 Volt refrigerator**: wire system for 24 Volt batteries and provide an inverter with 2,500 Watts peak power and up to 3,000 Watt-hours per day energy. The system will require a PV array of between 800 and 1,000 Watts peak output with a backup generator for "dark months." Depending on the location, more PV may offset the need for a generator provided winter sun hours are able to provide 100% of the daily load.

- **Full-time residence with all the electrical (high-efficiency) goodies, possibly including a home theater system, dishwasher, gas clothes dryer, central vacuum, washing machine, and electric refrigeration**: system may be wired for either a 24 or 48 Volt battery. Use 24 Volts for total daily energy consumption below 7,000 Watt-hours per day and 48 Volts for larger systems. Provide an inverter of between 2,500 Watts minimum and 4,000 Watts peak output or larger, depending on final peak loading. Will require a PV array of 2,000 Watts peak or higher depending on load. For off-grid systems a backup generator will be required.

Sunlight and PV Energy Generation

The amount of sunlight we receive varies from day to day, depending on clouds, rain, humidity, and smog, but also as a function of the seasons. In most locations, there will be a surplus of sunlight and resulting electrical energy production in the summer, with the opposite being true in the winter. It may be tempting to calculate the energy production of the PV panels using the worst sunlight hours of the year in an attempt to cover all of your energy requirements. However, this will increase the cost of the system and generate unusable, excess energy in the summer. Let's take a look at how this would work. Assume you live in the Rochester, New York area. Turn to Appendix 5, *Solar Illumination Map for North America (worst months)*. Find the Rochester area and note that the amount of solar illumination in sunlight hours per day for the worst months of the year is 1.6 hours. We can now plug in some numbers to calculate the size of PV array we require. (Watt-hours per day divided by sun-hours per day calculates the theoretical PV panel size. The theoretical size must then be derated as noted below):

4,000 Watt-hours/day ÷ 1.6 sun hours/day = 2,500 W PV Panel

Don't forget to derate the manufacturer ratings (assume a 20% derating factor):

2,500 real power rating ÷ 0.8 reality factor = 3,125 W manufacturer rating

A quick search of the web for PV panels reveals nothing even closely approaching that figure. The largest 12 V module you can find is from Sharp and is rated 123 W with a list price of $685.00. Kyocera has an 80 W model on sale for $375.00. This is starting to get complicated and expensive.

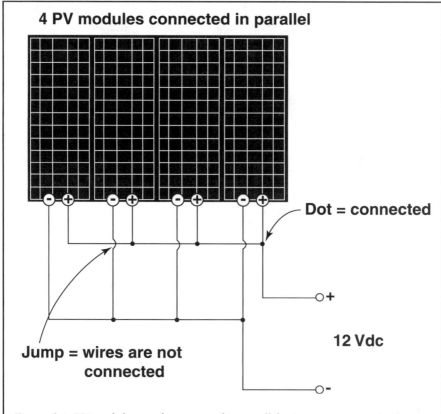

Figure 6-8. PV modules may be connected in parallel to increase wattage in the same way as connecting two small batteries in parallel increases their energy capacity.

PV modules are no different than batteries in the way they can be connected and their capacities increased. Having read Chapter 1.3, you will recall that batteries may be connected in *series* or in *parallel* to increase the voltage or current respectively; the same is true with PV cells. Assuming a PV array with 12 V output, modules can be grouped in parallel to increase the current and wattage rating:

3,125 W Array ÷ 123 W PV module = 25 modules

Or

3,125 W Array ÷ 80 W PV modules = 39 modules

With so little sunlight in the Rochester area in winter, this PV array will require a lot of modules to support 100% of the electrical load. Now let's take a look at the cost.

25 – Sharp 123 Watt modules x $685.00 = $17,125.00
39 - Kyocera 80 Watt modules x $375.00 = $14,625.00

Although these figures are **examples only**, they do reveal an interesting phenomenon. You would expect that since each PV array makes the same amount of power the cost would be the same. Welcome to the world of retail. This is no different than purchasing a car; it's best to shop around.

I will digress a moment to explain how to comparison-shop for PV modules. If you are the sort who cannot understand people who drive a few miles to buy tomatoes at another store because they cost ten cents less, just wait a minute. PV module comparison-shopping is easy and will save you more than enough money to pay for this book and possibly put your kids through college.

The energy-to-cost ratio allows for a very quick comparison of all module types. The calculation is quite simple: divide the cost of the module by the rated wattage. Using the PV modules from the example above:

Sharp 123 W modules @ $685.00 ÷ 120 W = $5.71 per watt
Kyocera 80 W modules @ $375.00 ÷ 80 W = $4.69 per watt

Depending on the warranty and on the installation cost for the additional modules required to make the same amount of power, the Kyocera modules are probably the least expensive based on the "$/Watt Factor." Now back to Rochester.

It takes quite a few modules to make 100% of the energy required to power a home when the winter sun hours are minimal. The opposite is true in the summer. Refer to Appendix 6, *Solar Illumination Map for North America (yearly average)*. You will find that the number of sun hours per day has nearly doubled. The effect is obvious: when the sun hours per day are peaking, the required number of PV modules in your array will be half the number required at the worst time of the year.

Many system owners and dealers use the monthly averages for one complete year to calculate your energy requirements. Using this approach, generation and consumption of electrical energy will never match on a day-to-day basis.

If you are using your PV modules in a summer cabin or cottage, you are in luck. Use the calculations for the best sun hours per day. If you visit the cottage for brief visits in the fall and winter, possibly add another panel or two.

For year-round homeowners, the choices are a bit more complex. You can spend the extra dollars and purchase all of the panels you require or consider a complementary renewable resource such as wind power generation. A third option, which tends to be the most common solution, is also available. Add a backup propane, gasoline, or diesel generator to help boost production in the worst of times. We will review all of these options in greater detail in later chapters.

The correct number of panels to make up your array will depend on more than just simply spending the money to purchase all you need. For starters, the weather seems to have a mind of its own. One year it's sunny all winter, the next you wonder if the sun will ever come out again. Charts like those shown in Appendix 5 and Appendix 6 are averages, which really means that you should be prepared for variability. Even if you purchase all the PV modules you can possibly mount on your roof, it is still a good idea to have a backup generator or alternate source, just in case. A thousand PV panels will not operate a single appliance if it's cloudy and dark!

Speaking of mounting panels on your roof, this is the next concern with the number of panels you select. You have several choices about where to mount your PV array. The roof is one of these places. The more panels you have in the array, the larger and heavier it becomes, placing limitations on some of the best mounting locations. If the panels are mounted on a sun-tracking unit, the limit is 16 to 20 modules per tracker.

What is the correct number of panels to purchase? There is no "correct" answer. If you have deep pockets, by all means purchase all the modules you require to offset the darkest winter days. If you are like most people and live on a tighter budget, start small and expand your system later as you can afford it. When you win the lottery or your great aunt leaves you some valuable stocks and bonds, splurge and purchase the additional twelve modules that

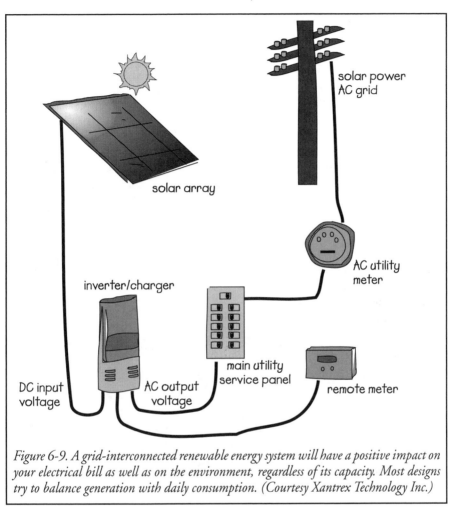

solar power
AC grid

solar array

AC utility
meter

inverter/charger

DC input
voltage

AC output
voltage

main utility
service panel

remote meter

Figure 6-9. A grid-interconnected renewable energy system will have a positive impact on your electrical bill as well as on the environment, regardless of its capacity. Most designs try to balance generation with daily consumption. (Courtesy Xantrex Technology Inc.)

you really need. The additional panels can always be added later, like those shown in Figure 6-6. Just remember to plan for the expansion at the start of the installation. This will lower future costs and headaches.

Step 2 – Ensure the Site Has Clear Access to the Sun

PV modules must face the sun to make electricity. As every romantic knows, the sun sets in a brilliant display of color in the west. My mother-in-law tells me that it does the same when it rises in the east, although I cannot personally attest to this statement. Where, then, do you face the modules? The quick answer is to point the panels directly at the sun. However, with the sun moving from east to west during the day and progressively higher in the sky as summer approaches, this might be a rather difficult task.

You have two options: mount the modules to a fixed location pointing *solar south* or install a tracking unit that automatically aims the panels directly at the sun. Either way, make sure the location you choose is not impeded by trees, buildings, or other obstructions. Site locations can be difficult to assess due to the seasonal variation in the sun's track, leaves on deciduous trees, and neighboring buildings. To help simplify the assessment of a site, consider purchasing or renting a Solar Pathfinder to eliminate any guesswork. This device works on the well-known principal observed by every schoolchild with access to a magnifying glass. As the sun tracks across the sky, light enters the magnifying and focusing dome of the device (see Figure 6-10). The concentrated sunlight falls on specially sensitized paper burning a copy of its path. Gaps in the marking indicate obstructions that will cause unacceptable shading of the PV panels.

Figure 6-10. The Solar Pathfinder is used to determine magnetic and solar south (using the data in Appendix 4) as well as to record the sun's track across the sky using the specially sensitized paper shown right. (Courtesy Solar Pathfinder)

Step 3 – Decide Whether to Rack or Track

Roof Mounts

Perhaps the simplest mounting system is to attach the panels to a solar-south-facing roof as shown in Figure 6-11. *Solar south* differs from magnetic south as a result of a phenomenon known as magnetic declination. Appendix 4 shows the correction of compass readings based on your geographic location. If you live in Florida, for example, there is no change between magnetic and solar south. If you live in Alberta, the error in the compass-magnetic versus solar south reading is so large that it will affect your system's energy production if you use fixed-mount PV arrays.

Another problem with fixed-mount arrays is their inability to change their angle of view throughout the seasons. The winter sun barely scrapes across the horizon in most of the northern United States and Canada. During the summer, the change in the earth's angle places the sun almost directly overhead. If your roof mount or other fixed mount cannot be adjusted for summer-to-winter sun angles, a good rule of thumb is to set the array at an angle from the ground equal to your latitude. Mounting racks, which have a summer/winter angle adjustment, should be lowered by fifteen degrees for summer. Likewise, raise the angle towards the vertical by fifteen degrees for winter.

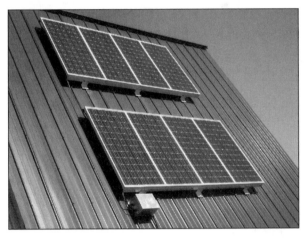

Figure 6-11. Attaching a PV array to a solar-south-facing roof is one of the least expensive mounting methods. The downside can be getting on the roof to brush off a foot of snow in the middle of winter.

Ground Mounts

If the thought of climbing onto your roof to wipe snow off the PV array has you going woozy, then ground or pole mounting is for you, a simple and inexpensive method of mounting the panels. The unit shown in Figure 6-13 was purchased, but you can also build a unit yourself. Select galvanized steel or aluminum to limit corrosion. When designing your mounting system, make sure to include a hinge assembly to allow for seasonal adjustment. It is possible to make a ground mount using preserved wood or cedar, but keep in mind that the panels will almost certainly outlast the wood.

A word of caution regarding ground mounting: PV modules are expensive and may have a tendency to walk away. Ensure that you have security bolts (or a good guard dog) to prevent theft. If possible, place the mounting legs in concrete footings to make sure that everything stays in its place. Snow is another problem. If you live in an area where "sweeping" the array requires

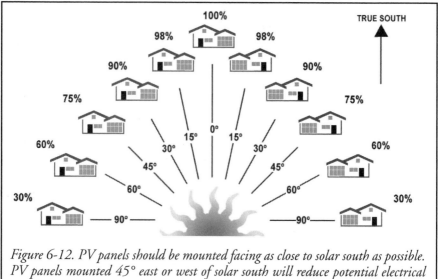

Figure 6-12. PV panels should be mounted facing as close to solar south as possible. PV panels mounted 45° east or west of solar south will reduce potential electrical output by 25%.

a snow blower rather than a broom, then ground mounting is not for you. Be careful to ensure that lawnmowers and playing children will not send rocks or other debris flying at the modules. Although it is very likely that no damage will result, there is no sense tempting fate.

Tracking Mounts

The tracking mount shown in Figure 6-14 is the most advanced means of pointing your PV panels at the sun and has the added benefit of being hyp-

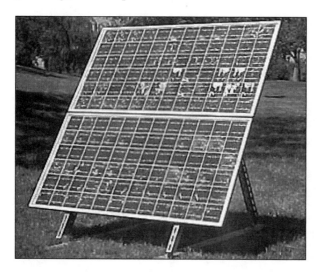

Figure 6-13. The commercial ground-mounting unit is ideal for smaller PV arrays and is also a good option where cost is a concern. (Courtesy Zomeworks)

notic to watch as it scans the sky. PV arrays comprising up to twenty panels are mounted on the tracker, forming a billboard-sized unit. (A sixteen-module array measures 16' x 8' or 4.9 m x 2.4 m.) The tracker is designed to move to an easterly location when it gets dark. At first light it follows the sun on its westerly track across the sky. At day's end, the unit returns to the easterly heading to repeat the process. Electronically controlled trackers can be fitted with a manual seasonal-adjustment device or with an automatic version.

Passively controlled trackers are also available. The Track Rack manufactured by Zomeworks (www.zomeworks.com) uses heat from the sun to operate a tracking mechanism. Although this mechanism is simpler than those used in active tracking units, the Track Rack model will go to sleep facing west and may require an hour or two of valuable sunlight before it will start tracking correctly on cold days.

Should you use a tracker? The debate rages on, but the following are considerations that may affect your decision:

- Trackers increase summer PV production by up to 50%. Winter production is improved only 10-20% due to the lower, smaller arc of the sun. If your energy consumption is higher in summer than in winter, perhaps due to the operation of a spa, swimming pool, or air conditioner, then tracking will be a benefit.

- The further north you are the less sense it makes to track in the wintertime. This is especially true along the United States/Canada border area.

Figure 6-14. Tracking units such as these increase summer electrical production by up to 50% but offer little improvement during the winter.

- If your site has a limited window of sunlight—less than six hours—then tracking will not greatly improve system performance.
- Trackers are not cheap. The cost of the tracker may be used to purchase a fixed rack and more PV panels, which might offset the loss in non-tracking production.
- Trackers add a degree of complexity to the system. The bits and pieces are just one more thing to have to maintain.

Regardless of which type of PV mounting system you decide to use, keep in mind that they take up a fair bit of area and make wonderful kites or sails in high winds. Ensure that proper mounting and foundation work has been undertaken in compliance with the manufacturer's installation instructions. If you want to harness the wind, don't use your PV panels.

Step 4 – Eliminate Tree Shading

Although you want to protect the south side of your house from the summer sun, the last thing you want to do is to shade even a very small portion of your PV array. Partial shading will cause a disproportionate reduction in electrical generation to the shaded area. Keep trees well clipped in the sunlight window (between the hours of 9am to 3 pm), keeping in mind both winter and summer sun tracking. If you are unsure about shading, consider evaluating your site using the Solar Pathfinder shown in Figure 6-10a.

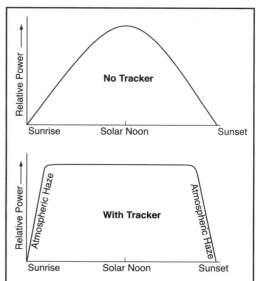

These graphs demonstrate the relative power output of a PV array equipped with and without a tracking unit. The top graph indicates that solar energy reaching a fixed array and its corresponding electrical power output increases slowly until the sun is perpendicular to the array at solar noon. Power output then decreases until sunset. With a tracking unit, solar energy rises very quickly as the tracker is always oriented towards the sun. The effect of dust and atmospheric haze and interference prevents power output rising to its maximum point immediately. Array power output will remain at its maximum point throughout the duration of the day, until just before sunset.

Step 5 - Consider Snow and Ice Buildup in Winter

If you live in a snowy area, take winter snow and ice buildup into account when considering PV location. Although PV angles at this time of year are almost vertical, snow and ice will stick to the array. If a coating of fresh, white, powdery snow is covering the array, it may take several days for it to fall off without brushing—and you may not want to brush off PV arrays if they're mounted on a second-story roof.

Step 6- Ensure That Modules Don't Overheat in Hot Weather

PV module electrical output fades (just like everyone I know) as the mercury rises. In order to ensure peak operation of the array, modules must not be seated directly on the roof surface. An air gap of 2-3" (5-8 cm) will allow cooling air to circulate under the array, providing maximum power output.

Step 7 – Decide on System Voltage

For off-grid systems, there are three commonly used system voltages: 12, 24, or 48. Although it is possible to select differing voltage and power levels, the industry uses the following guidelines.

System Size	PV/Battery Voltage
Up to 2 kWh per day	12 Volts
From 2 to 7 kWh per day	24 Volts
Over 7 kWh per day	48 Volts

The PV industry has standardized on modules with 12 or 24 Volt nominal outputs. As discussed above and as shown in Figure 6-8 we can interconnect PV modules in parallel to increase current flow and system wattage without changing the voltage of the panels. Figure 6-15 illustrates a mixture of series- and parallel-connected PV panels. In this arrangement there are two rows of four PV panels. Both rows are interconnected in series, increasing the system voltage. If you look carefully and pretend that each PV module is a battery, you can follow the series connection. The positive of one module connects to the negative of the next and so on. When each 12 volt panel is connected in a series grouping of four, the voltages are additive, resulting in a 48 Volt array.

Once both rows of panels are at the same system voltage, which in this example is 48 V, the array may be connected in parallel. Examine the wiring connection at the point indicated in Figure 6-15. The negative lead from each row is connected, as is the positive lead. Together, the two rows of 48 V arrays will deliver additional current flow (power) to the load.

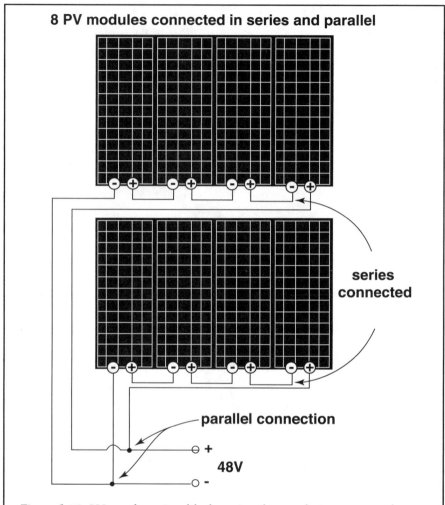

8 PV modules connected in series and parallel

series connected

parallel connection

+ 48V

-

Figure 6-15. PV panels are just like batteries: they can be interconnected in series to increase voltage and in parallel to increase the current of the array.

Step 8 - Locate the Array as Close to the Batteries as Possible

Have you ever connected 2 or 3 garden hoses together, turned on the tap, and wondered why the sprinkler was dribbling an anemic spray of water even though the pressure at the house was fine? What you have witnessed is resistance to flow. This phenomenon applies not just to garden hoses but to electrical circuits as well. The resistance to flow of electricity works in the same manner. Low-voltage (pressure) circuits require very large wires to carry the necessary current in the cable. It stands to reason that the further the PV array is from the battery or inverter, the greater the loss of electrical energy will be. Increasing the pressure or voltage of the electricity will help, but distance is a bad word when it comes to low-voltage direct current circuits. Keep the wire runs as short as possible and maximize wire size to prevent unnecessary power loss.

As a side note, this is the reason that the North American electrical grid transmits energy at very high voltages, reaching 750,000 volts in some areas. Even at this high voltage electrical losses through the system can be extreme. The province of Ontario has approximately 15,000 miles (24,000 km) of transmission grid. The electrical losses on this system are equal to the energy output of three large nuclear reactors! Distributed generation using off-grid or grid-interconnected photovoltaic systems places the consumption and generation of electricity at the same location, thereby eliminating transmission losses. At a couple of billion dollars apiece for nuclear generating plants, not to mention fuel disposal, decommissioning costs, and safety liability, PV panels mounted on the nation's roofs are a bargain.

Conclusion

PV systems are simple and reliable. Where a home or cottage is just a little too far from the electrical lines, there is no contest. It will make economic and environmental sense to use this technology without further consideration.

With electrical rates rising and governments committed to "greening the grid," there is little question that all of North America will be pushing PV in the same manner as California is now. According to Environment California, recent proposals before the state legislature would have 50% of all new homes running on solar energy within the next ten years. Although this goal is lofty and well ahead of any other jurisdiction in North America, we still have a long way to go before we catch the leaders, Japan and Germany.

Figure 6-16. The Conde Nast building in Manhattan is a magnificent example of what the future will look like for urban dwellers. This building, equipped with advanced energy-efficient designs, fuel cell electricity generation, and photovoltaic panels, pumps green electricity into the New York State grid. Over the coming years, commercial buildings, condominiums, and apartments will achieve an energy goal of almost net zero. (Courtesy Fox & Fowle Architects / Andrew Gordon Photography)

7
Electricity from the Wind

I have been told by a number of people that while women seem to like PV technology, men tend to gravitate towards wind turbines. Perhaps, for the men, the appeal lies in the big tools, concrete, propellers, and spinning stuff. Whatever the case may be, mankind has been capturing the wind for eons. All school children know about the quaint Dutch windmills of old. What they may not know is that citizens of Germany, Sweden, and Denmark are also amongst the world's foremost users of wind electricity generators.

Years of low fossil-fuel prices and certainty of supply convinced most North Americans that there was nothing to worry about.

Figure 7-1. This 10kW wind turbine is typical of larger home-based machines. With a blade diameter of 23 feet (7 m) and the right amount of wind, this unit can provide sufficient power for almost any home. (Courtesy Bergey Windpower).

The Europeans, on the other hand, have been worrying for years. Most European countries rely on imported oil and have high population densities. Conserving electrical energy has become a way of life there.

Fortunately we are waking up on this side of the pond and, as the alarm bells are ringing,, wind turbines are coming of age. Sites such as King Mountain in West Texas are good examples of how North Americans are catching up. Like everything else in Texas, this site is big, with 214 turbines installed. Together they provide an output of 278,000,000 Watts (278 MW), sufficient to power 80,000 super-sized Texas homes or 3% of Denmark's electricity requirements. These units are not toys.

It is doubtful if you will require this much power for your renewable energy-powered home, but this does indicate the scope of what wind power can do.

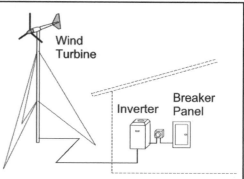

Home-Sized Wind Turbines

The market today includes many manufacturers of wind turbine systems. There is also a market

Figure 7-2. Home-sized wind turbines may be connected to the electrical grid in the same manner as small PV systems. (Bergey WindPower)

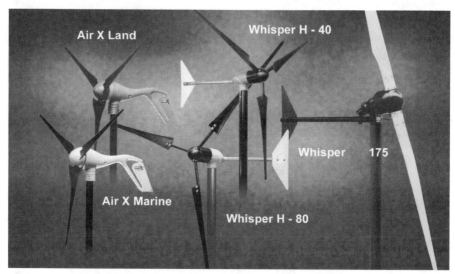

Figure 7-3. Wind turbines are available in a wide variety of sizes, designed to suit just about every application. (Southwest Wind Power)

Figure 7-4. Wind turbines are used extensively in this obvious off-grid installation. What better harmony than using the wind to sail the boat and make electricity at the same time. (Southwest Wind Power)

for rebuilt and used machines that were installed prior to rural homes being connected to the growing electrical grid in the early 1900s. The Resource Guide in Appendix 3 will provide you with a listing of the major manufacturers and rebuilders of new and refurbished equipment.

When people first see a wind turbine their reaction is to ask, "How big is it?" This can be a misleading question, since size in this case involves more than just the height of the tower, or rated electrical capacity.

Wind turbines require a more detailed assessment than PV modules when it comes to determining what "size" of turbine is required for your application. Let's begin by taking a unit apart in order to understand how they are made and how they operate.

The majority of home wind turbines are horizontal axis machines similar to the one shown in figure 7-5. The names of the major components are common to all wind turbines regardless of size. A tower structure supports the wind turbine and raises it high above the earth's surface. As the tower height is raised, the speed and smoothness of the wind increases: "the higher the tower, the greater the power." Typical tower heights range from 60 to 100 feet (18 to 30 m). The tower may be self-supporting or more commonly and less expensively have a series of guy wire cables to support it. A concrete pad that accepts the downward weight supports the tower and prevents it skidding sideways. The guy wires provide all of the support and are secured to earth or concrete foundations using anchors. Cable tensioning devices are installed on each guy wire to remove slack and prevent tower motion.

An important feature of the wind turbine is the rotor and blade assembly in the centerline of the turbine, consisting of the blades on the outside and

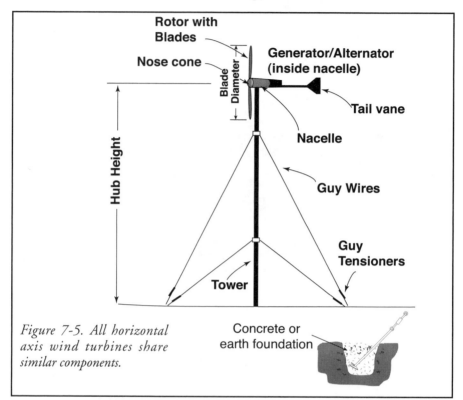

Figure 7-5. All horizontal axis wind turbines share similar components.

the generator/alternator on the inside. A nose cone is attached to the rotor to allow the wind a smooth surface to pass over as it travels through the blades. A nacelle or cover protects the generator and support components from the weather. Finally, a tail vane assembly is used to keep the rotor and blade assembly pointed into the wind. On some models the tail vane also provides a means of slowing the blades down in dangerously high winds by automatically performing a function known as furling. On models such as the Bergey Excel shown

Figure 7-6. The blades of a wind turbine are shaped into a form similar to an aircraft wing. Wind rushing past the blades causes aerodynamic lift that imparts a torque or rotational force powering the generator. (Bergey WindPower)

in figure 7-1, a manual furling winch is also provided, allowing the unit to be shut down during severe storm seasons.

In Figure 7-7 a close-up view of a small 1,000 Watt turbine assembly shows the rotor assembly with the blades removed from their sockets. The generator wiring is just visible at the rear of the rotor assembly. If you follow the wires from the generator, you will see them terminate at a black module known as the rectifier bridge,

Figure 7-7. This picture details the inside frame, rotor and generator of a small 1 kW wind turbine, and shows the blades and nacelle cover removed. (Bergey WindPower)

which converts the alternating current from the generator into direct current suitable for connection to the battery bank. The alternating current is supplied as 3 phase (which is similar to most industrial building services) at the top of the rectifier, while the direct current output is at the bottom of the rectifier. Following the positive and negative lead wires from the rectifier, you see they connect to a section of the tower assembly. As the turbine spins on the top section of the tower, it would quickly twist the power cables running down to the house. To prevent this, the wiring is fed to a "slip ring" assembly which acts as brushes, conveying the electrical power from the rotating "yaw tube" to the stationary tower head assembly. The tail vane mounts to the rear post, completing the assembly.

Wind Turbine Ratings

Now that you have been formally introduced to a typical home power-sized turbine, we need to clear up some of the details regarding size. Unlike the debate over certain biological issues, size in a wind turbine does count. However, the diameter of the wind turbine blades is the only size that is truly important. You may recall from Chapter 1.3, that in order to convert kinetic energy to electrical energy we have to capture as much of the "moving fluid" as possible. The blades catch the kinetic energy of the wind. A unit with a smaller blade area will capture a smaller amount of energy; it's that simple. Because the blades sweep a circular path we have to do a little bit of fancy mathematics to calculate the area they cover. Assume our turbine blades have a diameter of 10 feet (3 m), giving them a radius (½ the diameter) of 5 feet (1.5 m).

π x Radius^2 = Area Swept by Blades
3.14 x 25 = 78.5 Square Feet (7.3 m^2)

Machine manufacturers like to use power ratings that are not as well standardized as they are for PV modules. This makes comparison shopping between units difficult. Calculating the rotor swept area of different machines and comparing those values will provide an even keel for calculating power output. Keep in mind that rotor swept area is in a linear relationship with power. Simply stated, a doubling of the sweep area doubles the power output.

If the math is too much trouble, an optional means of comparing machine ratings is by looking at the "power curve" graph. The graph shown in figure 7-8 is for the large home-sized turbine shown in figure 7-1. This graph details the electrical power output in kilowatts based on a given wind speed in miles per hour. For example, at a wind speed of 24 miles per hour (38.5 kph), this unit outputs 6,000 Watts of electrical energy. An excellent way

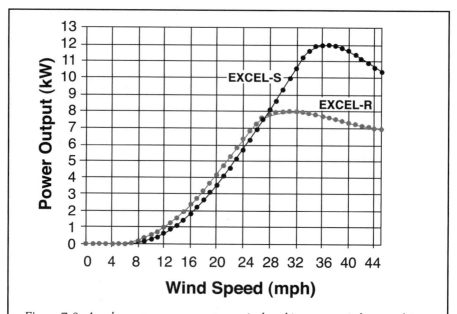

Figure 7-8. An alternate way to compare wind turbine output is by consulting the power curve rating. This chart shows the electrical power output for a given wind speed. (Bergey WindPower)

of cross checking manufacturers' figures is to compare both power curves at the same wind speed as well as the rotor swept area. The ratio between the two should be approximately the same: a machine with twice the rotor swept area should have a power curve with twice the power output at the same wind speed.

The power curve graph also details some other interesting facts concerning all wind turbines. You will notice that at low wind speed, below approximately 8 miles per hour (13 kph) the turbine output is almost zero power. Many dealers will talk about units that start in low-wind conditions or begin generating power earlier than others. This information is not factually supported by the physics of wind energy. The turbine has nothing to do with "starting to kick in earlier." Sure the unit might spin a bit in low-wind periods, but it will not output any energy. Energy is related to wind speed and you need speedy wind to make energy.

This brings up another important fact about wind energy. Wind speed is not linear to power. Removing the "tech talk," if we double the wind speed we do not get double the power. Take another look at the power curve in figure 7-8. Follow the Excel-S data line at a wind speed of 16 mph (25.5 kph) and we derive an output of approximately 2,200 Watts of power. Now if we double the wind speed to 32 mph (51 kph) we do not get 4,400 Watts as you might expect. Check the curve; our output is now up to a whopping 10,500 Watts!

What happened? This is a classic example of the non-linear relationship between wind speed and energy in the wind. In fact the theoretical relationship is cubic. This means that for a doubling of wind speed, we get an increase in power of eight times (2 cubed = 2^3 = 8)

Rating Jargon

The power curve also details information about how the unit behaves at various wind speeds. Referring to the Bergey model Excel-S power curve in Figure 7-8, you see a shape to the curve that is similar but not identical to the units produced by many other manufacturers. Relevant points on the curve are:

- **Start-up speed**. This is the wind speed when the rotor and blade assembly first begins to rotate. The start-up speed is irrelevant, as this low rotational speed will provide no usable power output.
- **Cut-in speed**. This is the wind speed when the generator actually begins to produce **usable power**. Cut-in speed is one of the most misunderstood terms in wind power generation. There is no power in slow wind; if the unit starts at 3 mph (5 kph) it will not be generating useful levels of power.

Likewise, if the turbine's cut-in speed is being touted as exceptionally low and superior to that of every other model, keep your money in your wallet.

- **Rated speed/power point**. The manufacturer's data gives the nominal power rating of the unit at a given wind speed. The Bergey Excel-S is rated 10,000 Watts and that occurs at approximately 31 mph (50 kph) wind speed. In practice, the rated speed/power point is of little value, as the machine will likely spend very little of its life at this wind level. Additionally, each manufacturer rates its turbines at different wind speeds, skewing the data values and making comparison difficult. Remember, the rotor swept area and blade diameter are the most important factors.

- **Peak or maximum power**. This is the maximum power the unit will generate. The Excel-S is capable of producing 12,000 Watts at 36 mph (58 kph) wind speed.

- **Furling speed**. The furling speed is the wind speed at which the unit enters a "self-protection" state. When wind speeds are too high, the turbine must be able to lower its rotational speed and resulting power output. In most small wind turbines, the tail and nacelle "hinge" so that the blade/rotor section turns out of the wind. In some models the rotor points upwards like a helicopter until wind speed subsides.

Ratings are important, but the most important consideration is to determine at what point on the power curve the turbine will work at your site. Don't become overly confused by the extra jargon that will become immediately unimportant the minute you switch the unit on.

Wind Resources in Your Area

The wind tends to vary greatly over the span of a year and depending on where you live. There is almost no wind in central Florida, but it never stops in Cape Cod. If a small change in wind speed causes a large change in electrical power output then we must be careful to capture the highest wind speed. Almost everyone I have talked to about installing a wind turbine tells me that they are on a hill or near a lake and have "great wind". But just what do they mean by "great wind"? A 9 mph (14.5 kph) breeze is pretty strong. At this speed it will make holding your umbrella a bit difficult or whip your tie around your face, so it must be a good wind turbine site. Think again. Looking at the power curve for the 10,000 Watt Bergey Excel-S model in Figure 7-8 you will find the power output is nearly zero.

You can't take chances installing a turbine and hoping it will work out. This is the surest way to electrical and possibly financial bankruptcy if there ever was one. Having someone tell you there is plenty of wind is just not going to work. A wind turbine is a big investment and a careful site study should be conducted before you commit your dollars and time.

There are two typical methods of determining the characteristics of the wind at a given site. The first method relies on wind maps prepared for the government; the second entails a site measurement study for a number of months or longer. However, before we discuss wind mapping and surveys we first need to review what is meant by wind assessment.

A wind turbine without wind is like the proverbial fish out of water. Wind makes a wind turbine operate and, within reason, the more wind the better. But what wind are we talking about? Wind requirements vary with the seasons and depend in part on the type of system you are running. Late fall and winter tend to be dark and rather tough on a PV system, so adding a bit of wind energy to boost things up is helpful then. On the other hand, a system which relies on wind as the main or only source of energy requires a steady supply of wind all year round, especially in the fall and winter when more time is spent indoors and energy needs tend to be higher.

Manufacturers publish tables that estimate the monthly or yearly energy production in kilowatt-hours at differing wind speeds when the turbine is installed on a tower of a specific height. An example taken from the Bergey Excel-S is shown in Table 7-1.

As the wind speed increases, so does the amount of energy produced. But why does the height of the tower have anything to do with energy production? Tower height relates indirectly to wind speed. Think of wind as water for a moment. If water were streaming over your wind turbine and it were close to the ground, trees, rocks, buildings, and other obstructions would cause

Tower Height	Average Wind Speed						
	8mph (13 kph)	9mph (14 kph)	10mph (16 kph)	11mph (18 kph)	12mph (19 kph)	13mph (21 kph)	14mph (23 kph)
60 ft (18 m)	330 kWh	480 kWh	670 kWh	870 kWh	1,110 kWh	1,350 kWh	1,610 kWh
80 ft (24 m)	430	620	840	1,100	1,370	1,670	1,960
100 ft (30 m)	490	700	950	1,220	1,510	1,820	2,130
120 ft (37 m)	550	780	1,050	1,340	1,650	1,970	2,280

Table 7-1. This chart details the relationship between wind speed, tower height, and monthly energy production in kWh for the Bergey Excel-S turbine. (Courtesy Bergey Windpower)

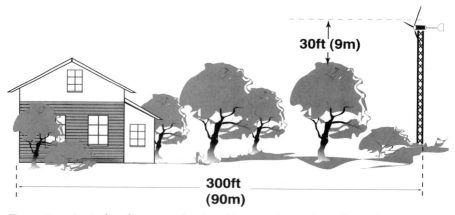

30ft (9m)

300ft
(90m)

Figure 7-9. A wind turbine must be placed in smooth air, clear of any obstructions.

the water to create rapids and generally slow down. As you moved above our "river bed" example, there would be fewer obstructions and the water would start to move much more smoothly. At some point, there would be minimal obstructions in the water's path and the flow would become very smooth. At this point we would have what is called laminar flow, which not only increases wind speed and energy production but also prevents "rough air" from stressing the turbine, ensuring longer life.

A general rule of thumb is to place the turbine on a tower of sufficient height that nothing within a 300 foot (90 m) radius is within 30 feet (9 m) of the height of the tower. For example, the house and trees shown in Figure 7-9 must be at least 30 feet shorter than the tower if they are within a 300 foot radius of the tower. Keep in mind that trees grow, so allow for this when determining tower height—or get pruning equipment to maintain this requirement.

The effect of tower height on turbine power cannot be underemphasized. At 8 mph wind speed, the Bergey Excel-S power output increases from 330 kWh to 550 kWh when tower height is increased from 60 feet to 120 feet. In other words, doubling the tower height increased power output by two thirds.

Before you decide to rush out and purchase a 700 foot (213 m) supertower, bear in mind that tower cost, foundation work, and wiring expenses will tend to throw cold water on your plans. A tower of 60 to 80 feet is a good start for smaller units under 1,000 Watts. Up to 120 feet is recommended for larger turbines, but work within your budget.

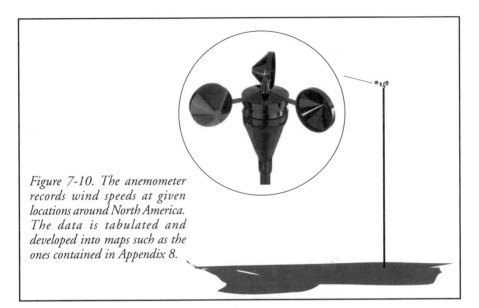

Figure 7-10. The anemometer records wind speeds at given locations around North America. The data is tabulated and developed into maps such as the ones contained in Appendix 8.

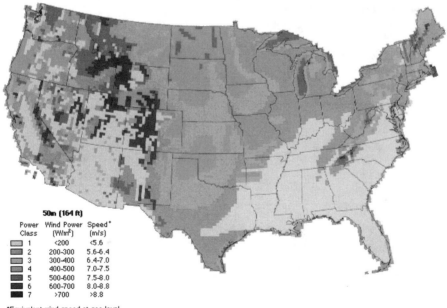

50m (164 ft)

Power Class	Wind Power (W/m²)	Speed* (m/s)
1	<200	<5.6
2	200-300	5.6-6.4
3	300-400	6.4-7.0
4	400-500	7.0-7.5
5	500-600	7.5-8.0
6	600-700	8.0-8.8
7	>700	>8.8

*Equivalent wind speed at sea level for a Rayleigh distribution.

Wind Energy Resource Atlas of the United States

Figure 7-11. This wind map of the United States shows annual average wind speed. Each area is broken down into different classes. A minimum power class for off-grid homes is Class 2. (Courtesy Bergey Windpower)

Wind Mapping

Table 7-1 shows energy output at varying wind speeds, which are yearly average wind levels measured at a standard 33 feet (10 m) height using a device known as an anemometer (see Figure 7-10). This device consists of a series of cups that are arranged so that greater wind speed causes the cups to spin faster. The spinning shaft is connected to a speed detection device, the output of which is fed to a data logger or recording instrument. The wind speed data is collected for long periods of time and then plotted using a mathematical "Rayleigh distribution" curve. The curve plots recorded wind speed according to frequency of occurrence. The highest point on the curve is the speed at which the wind blows the most often. When this data is recorded over a 12-month period, the result is the average annual wind speed.

Maps are available which detail wind speed during annual, seasonal, and monthly periods. These maps are compiled by and for the government for a variety of purposes such as crop planting, weather forecasting, and wind turbine installations. A typical wind energy map is shown in Figure 7-11, which details mean annual wind resources in the United States. A full overview map for North America is shown in Appendix 8.

Wind Atlases are available online at:

United States: http://rredc.nrel.gov/wind/pubs/atlas/
Canada: www.windatlas.ca/en/index.php

Wind speed maps are broken down into areas of speed and "power class." The typical speed rating is in meters per second, which can be converted to other units:

Wind Speed meters/second	Kilometers per hour	Miles per hour	Quality of Site for Wind Power	Power Class
4	14.4	8.9	Not Acceptable	1
5	18	11.2	Poor	1
6	21.6	13.4	Moderate	2
7	25.2	15.7	Good	3
7.5	27	16.8	Very Good	4
8	28.8	17.9	Excellent	5
8.5	30.6	19	Excellent	6
9.0	32.4	20.1	Excellent- HI	7

Table 7-2. Wind speed is usually provided in meters-per-second. Use this chart to convert to speeds in miles- and kilometers-per-hour.

Power class numbers that range from one to seven indicate relative energies in the wind based on average annual wind speed for that area. Power class 1 is considered unacceptable for wind turbine installations. Local geographic considerations such as large bodies of water, hilly or mountainous areas, and tree cover can change the average data up or down from the values indicated. Another consideration in assessing the accuracy of the mapping is the distance from the recording site to your location. Wind mapping is a reasonable method of determining relative wind speed, but it is not foolproof.

Site Measurement Study

Perhaps the best way to assess your site is to take your own site survey. This is recommended where a larger turbine will be installed for profit or if the wind mapping data for your location is questionable. Companies such as NRG Systems (www.nrgsystems.com) offer their Wind Explorer model on a 30 meter tower kit for approximately US $1,200. These units are battery-powered and contain a data logger that will record wind patterns and provide numerical information that will allow you or your dealer to calculate the potential energy contained in the wind.

Once you have determined the average wind speed for your area, you can apply this information to the turbine's power curve. Using the Bergey Excel-S as our model turbine, let's work through a simple energy calculation.

Assume that your site turns out to be a power class 2, with an average annual wind speed of 6 meters per second. If you consult Table 7-2 you will see that this wind speed translates into 13.4 mph (21.6 kph). Locate 13.4 mph on the wind speed or "x" axis of the graph shown in Figure 7-8, and correlate this to a power output of approximately 1,000 Watts. It would appear that we can now calculate the expected yearly energy output (energy = power x time):

a) *1,000 W output x 24 hr/day x 365 days/year = 8,760,000 Watt hours/year*
b) *8,760,000 Watt hours/year ÷ 12 months/year = 730,000 Watt hours/month*
c) *730,000 Watt hours/month ÷ 30 days/month = 24,000 Watt hours/day*

The energy production data provided by Bergey, shown in Table 7-1, shows a higher monthly energy output that what we have calculated above. Comparing average wind speed with instantaneous power output of the turbine **does not** take into account the effects of wind speeds that are higher or lower than the average. The relationship between wind speed and power output are not linear; wind speeds above the average will contribute significantly to increased power output. Likewise, below average wind speeds will have less impact on turbine power output.

As a general rule of thumb, using average wind speed to calculate power and energy output of any turbine will significantly underestimate the true power rating of the turbine.

Possibly the most interesting element in these calculations is the estimated power output per day. A large renewable energy powered home requires approximately 10 kWh of energy per day. The Bergey Excel-S provides more than twice this amount of energy, even when it is located in a "moderate" wind regime. Now make these calculations for a smaller, less expensive turbine. You may have to run through this calculation or compare estimated energy output charts a few times for different models, but the exercise will pay dividends by helping you purchase a machine that is neither over or undersized for your application.

Site Location and Installation

Another major difference between PV and wind-based systems is their size and height. You can a hang a few PV panels on almost any roof. Wind turbines, on the other hand, will attract attention for miles around. If those "miles around" happen to be forests or grazing lands, then only the cattle will pay attention. But don't even consider installing a wind turbine until you have all the I's dotted and T's crossed. Not everyone will share your enthusiasm for wind, and some may object simply out of fear.

The location of a wind turbine will generally be determined by your lot size, the cable distance to the main building, and tower construction. The best location, however, is where the winds blow strongest or wherever your pocketbook says to put it. Wind speed and smoothness (or laminar flow) will be better over unobstructed grazing land than in tall forested areas owing to ground surface smoothness. Tall trees in a wood lot or forest increase the "working" ground level to the tree height, which in effect reduces tower height. Follow the general site requirements in Figure 7-9 to ensure proper placement.

If the tower is freestanding, such as the one shown in Figure 7-1, it can be installed almost anywhere that will accommodate the concrete base. More often than not, the tower will be either a tubular or tripod lattice guyed tower as shown in Figure 7-12. Guyed towers are inexpensive and tend to be the most popular. The guy wires, however, necessitate considerably more room for installation. A good rule of thumb is to allow a minimum clearance radius of ½ of the tower height for guy wire placement.

A power feed cable is required to connect the wind turbine to the battery bank or inverter. An aerial cable may be run between the tower and house,

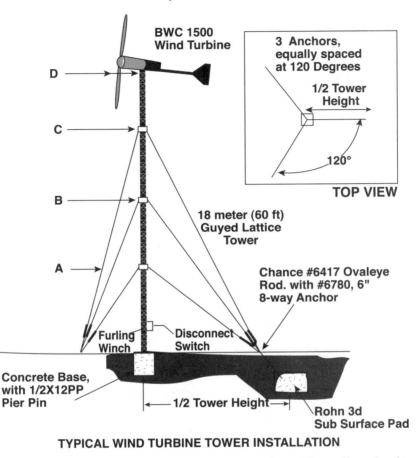

BWC 1500
Wind Turbine

3 Anchors,
equally spaced
at 120 Degrees

1/2 Tower
Height

120°

TOP VIEW

D

C

B

A

18 meter (60 ft)
Guyed Lattice
Tower

Chance #6417 Ovaleye
Rod. with #6780, 6"
8-way Anchor

Furling
Winch

Disconnect
Switch

Concrete Base,
with 1/2X12PP
Pier Pin

1/2 Tower Height

Rohn 3d
Sub Surface Pad

TYPICAL WIND TURBINE TOWER INSTALLATION

Figure 7-12. Guyed lattice and tubular towers are commonly used for small wind turbine installations. Their main drawback is the room required for guy wire clearance.

but more typically a direct burial cable or a cable in a PVC plastic conduit is used. The selection is based on preference as well as on obstacles in the cable path. It may be simpler to provide an overhead cable than to blast rock or dig up a paved laneway. Regardless of which method is chosen, there are specific electrical code requirements to ensure fire and shock safety. These items will be discussed in more detail in Chapter 13.

Zoning and building permits may be required depending upon your proximity to urban areas. Wind installations are installed almost exclusively in rural areas, where zoning restrictions are minimal and building permits are relatively easy to obtain. Regardless of whether the installation is rural or near an urban centre, however, the best advice is to plan things properly from the start.

A neatly drawn site plan showing property lines and setback distances is a good starting point. Building officials will almost certainly not know anything about wind turbines. This may help or hinder your cause. It may help if building inspectors are curious and willing to assist with something "neat." On the other hand their lack of knowledge may also cause problems. You may be asked whether a strong wind might blow the tower down or if aircraft will become tangled in the unit. It pays to do a bit of homework.

- Provide some pictures of local radio/TV antennas that are of similar or greater height. A wind turbine tower may seem odd, yet most people probably pass a 200 foot microwave tower just down the street without even noticing it.

- Ask your tower manufacturer to provide pre-engineered drawings of suitable installation guidelines. Your building inspector is used to walls and roof trusses, not towers. Pre-engineered drawings will provide some reassurance that the unit isn't going to crash through your roof during the next windstorm.

- If you have neighbors, be sure to get them onside first. A wind turbine shows a level of stewardship and concern for the environment that should be welcomed. But if George next door thinks it will kill all the birds and interfere with his TV signals and reruns of Baywatch, he might mount some opposition and you will have to convince him that these concerns are unfounded. Another common concern is noise. This is not a problem. When it's windy enough to activate the turbine, other sounds like trees blowing in the wind will tend to drown out the sounds of the turbine.

- Check for right of ways or legal easements, such as underground gas or telephone cables, deeded to your property. If you are unsure, check with your lawyer and show her your site plan. No one wants to pay a legal bill, but it is a lot less expensive than having to move your tower over a couple of feet.

- Height restrictions may come into play if you are in within a certain distance of an airport or if there are special zoning ordinances. Zoning variations should not be a problem, especially if there are radio towers or buildings in the township that are of similar height. Although aircraft interference is a minor issue, it is a good idea to check with your local FAA office in the United States or NavCan in Canada. Paint and lighting restrictions may be enforced in some areas, but usually only when the tower is over 200 feet (61 m) in the United States and 80 feet (24 m) in Canada. Both agencies will provide a letter indicating whether or not markings are required.

You have every legal and ethical right to install a wind turbine, just as you do if you want to plant a tall tree or erect a TV antenna. Problems with neighbors and building officials occur because of ignorance on their part or lack of planning on yours. You have the advantage of being enthusiastic and better informed. Take the time to transfer this knowledge and educate those around you. The results should lead to an amicable agreement between the concerned parties.

Getting Started: A Word about Safety

Once the wind turbine, tower, and site location have been selected, the next step is to start planning. Erecting even a small wind turbine is a job that can be dangerous and tax the skills of the most mechanically inclined, but don't let this dissuade you. Completing the work yourself, rather than hiring a professional, is satisfying and will save on installation costs.

There are three common methods of erecting a wind turbine and tower. The first uses a tilt-up arrangement that allows the tower to lie flat on the ground during installation and servicing. When the tower is ready for erection a hand or electric winch pulls the tower to a vertical position. This system is primarily used with smaller wind turbines (<1kW) and shorter tower sections. Small winch-up towers are guyed, while larger "tubular" towers are becoming available which are elegant tapered poles that can be hinged to allow lowering for service.

The second type of installation involves building the tower on the ground and then using a crane to erect the tower in one piece. The advantage of this system is that no one has to leave the ground or work up on the tower. The disadvantage is that servicing requires taking a mortgage on the first-born to pay for the crane. Alternatively, you can climb the tower to perform the necessary work.

The third method requires the use of a "gin pole" and some tower climbing theatrics that would rival the gymnastics of Cirque du Soleil performers. A gin pole is a boom or pipe that is bolted to the top of an installed tower section and extends upwards to where the next section is to be placed. A ground crew using a pulley mounted to the top of the gin pole raises the next tower section, while someone who is strapped to the tower gingerly connects the successive sections. Notwithstanding my personal bias against them, I believe the term "gin pole" reflects the need for such fortification after successfully completing this death-defying act.

I cannot emphasize enough how difficult working on a tower can be. Climbing a long extension ladder is nothing compared to climbing and working on a tower. A ladder always has some horizontal slant, making this more akin to walking up steep stairs. Climbing a tower with work boots as well as safety climbing and fall protection gear and a bag of tools is great provided you spend your weekends rappelling El Capitan. It can be very tiring and dangerous work. I would stick with the tilt tower or mortgage the first-born and hire a crane before resorting to the gin pole system.

At some point it may be necessary to actually climb the tower. Before you get to the top and realize you are too scared to let go and actually do some work, let's discuss climbing gear.

There are too many people who think safety equipment isn't cool or even necessary. The smallest wind turbine tower for home use is in the range of 60 feet (18 m). Falling from that tower would be about the same as falling from a five-storey balcony; it is unlikely that you would survive. Visit your local safety supply store and for a few dollars purchase an approved full-body/climbing harness. The Occupational Safety and Health Administration in the United States and The provincial workplace health and safety boards in Canada certify climbing gear suited to this work. Do not purchase a "safety belt" or other non-approved device. One slip off the tower rung and you may be dumped upside down and slide right out of your safety belt.

A safety harness is of no value if it isn't connected to something, preferably you and the tower. A short section (< 6 ft/2 m) of nylon rope may be attached to the harness and then successively clipped and unclipped to the tower with a snap hook as you climb. While steel snap hooks and approved nylon rope or lanyard will work, there are drawbacks to this system.

The first problem is that constantly clipping and unclipping the hook impedes your progress, and

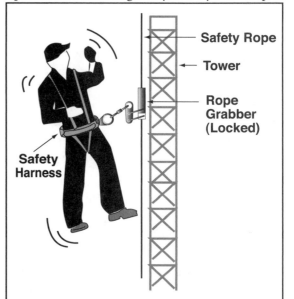

Figure 7-13. Falling from a 100 ft wind turbine tower is not something most people find enjoyable. A safety harness and rope grabber, such as this model shown here, makes climbing and servicing a much safer task.

there is a tendency to not use the clip until reaching the tower top. Climbing the small round legs of a tower can be difficult and slippery, and during the descent you may be quite tired after spending time at the top. Safety is just as important on the way down as it is on the way up.

The second problem involves the length of lanyard required between the tower and safety harness. A lanyard that is 6 ft (2 m) long will allow you to drop that same distance before breaking your fall. This may be enough of a drop to cause serious back or muscle injury with a difficult recovery. Also, a lanyard of steel wire will not give, causing an abrupt stop and causing further injury.

The best value for your money and life is a device from Flexible Lifeline Systems (www.fall-arrest.com). These units comprise a long nylon rope or lifeline attached to the top and bottom of the tower and a "rope grabber" that is attached to the rope and then clipped to the full body harness. When you climb up the tower, the rope grabber slides along, allowing easy ascent.

Should you fall, your body weight causes a cam inside the rope grabber to pinch the lifeline and break your fall, as shown in figure 7-13. During a normal descent, the rope grabber must be lifted and slid down the lifeline one section at a time through the descent.

This system will limit your freedom of motion at the top of the tower, as the lifeline is affixed to the tower top section. To provide a bit more freedom, a short lanyard may be attached between the rope grabber and the safety harness.

In addition to protecting yourself from a fall there are a few other points of safety to be considered.

- Always carry tools in a fitted and sealed pouch. You will need both hands free while you're climbing the tower structure.
- Wear leather gloves and non-slip work boots. Tower rungs are very rough and can be slippery from dew, leaves, pollen and dust.
- When you're climbing the tower, only one hand or foot should be off the tower at a time. For example, release a hand and reach for the next highest tower rung. Make sure you are gripping the rung before lifting a foot to climb to the next rung.
- Always immobilize the turbine by apply the brake switch and/or furling the machine. Never work on a windy day.
- Wear a hard hat. A light breeze can easily cause the turbine to pitch around or "yaw," striking you in the head.
- Never work on the tower alone. Make sure a buddy is on the ground to assist when needed.

- Make sure that ground persons are not standing in the way of errant nuts and bolts. A bolt dropped from 100 feet (30 m) is not much different from a bullet.

Being cheap or stupid when it comes to safety is a no-win proposition.

Tower Foundations and Anchors
Self-Supporting Towers

The self-supporting tower shown in figure 7-1 is built on a single pad of concrete reinforced with metal rebar. Alternatively, three holes may be drilled and filled with a rebar and concrete mix, providing the necessary pads. It is not possible to cover the design of freestanding tower foundations as there are too many variables to consider, including tower height and weight, wind load, and soil type.

The turbine manufacturer is often able to provide foundation plans that have been pre-approved for most local soil conditions.

Guy Wire Supported Tower

Guy wire-supported towers such as the unit shown in Figure 7-12 are the most common design owing to their lower cost. Guyed towers can be tilt-up or installed with a crane or gin pole arrangement. Smaller turbines, typically less than 8 ft (2.5 m) diameter, can be mounted on a pipe tower. These towers use very inexpensive 2.5 " OD steel pipe. The turbine manufacturer supplies the guy wire and hardware kit, and you purchase the pipe locally.

Lattice towers such as the Rohn product line manufactured by Rohn Industries (www.rohnnet.com/ROHNNET/rohnnet2004/html2004/index.html) are more costly than pipe towers but offer the advantage of fewer guy wires and the option of climbing the tower when servicing is required.

The tower mast sits on a compressive foundation known as a pier. The function of the center pier is to prevent the tower from pressing itself into the ground or slipping sideways. It offers no vertical support to the tower.

The guy wires are connected to anchors that are spaced 120 degrees apart from each other as noted in the top view of Figure 7-12. Pipe towers, being structurally weaker, will generally require spacing of 90 degrees. The anchor is placed a minimum of ½ of the tower height away from the center pier.

The pier is most often built with concrete, although it is possible to drill a hole in exposed bedrock and mortar a tower support pin into the hole. It is not necessary to create forms to hold concrete provided the soil will not

Figures 7-14 a, b, c. The successive steps in pouring a center pier to hold a guy wire-supported tower include: a) forming the pier excavation; b) pouring the concrete; and c) aligning the center pin and smoothing the surface. (Courtesy Bergey Windpower)

cave in when pouring begins. Figure 7-14a shows a simple hole dug into the ground, with 2 x 4 inch lumber used to create a "finished" look to the surface of the concrete. The hole is lined with rebar that helps prevent the concrete from cracking. Rebar is sold in lengths that can be cut with a hacksaw and held together with wire to form a metal "cube". The rebar cube should be supported so that it is completely immersed in the concrete.

Provided your excavation is approximately 1 cubic yard or less, it is possible to hand mix the concrete, although most prefer to have a ready-mix truck deliver and pour the pier and anchors in one shot as shown in Figure 7-14b.

A metal rebar pin is shown protruding from the center of the pier in Figure 7-14c. The pin, cut just above the surface of the concrete once it has set, accepts the base of the tower. The sole purpose of the pin is to prevent the tower from sliding off the pier.

Anchors

Smaller wind turbines less than 60" (1.5 m) may be anchored using "screw in place" augers. As the wind turbine size increases, the augers must be encased in concrete as shown in Figure 7-15. The Chance company (www.abchance.com) is a major supplier of anchors to the electrical utility market. The company Web site provides a question and answer support service which can be used to discuss soil types and matching anchor designs. Screw-in anchors may be used in dense clay, sand, gravel, and hard silts. When in doubt, stick with concrete. A hole dug 4 ft (1.2 m) deep (or below frost line) by 2.5 ft (0.8 m) in diameter must be filled with 2 ft (0.6 m) of concrete to cover the auger helix and will provide sufficient holding strength. It is important to ensure that the anchor is fixed below the frost level of the local area or the an-

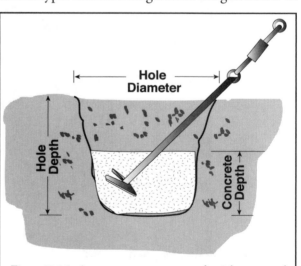

Figure 7-15. A guy support auger may be either screwed directly into suitable soil or cemented in place where soil is loose.

chor may be subjected to "jacking." This is the tendency of buried objects to be pushed towards the surface during the freeze-thaw cycle. Ensure that the concrete hole is lined with rebar to support the concrete.

The anchor must exit the ground at an angle of approximately 45° and point directly towards the center of the tower. This will ensure that the bending force placed on the anchor by the guy wires is equal. If the angle is too great or small, the anchor will have a tendency to bend up or down, reducing its strength over time.

Guy Wires

Guy wires may be supplied as part of the tower kit or they can be purchased in bulk rolls for custom fitting. Use extra-high-strength (EHS) steel cable for supporting the tower. Refer to the tower manufacturer's data sheets for the diameter of guy cable required for your tower and turbine size. A very conservative estimate is ¼" EHS steel cable for units of up to 3 kW output on a 100 ft (30 m) tower.

The length of guy wire required for a tower will depend on the tower height and the distance from the anchor. Table 7-3 provides the location of the guy wires for levels A, B, C and D on the installation shown in Figure 7-12.

The guy wire length required to run from the various heights to the anchor can be calculated using a bit of grade 9 trigonometry called the Pythagorean theorem. As my classmates and I were in a deep slumber during this lesson, we should take a moment to review the calculation.

Tower Height ft / (m)	Height to "A"	Height to "B"	Height to "C"	Height to "D"
40 / 12	17 / 5	35 / 11		
60 / 18	27 / 8	55 / 17		
70 / 21	22 / 7	44 / 14	65 / 20	
80 / 24	25 / 8	50 / 16	75 / 23	
90 / 27	28 / 9	57 / 15	85 / 26	
100 / 30	24 / 7	48 / 14	72 / 22	95 / 29
110 / 34	27 / 9	53 / 16	79 / 24	105 / 32
120 / 37	29 / 9	57 / 17	86 / 26	115 / 35

Table 7-3. The number and height of each guy wire is determined by the overall tower height.

Flip back to Figure 7-12 and you will see that the tower and ground form two sides of a triangle, with the angle at the ground/tower intersection point being 90°. The right angle formed by the two sides and the hypotenuse created by the guy wire form a right angle triangle. Assume that we want to find the length of the guy wire at position "B" for a 60 ft (18 m) tower. Refer to Table 7-3 and you will see that a 60 ft tower requires a guy wire at position "B" of 55 ft (17 m) above grade. The guy radius is ½ the tower height or 30 feet (9 m). With these data we can make the following calculation:

$$(Tower\ Guy\ height^2 + Guy\ Radius^2) = Guy\ Wire\ Length^2$$

Therefore:

$$\sqrt{(Tower\ Guy\ Height^2 + Guy\ Radius^2)} = Guy\ Wire\ Length$$
$$= \sqrt{(55^2 + 30^2)}$$
$$= \sqrt{(3,025 + 900)}$$
$$= 62.5\ ft\ (19\ m)$$

This length provides the exact distance between the tower and the anchor. An allowance of an additional 5 ft (1.5 m) should be provided to allow for misalignment and affixing the guy to the tower and anchor brackets.

Once the guy wire lengths are determined, carefully unroll the guy cable in a straight line along the ground. Even the slightest kinking will render the guy wire useless, so use caution. And speaking of caution, be sure to wear proper eye protection in case the cable whips around and hits you.

The guy wires may be attached to the tower and anchor using "U" bolts or preformed cable grips. Either method is adequate, although most guy cable suppliers will recommend the cable grips because they can easily be adjusted.

Larger wind turbines require the guy wires to be tightened with turnbuckles, as shown in Figure 7-16. After tightening the cables, thread a safety chain or piece of guy wire cable through the turnbuckle to prevent inadvertent loosening.

Smaller units less than 60" (1.5 m) in diameter do not require turn-

Figure 7-16. Guy wires for larger installations are connected to a turnbuckle that allows the cables to be tightened as required. A safety chain or piece of guy wire is threaded through the turnbuckle to prevent unintentional loosening.

buckles. These guy wires may be tightened by pulling them taught using a cable puller or come-along prior to applying the guy wire clamps, as shown in Figure 7-17.

Electrical Supply Leads

Small turbines output direct current at 12, 24, or 48 Volts, while large units supply 3-phase alternating current, which is rectified at a remotely located "controller box." In order to transfer this power down the tower, the turbine will have a connection terminal or be supplied with a

Figure 7-17. Guy wires can be tightened using a come-along or cable puller. Smaller units less than 60" (1.5 m) in diameter do not require turnbuckles. (Courtesy Bergey Windpower)

short section of "flying leads." A direct current machine will be supplied with two wires, one positive and one negative. A machine which outputs 3-phase alternating current will have three wires without any polarity or connection concerns.

The connections between the electrical leads from the wind turbine and the tower wires should be located in a weatherproof junction box mounted near the top of the tower. The tower-to-house lead wires should be weather-resistant flexible armored cable (Teck is one trade name) or else they should be enclosed in conduit. The power lead wires or conduit may be strapped to one of the tower legs using either ultraviolet- protected tie wraps or cable clamps. Wire size and selection will be discussed in Chapter 13.

Figure 7-18. The tower and turbine are ready to be lifted by the crane. Note the tower bottom plate center hole and mating pier pin at lower left. (Courtesy Bergey Windpower)

Erection of the Tower with a Crane

Pre-Lift Check List

At this point, the tower should be flat on the ground or slightly raised and supported along its length if the turbine is already installed. We are now ready to prepare for final inspection prior to lifting.

- Guy wires (if used) are attached to the tower and secure. Guys should be run down to the tower. Ensure that cables are in order from lowest to highest and are not twisted.
- Turbine is wired into the junction box at the top of the tower. Connections are made with tower lead wires.
- Tower lead wire is secured to the tower and is sufficiently free to allow tower movement when aligning base plate with pier pin.
- Turbine is securely mounted to tower and bolt torques tested.
- Tower bolts are checked for security and torque.
- Turbine brake and furling are enabled.
- Turbine is tied with rope to prevent it from yawing and hitting the crane or lift cables.
- Check, check, and re-check that all hardware and tools are ready and close

Figure 7-19. The crane lifting straps are installed at 75% of tower height and wrapped through the rungs to prevent them from sliding up the tower. Note the support cradle used to hold the tower off the ground, allowing turbine installation before the arrival of the crane. (Courtesy Bergey Windpower)

at hand. The crane service charges by the hour, even when you have to run to town to pick up a couple of last minute ten cent washers.

• Check the weather for a still, clear day and call for the crane.

The Lift

Once the crane is ready, the tower lifting straps are threaded through the tower rungs at 75% of the tower height. The tower is then slowly lifted, making sure that the guy wires, power feed cable, and turbine assembly do not tangle in or hit the crane structure.

If the turbine was not installed at ground level, lift the tower to a comfortable working height and install the unit. (It is preferable to have this step completed prior to the arrival of the crane, as this time is billed at a rate similar to that of a New York lawyer.)

Once the tower is located over the center pier, the guy wires may be secured to their respective anchors as described above and detailed in Figures 7-15 through 7-17. At this point, the leveling of the tower and

Figure 7-20. The tower and turbine are slowly lifted into position, with care being taken not to tangle the guy wires or cause the turbine body or blades to strike the crane. (Courtesy Bergey Windpower)

Figure 7-21. Once the tower is secured in position, the lifting straps and anti-yawing ropes are removed. This is also a good time to appreciate the view and congratulate yourself on a job well done.(Courtesy Bergey Windpower)

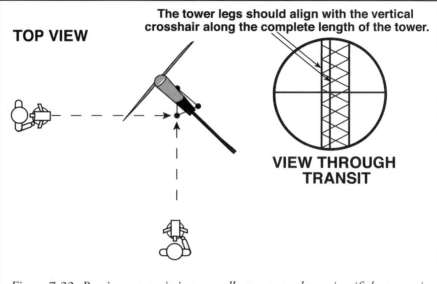

TOP VIEW

The tower legs should align with the vertical crosshair along the complete length of the tower.

VIEW THROUGH TRANSIT

Figure 7-22. Renting a transit is an excellent way to determine if the tower is properly leveled.

Figure 7-23. At least one tower post (more in sandy conditions) should be connected to a grounding rod driven 8 feet (2.5 m) into the soil. (Courtesy Bergey Windpower)

tightening of the cables is only approximate. Once the tower is secure, you need someone to climb the tower, remove the crane lifting straps, and untie the anti-yawing ropes securing the turbine. Use caution when removing the anti-yawing ropes to ensure that the turbine doesn't yaw and strike you.

The crane may now be packed up and sent on its way—just as soon as you've written a big check.

Hinged Tower Lifting

There are numerous designs of hinged towers on the market today, including guyed or tubular freestanding models. The advantages of a hinged tower are obvious, as they allow all of the installation and service work to be done safely from the ground. In addition, a crane is not required to perform the lift.

Hinged towers employ a hand-operated or power winch that acts on a levered section of the tower. The anchor foundations are slightly larger, but the lift is the same as it is with a crane.

Final Assembly

After the tower is raised, or when the crane has left (or stopped billing), set the tower vertical by using a transit sighting scope as detailed in Figure 7-22. A transit may be rented from a local tool company. These units are often seen on sites where construction workers are building roads or checking various angles. A basic unit comprises a tripod stand with an optical sighting scope similar to those used on rifles. The tripod is placed between one and two tower heights away from the center pier and leveled using the integrated water bubble levels and adjustable legs. The vertical crosshair in the scope is then aligned with one tower leg, close to the ground. The scope is then moved vertically up and down, ensuring that the tower leg and crosshair remain in parallel through the entire length.

Guy wires may be adjusted tighter or looser using the turnbuckles to compensate for error. The transit is then moved 90° and checked again from this alternate position. It may be necessary to move back and forth a few times to get the alignment perfect.

If you do not have access to a transit, it is possible to use a standard carpenter's level and check for tower alignment at the bottom section. You can then site up the tower, placing your eye as close to the tower leg or pipe as possible. Bends in the structure will be obvious and can be adjusted by aligning the guy wires as described.

Once the tower is aligned, tighten the guy wires until there is little slack in the cables. A good shortcut for testing cable tension is to strike the cable with a hammer. The cable should sound a guitar-like note and a "wave" will be seen running quickly up and then back down the wire. Don't forget to attach the turnbuckle security chain as outlined in Figure 7-16.

The last step in the installation process is to ground the tower and guy wires. In clay and other moisture-retaining soils, a single 8 ft (2.5 m) electrical grounding rod should be driven into the ground as close to or through the tower pier as shown in Figure 7-23. In sandy soils, at least two grounding rods should be installed, each at a different tower leg, to ensure proper lightning protection qualities.

Figure 7-24. Each set of guy wires should be grounded using a set of U-shaped galvanized steel clamps and attached to the guy anchors. Use smooth bends and as short a run as possible.

Figure 7-25. A job well done! A Whisper 175 reaches for wind as the PV array complements the production of energy for this very rural off-grid home. (Courtesy Southwest Windpower)

Each of the guy wires sharing a common anchor should be connected together using a minimum 0-gage grounding wire and galvanized U-shaped bolts as detailed in Figure 7-24. The end of the ground wire may be connected either to the guy anchor or to a separate 8 ft grounding rod.

For both the tower leg and guy wire grounding, be sure to use as short a run of wire as possible and smooth, even bends. Lightning hates sharp corners and may completely miss the grounding rod if severe wire bends are used.

Recycling a Jake

A Homeowner's Story *by Josée Guénette*

We purchased our Jake in 2003, three years into the construction of our owner-built log home and a full eighteen months before we would be ready to use it. At that time, I was not really aware that the Jacobs wind turbine, also known as "The Jake," was considered the gold standard in the industry; but my husband was, and he jumped at the opportunity to buy one second-hand, even if it needed a bit of work!

A family in Lanark, Ontario had been relying on this Jacobs turbine to supply their household electricity for many years, but an unexpected grid extension, corresponding with a much needed maintenance cycle, helped them decide to sell not only the turbine they were using, but also a second one they had purchased and not yet commissioned. We bought the largest of the two, a 2.4 kW turbine which dated back to 1958. The current owner had purchased it from a farmer in Saskatchewan. Our deal was Cash & Carry—we paid for the turbine where it stood, including everything up to

DJ MacIntyre of Le Boise' Alternatives (www.leboise.com) nearing completion of the 2.4 kW, 1958-era Jacobs wind turbine installation.

and including the wires into the battery bank. Moving it was our responsibility, and the next challenge!

The dismantling and re-commissioning of the Jake were feats of assiduous project management, and our success was due largely to planning and family involvement. There were conference calls, tool lists, battle plans in triplicate, personalized tool kits, troop deployment diagrams, and tight schedules. Oh, and a nice picnic lunch, to keep everyone in good spirits.

It took about six hours to get our Jake out of the sky, disassemble the tower, pull up the 4/0 wiring and send it all to our home in La Pêche, Québec. The turbine did need some work: the insulation on the windings was worn, the pickup rings had fused, the original sitka spruce blades needed some paint. During the repairs to the turbine, we reassembled the tower on the ground. It took a lot longer to reassemble it than to take it apart!

But now, triumphantly refurbished and repainted, our Jacobs turbine stands 90 feet in the air and is largely responsible for the electricity in our home. It recharges a NiCad battery bank which was salvaged from a decommissioned military installation. Considering that we started this adventure by decreasing our electricity use where practical, our system is perhaps a good example of Reduce, Reuse and Recycle!

The "Jake" is just about ready to fly, looking pretty sharp in a new coat of paint.

DJ is feeling pleased now that the Jake is flying again at its new home in La Pêche, Quebec, Canada.

8
Micro Hydro Electricity Production

If you are lucky enough to have a stream, river, or waterfall on your property it is well worth considering micro hydro electricity production. While PV is undeniably the simplest renewable energy source to install and maintain, micro hydro systems (units under 1 kW capacity) can be the least expensive. The density of water is considerably higher than that of wind. This increase in density allows a hydro turbine to be many times smaller than a wind turbine of similar output.

Smaller turbines and fairly simple installation help keep the cost to a reasonable level. In addition, hydro sites are generally able to supply electricity on a 24–hour-a-day basis. A smaller amount of power supplied over a longer period equates to a larger amount of energy. For example, a typical PV array is rated at 1,000 Watts of output power. Over a five-hour period of sunlight, this equates to 5 kilowatt-hours of energy. A small 200 Watt hydro turbine operating 24 hours per day produces approximately the same amount of energy. On the down side, suitable hydro power sites are not as common as either PV or wind sites.

Understanding the Technology

As we discussed in Chapter 1.3, hydro electricity is produced from the kinetic energy in moving water under the force of gravity. The pressure or head and the flow determine the potential energy in water. The vertical distance that water must fall under the influence of gravity determines the head pressure. Head is measured either in feet or in units of pressure such as pounds per square inch (psi) or kilopascals. (kPa). It is interesting to note that water standing in a vertical tube will exert a known force at the bottom of the tube at the ratio of 1 psi (6.9 kPa) per 2.31 vertical feet (27.7 inches / 0.7 m). This allows head to be easily converted back and forth between vertical height and pressure as necessary.

Flow is the quantity of water flowing past a given point in a given period of time and is expressed as a volume. Typical units are gallons of water per minute (gpm) or liters per minute (lpm).

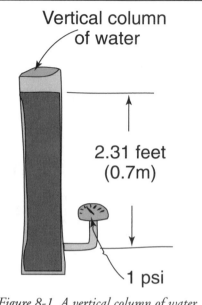

Figure 8-1. A vertical column of water 2.31 feet high (0.7 m) exerts a pressure of 1 PSI (6.9 kPa).

Figure 8-2. This turgo wheel turbine and generator are suited for high head and low flow applications. This unit is installed in the Rocky Mountains of British Columbia and provides an output of 500 Watts. (Courtesy Energy Systems and Design)

When the flow and head are integrated mathematically the result is power, expressed in horsepower or Watts. Once the units of waterpower are converted to Watts by the turbine's generator, we are back to common electrical units of measure. Because head and flow are interrelated, a high head site with a low flow produces the same amount of power as a site with a low head

Figure 8-3. A low head turbine such as this propeller-based model requires high water flow. With 3 feet (1 m) of head, this unit produces 200 Watts of power. (Courtesy Energy Systems and Design)

and a high flow. For example, a site with 100 feet of head and 2 gallons per minute of flow has the same potential energy as a site with 2 meters of head and 100 gallons per minute of flow. This is not semantics. Most sites are located in an either/or situation. A high head site is generally located in a very hilly or mountainous region, where the water might flow through sev-

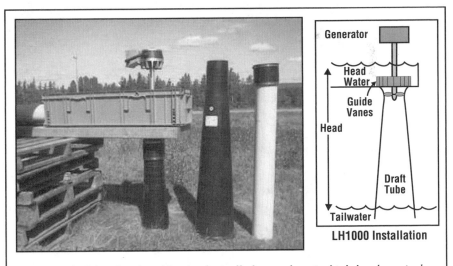

Figure 8-4. A low head turbine is physically larger than its high head cousin but retains many of the same components. (Courtesy Energy Systems and Design)

eral hundred to several thousand feet (100 m to 1000 m) of plumbing line. Fortunately a high head site requires little flow, so that the pipe and turbine in this instance are fairly small.

The opposite is true of low head sites. Commonly known as "run of river" designs, low head sites can be sized to operate from 2 to 10 feet (0.6 m to 3 m) of head. What we gain in lower head we pay for in more water flow and a larger turbine; nobody rides for free.

Water turbines come in two broad categories and dozens of subcategories that we do not need to worry about. The main consideration with respect to the turbine selection is whether there is a large flow of water and little vertical drop or vice versa.

All turbines have a number of components in common, as do the civil works of the hydro site. Figure 8-4 shows the workings of a low head, high flow hydro turbine manufactured by Energy Systems and Design of Canada (www.microhydropower.com). As the low head turbine must have a large flow, the intake is formed using the large gray plastic "flume." Water enters the turbine heart and passes through a set of guide vanes. These vanes are stationary and are used to force the water to hit the runner blades at the correct angle. In this low flow turbine, the runner blades are shaped like a large boat propeller. As the water falls through the vertical head, the runner blade captures the kinetic energy and causes the generator to rotate, producing electricity. The draft tube forms the exit path for the water. The tapered draft tube acts as a siphon, helping to draw water down the tube and away from the runner blades, increasing turbine efficiency. To increase the head and resulting output power of the turbine, a series of draft tube extensions can be added.

High head turbines such as the turgo wheel unit shown in figure 8-5 are considerably smaller owing to the low flow of water required to turn the runner blade. Increased head and resulting water pressure make up for what is lost in flow. For example, a turbine installed with 300 feet (91 m) of head will have a water pressure at the runner nozzles of 130 psi (900 kPa).

300 feet head ÷ 2.31 feet per psi = 130 psi nozzle pressure

Figure 8-5 shows a water intake pipe plumbed to a series of two shutoff valves prior to entering the turbine. These valves may be used for turning off the turbine during maintenance or when insufficient water flows to operate the system during dry periods. After leaving the shutoff valves, water is directed through a nozzle prior to striking the turgo-style runner blade. Nozzle openings can be changed to modify the water flow and resulting output power of the turbine.

Figure 8-5. High head turbines tend to be quite small owing to the high water pressure and resulting low flow required to operate the tiny runner blade. (Courtesy Energy Systems and Design)

The high-pressure water jet strikes the runner blade, imparting a turning effort or torque to the generator shaft, producing electricity. The water then exits the turbine heart by simply dropping away from the open-bottom casting to return to the tailrace downstream.

The civil works comprise the dam, intake, penstock, and tailrace sections of the hydro site. Figure 8-6 details the major components required for a high head site. A dam structure provides a means of inserting an intake pipe and strainer midway between the riverbed and river surface. If water levels are sufficiently deep it may be possible to eliminate the dam and simply install the intake in a submerged strainer or crib.

The dam's outlet pipe is fed to a stop valve that is used to drain the penstock system during servicing or in the event of a broken pipe. Penstock pipes are often buried to prevent mechanical damage and chewing by animals. The penstock is also provided with an air valve at the top and a pressure relief

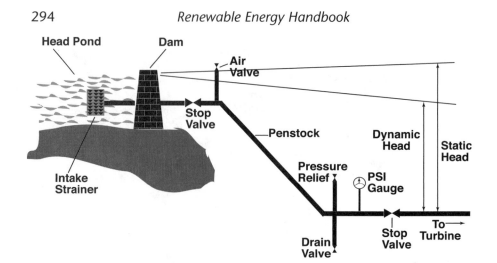

Figure 8-6. The dam, water intake, penstock, and support components are collectively known as the civil works.

and drain valve at the bottom. In addition, a second turbine stop valve and pressure gage are also provided.

When the turbine stop valve is closed, water will flow stop flowing and the penstock will remain filled with water to the height noted as the "static" or "not moving" head level. As discussed earlier, the static head can be directly translated into a pressure by dividing head in feet by 2.31 to produce psi. The pressure gage will record this static pressure when the turbine stop valve is closed.

As the turbine stop valve is opened, water will begin to flow down the penstock to the turbine. As this occurs, the pressure on the psi gauge will drop as a result of pressure loss due to friction in the penstock pipe. The resulting pressure drop will lower turbine output accordingly. To circumvent this problem, a correspondingly large penstock pipe may be used. Alternatively, lower friction polyethylene SDR pressure-rated pipe may be chosen over PVC material. Pipe friction losses are noted in Appendix 9a and 9b.

If you have ever been in a home where suddenly turning off a water tap causes the whole house to shake, you have experienced the effects of water hammer, which occurs when moving water is suddenly stopped. Moving water contains kinetic energy (the energy of motion) that must be removed when the movement stops. If the turbine stop valve is suddenly turned off, the entire weight of water in the penstock has to find someplace to let off this energy. The effect is not much different from a freight train slamming into a solid wall: something has to give. Usually it's the penstock that gives. If the water hammer is severe enough, the penstock pipe will absorb the energy and possibly split or explode. To help stop damage to the penstock a pressure relief valve is located near the bottom. Should pressure build due to water hammer, the pressure relief valve is sized to open before the pipe bursts.

An air valve is provided at the topmost section of the penstock to eliminate any air trapped within.

Low head or run of the river sites are generally simpler but still require a dam, intake, and penstock pipe feeding an flume (water delivery box) as shown in Figure 8-3.

What Is a Good Site?

In order to determine if you have a suitable hydro site, it is necessary to perform an evaluation of the available resources. In assessing a particular site, the variables to consider are:

- Water head level
- Flow at the turbine intake
- Penstock design issues
- Electrical energy transmission distance

• Water Head Level

Head may be measured using several techniques. One method is to hold one end of a garden hose at the same level as the proposed intake from the stream.

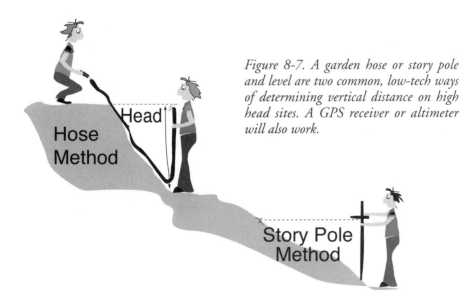

Figure 8-7. A garden hose or story pole and level are two common, low-tech ways of determining vertical distance on high head sites. A GPS receiver or altimeter will also work.

A first person ensures that the hose remains filled with water. The other end of the hose is routed downhill, causing water to flow out. If a second person slowly lifts the downstream end of the hose, water will stop flowing once the upstream and downstream ends are at equal heights. Measure the distance from the ground to the point where the low-side hose stopped leaking water. This is the head measurement for the first section. This is repeated with the first person now holding the upstream end of the hose at ground level where the second person was standing. The process is repeated until the final head measurement is completed at the desired intake location for the turbine.

A variation on the hose method is to use a measuring stick or "story pole" and a carpenter's level, working uphill. The story pole is placed at the desired turbine intake location. A carpenter's level is placed against the pole at any convenient height. By sighting along the level, a spot will be noted ('x' in Figure 8-7) which denotes a point of equal elevation. The story pole is moved to this spot and the process repeated until the stream intake location is reached.

A third and much simpler method is to use a global positioning system receiver or accurate altimeter if one is available.

• Flow at the Turbine Intake

The easiest method to measure small flows is to channel the water into a pipe and, using a temporary dam of rocks, plywood or whatever, fill a container to a known volume. By measuring the time it takes to fill the container you will be able to calculate the flow rate.

For higher flow rates the weir method is more

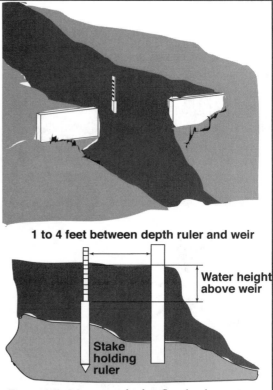

Figure 8-8. Measuring higher flow levels requires the use of a temporary dam or weir installed in the brook, stream, or river.

versatile. This technique uses a rectangular opening cut in a board which is set into the stream much like a dam, as detailed in Figure 8-8. The water is channeled into the weir and the depth is measured from the top of a stake that is level with the edge of the weir and positioned several feet upstream.

The resulting measurement can then be applied to the data contained in Table 8-1 that converts the weir measurement to U.S. gallons per minute. Multiply gallons per minute by 3.8 to determine liters per minute.

TABLE 8-1 - WEIR MEASUREMENT TABLE

Table shows water flow in gallons/minute (gpm) that will flow over a weir one inch wide and from 1/8 to 10-7/8 inches deep.

Inches		1/8	1/4	3/8	1/2	5/8	3/4	7/8
0	0.0	0.1	0.4	0.7	1.0	1.4	1.9	2.4
1	3.0	3.5	4.1	4.8	5.5	6.1	6.9	7.6
2	8.5	9.2	10.1	10.9	11.8	12.7	13.6	14.6
3	15.5	16.5	17.5	18.6	19.5	20.6	21.7	22.8
4	23.9	25.1	26.2	27.4	28.5	29.7	31.0	32.2
5	33.4	34.7	36.0	37.3	38.5	39.9	41.2	42.6
6	43.9	45.3	46.8	48.2	49.5	51.0	52.4	53.9
7	55.4	56.8	58.3	59.9	61.4	63.0	64.6	66.0
8	67.7	69.3	70.8	72.5	74.1	75.8	77.4	79.1
9	80.8	82.4	84.2	85.9	87.6	89.3	91.0	92.8
10	94.5	96.3	98.1	99.9	101.7	103.6	105.4	107.3

Example of how to use weir table:
Suppose depth of water above stake is 9 3/8 inches. Find 9 in the left-hand column and 3/8 in the top column. The value where they intersect is 85.9 gpm. That's only for a 1-inch weir, however. You multiply this value by the width of your weir in inches to obtain water flow.

Measuring the flow at different times of the year helps you to estimate your minimum and maximum useable water flows. If the water source is seasonably limited, you will have to depend on a hybrid system or fossil-fuel generator to make up the loss of hydro energy.

You must also keep in mind that a reasonable amount of water should be left in the natural stream path to support life in that section of the stream. An intake screen will ensure that fish and other animals will not be driven into the turbine's intake system. When you have removed the energy, the water will be returned to the stream with no loss in total water flow. Such an arrangement not only provides you with renewable energy, it is also environmentally responsible.

• Zero Head Sites with Good Flow

After carefully explaining above that head and flow combine to make power, I am going to recant that statement enough to allow for an interesting class of zero head turbines. Zero head turbines rely completely on the flow or run of the river to extract their energy. There is currently only one commercially available unit of this class, sold by Jack Rabbit Energy Systems (www.jackr abbitmarine.com) and illustrated in Figure 8-9.

Zero head units were originally designed to be pulled behind slower-moving sailboats as a means of charging batteries in these seriously off-grid systems. It stands to reason that if the turbine stops moving and the water doesn't, the same power will be generated.

Installing a unit in a culvert or fast-moving stream will start the outboard motor type of propeller spinning. Throw a ping-pong ball into the stream. If you can keep pace with the ball at a walk the site is too slow. A brisk walk to partial jog indicates a power output of 6 amps at 12 volts potential (1,700 Watt-hours per day). If you need to increase your speed to a jog, the power output will be 9 amps at 12 volts (2,500 Watt-hours per day).

Figure 8-9. Zero head turbines such as this model from Jack Rabbit Energy Systems (www.j ackrabbitmarine.com) feature special low-speed generators that can be mounted in fast-moving streams.

These units are ideal where the stream will not freeze or for summer cottage installations.

• Penstock Design Issues

All hydro systems require a pipeline. Even systems that operate directly from the dam require at least a short plumbing run or flume. It is important to use the correct type and size of pipe to minimize restrictions in the flow to the nozzle(s) or turbine runner wheel. More power can be obtained from the same flow with a larger pipe which has lower losses. Therefore, pipe size must be optimized based on economics. As head decreases, efficiency of the system decreases; it is important to keep the head losses low.

The pipe flow charts shown in Appendix 9 show us that 2" (50 mm) diameter polyethylene pipe has a head loss of 1.77 feet (0.5 m) of head per

100 feet (30 m) of pipe at a flow rate of 30 gpm (113 lpm). This is 17.7 feet (5 m) of loss for 1000 feet (300 m) of pipe. Using 2" (50 mm) PVC results in a loss of 1.17 feet (0.36 m) of head per 100 feet of pipe or 11.7 feet (3.6 m) per 1000 feet.

Polyethylene comes in continuous coils because it is flexible (and more freeze resistant than steel). PVC comes in shorter lengths and has to be glued together or purchased with gaskets (for larger sizes).

• Additional Pipeline Construction Details

At the inlet of the pipe, a filter should be installed. This can be a screened box with the pipe entering one side or a section of pipe drilled full of holes and wrapped with screen. If the holes are sufficiently small, the screen may not be necessary. Make sure that the filter openings are smaller than the smallest nozzle used on high head machines. This prevents debris from blocking the nozzles and restricting power output.

The intake must be above the stream bed so as not to suck in silt and deep enough so as not to suck in air. The intake structure should be placed to one side of the main flow of the stream so that the force of the flowing water and its debris bypasses the intake. This also ensures that fish will not become entrapped in the intake system. Routinely clean the intake of any leaves or other debris.

If the whole pipeline doesn't run continuously downhill, at least the first section should so that the water can begin flowing. A bypass valve may be necessary. This should be installed at a low point in the pipe.

For pipelines running over dams, the downstream side may be filled by hand. Once filled, the stop valve at the turbine can be opened to start the flow using a siphon action. If full pressure is not developed, a hand-powered vacuum pump can be used to remove air trapped at the high point.

At the turbine end of the pipeline a bypass valve may be necessary to allow water to run through the pipe without affecting the turbine in order to purge the line of air or increase the flow to prevent freezing.

A stop valve should be installed upstream of the nozzle and a pressure gauge should be installed upstream of the stop valve so both the static head (no water flowing) and the dynamic head (water flowing) can be read.

The stop valve on a pipeline should always be closed slowly to prevent water hammer (the column of water in the pipe coming to an abrupt stop). Water hammer can easily destroy your pipeline and for this reason you may wish to install a pressure relief valve just upstream of the stop valve. Water hammer can also occur if debris clogs the nozzle.

• Electrical Energy Transmission Distance

Let us assume we have a 12 Volt battery system and the transmission distance from the turbine to the battery bank is 200 feet (60 m) for the round trip. Because of electrical resistance losses in the transmission cable we will require a slightly higher voltage at the generator to charge the batteries. If our turbine is providing 300 Watts of energy, this will translate into a current of approximately 20 Amps at 15 Vdc.

No electrical conductor is perfect, and a decision must be made about how much energy should be lost in the wiring. Systems with lower losses require larger, more expensive cable than those with slightly higher losses. An alternative is to raise the battery voltage to 24 or even 48 Volts. Remember that a higher voltage will result in lower current.

A good starting point is to work with a 10% loss. This yields a 30 Watt loss out of the original 300 Watt production assumption. The formula for conductor resistance loss when wire resistance is not known is:

$$
\begin{aligned}
\textit{Watts of loss} \div \textit{Current}^2 \quad &= \quad \textit{Wire Resistance in Ohms} \\
&= \quad \textit{30 Watts loss} \div 20^2 \textit{ Amps} \\
&= \quad \textit{30 Watts loss} \div 400 \\
&= \quad \textit{0.075 Ohms of Wire Resistance}
\end{aligned}
$$

This is the calculated wire resistance that will produce a 10% electrical transmission loss. The wire resistance chart in Table 8-2 shows losses per 1000 feet of cable, while our sample installation requires only 200 feet round trip:

1000 feet ÷ 200 feet x 0.075 ohms = 0.375 ohms per 1000 feet.

Table 8-2 indicates that 6-gauge wire has a resistance of 0.40 ohms per 1000 feet:

200 feet ÷ 1000 feet x 0.40 ohms = 0.08 ohms.

Increasing the wire size further reduces the losses in the same manner albeit at a higher cost.

We must also take into consideration that the power loss within a cable is shared between the current and the voltage. Voltage drop in the wire is equal to:

$$
\begin{aligned}
\textit{Current flow in Amps x resistance of wire} \quad &= \quad \textit{Voltage drop in wire} \\
&= \quad \textit{20 Amps x 0.08 Ohms} \\
&= \quad \textit{1.6 Volts drop}
\end{aligned}
$$

Wire Gauge	Diameter (Inches)	Ohms per 1000 Feet	Ohms per Kilometer
0000	0.460	0.05	0.16
000	0.410	0.06	0.20
00	0.364	0.08	0.26
0	0.324	0.10	0.32
2	0.258	0.16	0.52
4	0.204	0.25	0.83
6	0.162	0.40	1.32
8	0.128	0.64	2.10
10	0.102	1.02	3.34
12	0.081	1.62	5.31
14	0.064	2.58	8.43
16	0.051	4.10	13.39
18	0.040	6.52	21.33

Table 8-2. Copper wire resistance loss chart measured in ohms per 1000 feet or ohms per kilometer.

Therefore if the battery voltage is 13.4 the hydro generator will be operating at 15.0 Volts. Keep in mind that it is always the batteries that determine the system voltage. That is, all voltages in the system rise and fall according to the battery's state of charge, as discussed in Chapter 9, "Battery Selection and Design."

Site Selection Summary

After having determined the head, flow, penstock, and electrical transmission issues, the question still remains: will this site be suitable for a micro hydro installation? To make this decision, it is now necessary to refer to the manufacturer's data sheets to determine how well the site will work with the available turbine selection. Hydro turbines, like wind turbines, come with a power curve data sheet, but instead of relating wind speed to power, hydraulic units relate flow and head to generator output power

The power curve for this model of turbine is based on the assumption that a minimum of 2 feet (0.6 m) must be provided before the unit develops useable power of 100 Watts output (or 2.4 kWh of energy per day). Even at this level of power output, the unit will require 500 gpm (1,900 lpm) of water flow. On the other hand, a high head site will require much less water flow but correspondingly higher head pressure levels.

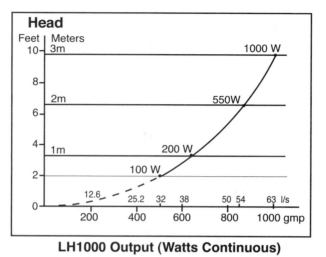

Table 8-3. Output of a low head turbine rated 1 kW at various flow and head levels. (Energy Systems and Design)

It will take a bit of juggling to determine which turbine will work best for your site based on the manufacturer's data. As with any complex technology, a dealer who has experience installing these units will be invaluable. Additional turbine-specific installation data is available from the various manufacturers listed in the resource guide in Appendix 3.

Splashing About with Micro Hydro

Micro hydro systems are a little bit more difficult to understand, design, and install than PV or wind-based systems. Offsetting the complexity issue is the fact that hydro systems can generate more power and energy than can other forms of renewable energy. During the early part of the industrial revolution water power was used to operate mills directly, and within a few years those same waterwheels and turbines began generating electricity.

Water has the advantage of being much denser and having higher energy content than wind. Because of this, a hydro turbine of the same capacity as one powered by wind will be many times smaller. The turbine shown in Figure 8.1-1 will provide power 24 hours per day for battery charging and inverter systems.

With sufficient water resources, it is possible to power any home with a direct-to-alternating current turbine. Although these systems are still considered micro hydro, they operate on the exactly the same principle as mega-hydro systems such as Niagara Falls or Hoover Dam. Water falling through a vertical drop or "head" strikes the turbine blades causing a shaft connected to an alternating current electrical generator to spin. The power from the generator is fed via a control system known as a load shedding governor to the house electrical panel.

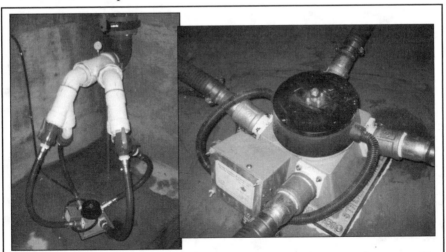

Figure 8.1-1. Because water has a higher energy density than any other renewable energy source, it requires very small turbines to make a considerable amount of energy. This four-jet turbine with sufficient water resources can power any battery/ inverter-based off-grid home. (Courtesy EA Energy Alternatives Ltd, www.energyalternatives.ca)

Figure 8.1-2. With sufficient water resources, it is possible to power any home with a direct-to-alternating current turbine. Although these systems are still considered micro hydro, they operate on the exactly the same principle as mega-hydro systems such as Niagara Falls or Hoover Dam. (Courtesy Canyon Turbines Inc.)

With a given amount of water flow through the turbine and a given amount of electrical load connected to the generator, the turbine speed and resulting power frequency will remain fixed. However, electrical loads are never fixed. When a refrigerator switches on a load is placed on the turbine, causing it to slow down and lower the frequency of the generated power. With a direct-to-alternating system, electrical frequency must be carefully controlled and maintained at 60 cycles per second or 60 Hz, as this is the frequency at which most appliances must operate. The slowing of the turbine and generator is therefore not acceptable.

Figure 8.1-3. A load shedding governor is designed to monitor the generated power and rapidly remove or apply some auxiliary electrical load to compensate for the varying electrical frequency resulting from changing household loads. Using this technique, the load shedding governor can maintain the fixed electrical load required by the turbine and provide a constant frequency of electrical power to the home. (Courtesy Canyon Turbines Inc.)

A load shedding governor is designed to monitor the generated power and rapidly remove or apply some auxiliary electrical load to compensate for the varying electrical frequency resulting from changing household loads. The auxiliary load is typically an electrical water heater which is either mounted in the domestic water system of the home or installed in the flowing river to dissipate excess electrical energy. Using this technique, the load shedding governor can maintain the fixed electrical load required by the turbine and provide a constant frequency of electrical power to the home.

The water intake is typically located in a small creek or stream at a point with a considerably higher elevation that the hydro turbine. Higher vertical head yields more energy while using less water than a low-head site. Although this requires longer penstock runs, it reduces the size of the pipe and turbine.

Figure 8.1-4. The water intake is typically located in a small creek or stream at a point with a considerably higher elevation than the hydro turbine. A self-cleaning strainer prevents debris, fish, or other materials from entering the system and causing failures due to mechanical damage or blockages. This intake filter provides the water power to operate a 12 Volt, 30 Amp micro hydro system that powers a provincial park ranger station on the coast of British Columbia. (Images 8.1-4 through 8.1-22 all Courtesy Peter Talbot)

The intake is designed to extract only a small portion of the river flow, particularly if there is marine life. A self-cleaning strainer prevents debris, fish, or other materials from entering the system and causing failures due to mechanical damage or blockages.

Figure 8.1-5. Where there are no roads, it can take extraordinary measures to deliver the penstock. Here a helicopter delivers 50 ft (15 m) sections of DR17 poly pipe 1,500 ft (457 m) up a mountain.

Figure 8.1-7. This view shows the installed penstock running through a mountain forest. As the penstock runs down the mountain, water pressure increases, necessitating the conversion from poly to steel pipe, as in this 250 PSI (1723 kPa) section.

Figure 8.1-6. Moving long sections of steel or poly pipe uphill can be demanding work. The job is made easier with a special winch attached to a gasoline-powered chain saw.

Water is then directed down the penstock pipe, building pressure under the influence of gravity. At the power house, the high pressure water strikes the turbine blades, causing the drive shaft to spin, operating the electrical generator.

Figure 8.1-8. High pressure water strikes the turbine blades, causing the drive shaft to spin, operating the electrical generator. This three-phase induction motor is used as a generator, along with a step-down transformer to provide battery charging for an inverter-based system.

Figure 8.1-9. This beautifully designed power house contains the turbine and generator of a mid-sized hydro generating station. In this view, the penstock can be seen entering the building on the right. Water leaves the turbine and exits the power house through the rectangular opening in the concrete foundation.

Direct current generators connect to a battery bank and inverter system in exactly the same way as a wind turbine would be interconnected to an off-grid home.

Alternating current systems power the house with 120/240 Volts A.C. in the same manner as a regular grid-connected home.

Tantalus View Retreat

Janice and Warren Brubacher have enjoyed and nurtured their homestead in the Coast Mountains of British Columbia for over ten years. Unfortunately, the first few years required the use of a small gas generator to power their home, which was not in keeping with their desire for peace and quiet and their wish to live in harmony with nature.

The Brubachers decided to see if the small stream on the property would have sufficient water flow to power their home. They contacted Peter Talbot of Home Power Systems (www.homepower.ca) to help with this assessment.

Now, several years after the successful installation, the family is living in a beautiful, pristine location with their home powered by the fresh mountain rain.

The family now runs two home-based businesses. Costal Cedar Creations is Warren's wood artistry business, where he teaches the art of log home building and produces cedar carvings and one-of-a-kind log furniture and structures. Janice, who is an avid gardener and fabulous cook, operates the guest house inn known as Tantalus View Retreat. Both businesses are on the Web at www.tantalusviewretreat.com (retreat@telus.net).

Figure 8.1-10 Janice and Warren Brubacher welcome people to partake in the views, clean mountain air, and country cooking all offered in their micro hydro-powered resort.

Figure 8.1-11. Relaxing in Warren's beautiful handmade cedar furniture and enjoying the view of the vast Coast Mountains has to remind people that we are part of nature and not master of it.

Figure 8.1-12. Warren demonstrates the water intake for the micro hydro system. The intake is located 86 ft (26 m) above the turbine.

Figure 8.1-13. The penstock runs through the forest to the intake strainer and sand settling box. The penstock is insulated to prevent freezing where the pipe crosses a creek.

Figure 8.1-14. There is almost no sound emanating from the turbine housing, even if you're standing right beside it.

Figure 8.1-15. An Energy Systems and Design turbine generates 24 Volts at 80 Amps (approximately 2 kilowatts) of power or 48 kWh per day during periods of maximum water flow. This is enough energy to power almost any home.

Figure 8.1-16. The discharge or tail race drops into a creek after a run through 550 ft (167 m) of 4" (100 mm) plastic poly pipe.

Figure 8.1-17. Janice is seen preparing a morning feast in the kitchen.

Figure 8.1-18. The kitchen of the Brubacher home is typical, using a combination of electric and gas appliances. Propane is used to heat the hot tub, as well as for the clothes dryer and cook stove.

Figure 8.1-19. The living room area has a television, VCR, and stereo system, common fare for any home.

Figure 8.1-20. Warren checks out the hot tub, which is heated with propane but uses the hydro system to operate the pump and pressure jets.

Figure 8.1-21. The organic gardens provide Janice with the raw materials for her cooking creativity.

Figure 8.1-22. The couple's home blends with nature at every turn, from the clean micro hydro power to Warren's wonderful woodworking.

9
Battery Selection

Why Use Batteries?

Why use a battery bank in the first place? This is a reasonable question with a simple answer. If you want to go off-grid now, there is no other way to store electricity as easily and economically, *period*. While it is possible to store electricity by other means, the technology and cost benefits are completely uneconomic.

What about Fuel Cells?

Automobile manufacturers are starting to use fuel cells, so why not install some of these? Fuel cells are wonders of technology and work well in specific applications, but home use isn't one of them. Besides, fuel cells don't store energy; they simply convert it from one form to another.

Electrical energy from photovoltaic panels and wind turbines combines with water in a device known as an electrolyzer. The electrolyzer generates hydrogen and byproduct gas oxygen. The hydrogen is compressed and stored in a pressurized cylinder until it is required by the fuel cell, where it recombines with atmospheric oxygen, generating electrical power and waste heat. The poor overall efficiency and the expense of this process mean that fuel cells cannot match the economics of today's industrial deep-cycle battery.

Industry fuel cell manufacturer Ballard in conjunction with Coleman produced a small home-sized unit to act as a battery-backup system for computer and office equipment. Unfortunately, safety and supply issues with hydrogen gas as well as capital costs have killed that project for now. If we revisit this in another ten to twenty years the story *might* be a little bit

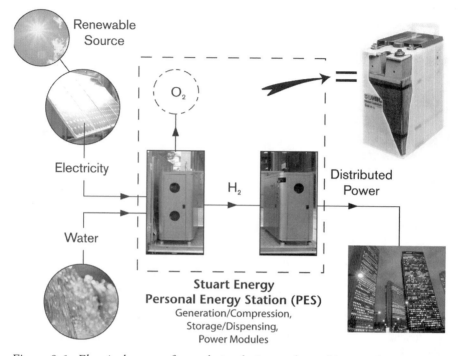

Renewable
Source

Electricity

Water

O_2

H_2

Distributed
Power

**Stuart Energy
Personal Energy Station (PES)**
Generation/Compression,
Storage/Dispensing,
Power Modules

Figure 9-1. Electrical energy from photovoltaic panels combines with water in an electrolyzer device which generates oxygen and hydrogen. The hydrogen must be compressed and stored until it is required. Feeding the hydrogen gas to a fuel cell generates electrical power and waste heat. The poor overall efficiency and the expense of this process mean that fuel cells do not yet match the economics of the industrial deep-cycle battery.

different. However, although fuel cells sound exciting the battery industry is also developing advanced methods of energy storage with lower pricing. The jury is out on which technology will prevail in this particular application.

Battery Selection

There are thousands of batteries available, and the more you look the more selection and complexity you will find. Car, truck, boat, golf cart, and telephone batteries are a few examples that come to mind. Then, of course, there are the "special" kinds such as NiCad, nickel metal hydride, and lithium ion, to name a few. What is the correct choice for our renewable energy system?

First of all, let's eliminate a few of the legends and quick fixes that abound:

• Used batteries of almost any size and type are not the answer. It doesn't matter how good a deal you got, the batteries are cheap because there is a problem with them. All batteries die after a given life span or if incorrectly serviced. Sure, they might work for a year, maybe even two or three, but the lower charge capacity of older cells makes this a sour deal.

- Quality battery costs are high. A golf cart battery of the same capacity as a high-quality "deep-cycle" battery may cost less, but the reduction in life span is not worth the hassle except for very small electrical loads or short grid interruptions. Changing batteries is backbreaking and dangerous work. You only want to do this once or twice in your lifetime, so stick with the best.

- Fancy batteries such as NiCad or lithium ion work well in cell phones and camcorders, but have you ever bought one of these babies? To get a lithium ion battery big enough to operate your renewable-energy system would probably cost as much as your entire house. Be thankful you can have one in your cell phone and leave it at that.

The only type of battery to consider in a full-time renewable-energy system is the deep-cycle lead-acid industrial battery manufactured with liquid or gelled "maintenance-free" electrolyte. These batteries have been around a long time and have been engineered and re-engineered so that they offer the best value for your money. Good-quality liquid electrolyte industrial batteries should last between fifteen and twenty years with a reasonable amount of care. In addition, the batteries are recyclable, and many companies offer trade-in allowances for their worn-out models, which also eliminates an environmental waste issue. The same is not true for NiCad or other exotic blends, which are considered hazardous waste.

How Batteries Work

The typical deep-cycle electrical storage battery, like a car battery, uses a lead-acid composition. A single-cell batter uses a plastic case to hold a grouping of lead plates of slightly different composition. The plates are suspended in the case, which is filled with a weak solution of sulfuric acid, called *electrolyte*. The electrolyte may also be manufactured in a gelled form, which prevents spillage. (Batteries of this type are often sold as "maintenance free"). The lead plates are then connected to positive and negative terminals in exactly the same manner as the AA and C cells described in Chapter 1.3. A single-cell with one negative and one positive plate has a nominal rating of 2 V.

Connecting an electrical load to the battery causes sulfur molecules from the electrolyte to bond with the lead plates, releasing electrons. The electrons then flow from the negative terminal through the conductors to the load and back to the positive terminal. This action continues until all of the sulfur

Figure 9-2. These models from Rolls Battery Engineering (www.surrette.com) are typical of long-life renewable-energy-system batteries. The two batteries in the foreground have the same voltage rating (6 Vdc). The model on the right has twice the capacity of the model on the left. (Courtesy Surrette Battery Company)

molecules are bonded to the lead plate. When this occurs, it is said that the cell is discharged or dead.

As the cell is discharged of electrical energy, the acid continues to weaken. Using a device called a hydrometer (see Figure 9-4), we can directly measure the strength or specific gravity of the battery electrolyte. A fully charged battery may have a specific gravity of 1.265 (or 1.265 times the density of pure water). As the battery discharges, the specific gravity continues to drop until the flow of electrons becomes insufficient to operate our loads.

When a regular AA or C cell is discharged, the process is irreversible, meaning the cell cannot be recharged. Discharging a deep-cycle battery bank is reversible, allowing us to put electrons back into the battery, thus recharging it.

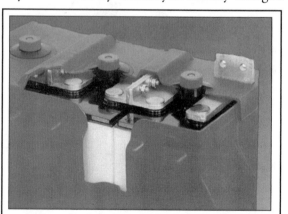

Figure 9-3. All lead-acid batteries comprise lead plates suspended in a weak solution of sulfuric acid. The size of the plate and the acid capacity directly affect the amount of electricity that can be stored. Each cell of the battery can be interconnected with others, increasing capacity and voltage. (Courtesy Surrette Battery Company)

Forcing electrons into the battery causes a reversal of the chemical discharge process described above. When a photovoltaic panel is placed in direct sunlight, it generates a voltage. When the voltage or pressure at the PV panel is higher than that of the battery, electrons are forced to flow from the panel into the cell plate. Electrons combine with the sulfur compounds stored on the plate, in turn forcing these compounds back into the electrolyte. This action raises the specific gravity of the sulfuric acid and recharges the battery for future use. Although there are many types of battery chemistry, the charge/discharge concept is similar for all types.

Depth of Discharge

A deep-cycle battery got its name because it is able to withstand severe cycling or draining of the battery. A car battery can only withstand a couple of "Oops, I left my lights on. Can you give me a boost?" mistakes before it is destroyed, whereas a deep-cycle battery may be subjected to much higher levels of cycling.

Figure 9-5 graphs the relationship between the life of a battery in charge/discharge/charge cycles and the amount of energy that is taken from the cell. For example, a battery that is repeatedly discharged completely (100% depth of discharge) will only last 300 cycles. On the other hand, if the same battery is cycled to only 25% depth of discharge, the battery will last 1,500 cycles. It stands to reason that a bigger battery bank will provide longer life, albeit at a higher cost for the added capacity. A good level of depth of discharge to shoot for is a maximum of 50%, with typical levels of between 20% and 30%.

A grid-interconnected battery system spends the majority of its life filled to capacity waiting for the next utility failure. Provided power outages are not an everyday occurrence, frequent cycling should not be a major concern. On the other hand, depth of dis-

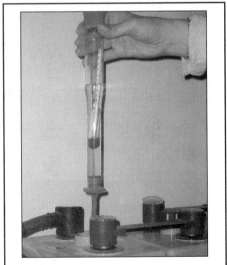

Figure 9-4. The hydrometer measures the density or specific gravity of fluids such as the electrolyte in this cell. The higher the reading, the more electrons are stored in the cell, indicating a higher state of charge.

Depth of Discharge %	Specific Gravity @ 75° F (25° C)	Cell Voltage
0	1.265	2.100
10	1.250	2.090
20	1.235	2.075
30	1.220	2.060
40	1.205	2.045
50	1.190	2.030
60	1.175	2.015
70	1.160	2.000
80	1.145	1.985
90	1.140	1.825
100	1.130	1.750

Table 9-1. Using the hydrometer shown In Figure 10-4, it is possible to accurately determine the state of charge, specific gravity, and voltage of each cell in a lead-acid battery bank.

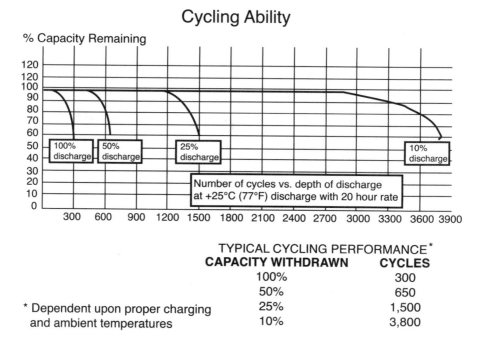

Cycling Ability

TYPICAL CYCLING PERFORMANCE*	
CAPACITY WITHDRAWN	**CYCLES**
100%	300
50%	650
25%	1,500
10%	3,800

* Dependent upon proper charging and ambient temperatures

Figure 9-5. This graph relates the life expectancy in charge/discharge/charge cycles to the depth of discharge level. Grid-interconnected systems do not generally cycle deeply or frequently unless the electrical utility is particularly unreliable in your area.

charge considerations may become a major issue where grid interruptions are frequent. When a lengthy power outage occurs, significant battery-capacity depletion can result from inexperienced operation. Novice operators should consider battery-metering equipment as discussed later in this chapter.

Operating Temperature

A battery is typically rated at a standard temperature of 75°F (25°C). As the temperature drops, the capacity of the battery drops as a result of the lower "activity" of the molecules making up the electrolyte. The graph in Figure 9-6 shows the relationship between temperature and battery capacity. For example, a battery rated at 1,000 amp-hours (Ah) at room temperature would have its capacity reduced to 70% at -4°F (-20°C), resulting in a maximum capacity of 700 Ah. If this battery is to be stored outside where winter temperatures may reach this level, the reduction in capacity must be taken into account. Installing batteries directly on a cold, uninsulated cement floor will also cause a reduction in capacity for the same reason. Be sure to use a wooden skid or other frame to allow room-temperature air to circulate around the battery, maintaining an even temperature.

Freezing is another concern at low operating temperatures. A fully charged battery has an electrolyte-specific gravity of approximately 1.25 or higher. At this level of electrolyte acid strength, a battery will not freeze. As the battery becomes progressively discharged, the specific gravity gradually falls until the electrolyte resembles water at a reading of 1.00.

For cold-weather applications you may consider a Nickel Metal Cadmium or NiCad battery. Although more expensive than standard lead-acid batteries, the NiCad cell is relatively unaffected by extremes in temperature, particularly where deep-discharge cycles and cold weather threaten to destroy lead-acid varieties.

Battery Sizing

A single cell does not have enough voltage or capacity to perform useful work. As discussed in Chapter 1.3, single cells may be wired in series to increase voltage and/or in parallel to increase capacity.

Batteries are just big storage buckets. Pour in some electrons and the "buckets" will fill with electricity. If cells are wired in parallel, the capacity is increased. If the cells are wired in series, the voltage or pressure rises.

Batteries may be purchased in several voltages. Individual cells have a nominal rating of 2 V for lead-acid and 1.2 V for NiCad brands. The lead-acid battery shown in the background of Figure 9-2 has two cells, each of

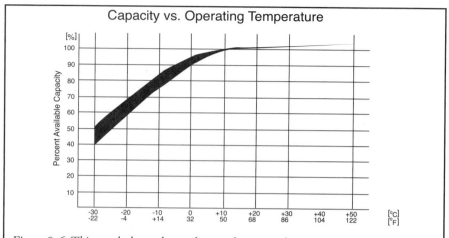

Figure 9-6. This graph shows the resulting reduction in battery capacity as the ambient temperature is lowered. A thermometer should be used to correct the specific gravity reading of very cold or overly hot electrolyte.

which can be identified by the servicing cap on the top. Using interconnecting plates inside the battery case, the cells are wired in series, providing 2 + 2 volts = 4 V total capacity. The batteries in the foreground have three cells, providing an output of 6 V. Using the same approach, a car battery contains six cells, providing a nominal 12 V rating. Typical battery voltages for grid-interconnected renewable energy systems are 24 V and 48 V.

It stands to reason that the more energy consumed, the larger the storage facility required to hold all the electricity. While the capacity of buckets is given in gallons or liters, the capacity of batteries is rated using watt-hours or amp-hours, units of electrical energy. We discussed earlier that a battery's voltage tends to be a bit elastic, with the voltage rising and falling as a function of the battery's state of charge (see Table 9-1). Additionally, when the battery is charged from a

Figure 9-7. For applications where the battery bank must be stored outdoors in extremely cold locations, consider the more expensive yet longer life NiCad style. (Courtesy Saft Battery Company)

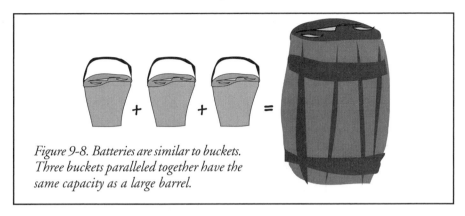

Figure 9-8. Batteries are similar to buckets. Three buckets paralleled together have the same capacity as a large barrel.

source of higher voltage such as a PV array in full sun, the voltage increases further. It is not uncommon for a battery bank with a nominal 24 V rating to have a voltage reading of 30 V when undergoing equalization charging.

This fluctuation of battery voltage makes it very difficult to calculate energy ratings. You will recall that energy is the voltage (pressure) multiplied by the current (flow) of electrons in a circuit multiplied by the amount of time the current is flowing. The problem with this calculation when it is applied to a battery is deciding which voltage to use.

Energy (watt-hours) = Voltage x Current x Time

Battery manufacturers are a smart bunch. To eliminate any confusion with ratings, they have simply dropped the voltage from the energy calculation, leaving us with a rating calculated by multiplying current x time.

Battery Capacity (amp-hours) = Current x Time

The assumption is that there is no point in trying to shoot a moving target. If we really want to look at our energy storage in more familiar watt-hour terms, we can simply multiply the amp-hour rating by the nominal battery-bank voltage.

Let's look at how this relates to the storage level our battery must have in order to run the essential loads of our home when the grid goes south. If you haven't already done so, it will be necessary to complete the *Energy Sizing Worksheet* in Appendix 7. This form will determine the amount of electrical energy you require in watt-hours per day to operate these essential loads.

For our example, we will assume an energy consumption of 4,000 watt-hours (4 kWh) per day for a 24-hour blackout. This is a reasonable amount of energy for grid-interconnected system users who wish to ride through power failures. In this case, it will be necessary to provide a separate wiring circuit for essential loads such as house lights, a refrigerator, and a few plugs to receive power during a grid interruption. Large or inefficient loads such as electric heating or air conditioning would remain off during the power failure. A detailed discussion of this system configuration follows in Chapter 13.

Remember that we have to take into account the usable amount of energy stored in the battery bank, not the gross or rated amount. A maximum depth of discharge is typically 50%, with less being recommended.

4 kWh capacity required x battery derating factor of 2 = 8 kWh capacity

Some manufacturers will rate the capacity of their batteries in watt-hours or kilowatt-hours; however, as we discussed earlier, most choose to use amp-hours. To convert watt-hours (kWh) to amp-hours (Ah), divide the desired battery rating by the nominal battery voltage required for your system, typically 24 or 48 Volts (assume 24 V).

8 kWh capacity (8,000 Wh) ÷ 24 V = 333.33 Ah capacity
≈ 333 Ah capacity

A quick scan of the resource guide data in Appendix 3 indicates that there are no batteries available with exactly 333 amp-hours of capacity. Remember, however, that batteries can be connected in series to increase voltage and in parallel to increase capacity. Two sets of 150 Ah batteries wired in parallel provide 300 Ah of capacity. This is a bit below the calculated rating, but it's close enough. We could also consider 2 sets of 200 Ah batteries, offering 400 Ah. A bit of juggling may be required, but either way we're in the ballpark.

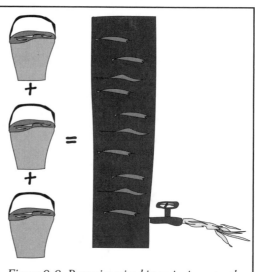

Figure 9-9. Batteries wired in series increase the voltage or pressure. In the same manner, a tall tank will shoot water further than a short tank due to the higher pressure.

As discussed earlier, under-sizing the battery can lead to over-discharging and shortening the battery's life. If you have to estimate and select sizes, it is always in your best interests to round up. Battery capacity is like money in your bank account: it never hurts to have extra on hand.

Each of the two rows of eight batteries shown in Figure 9-10 is shown in the schematic in Figure 9-11. The columns comprise 3 batteries. Each battery has two cells, as you can see from the two servicing caps on their cases. The nominal voltage of each battery is 4 V (2 cells x 2 Volts/cell). The batteries are wired in series by connecting successive "-" to "+" terminals, making the voltages additive. Thus, six batteries wired in series multiplied by 4 V per battery provide a total bank of 24 V.

Each of the batteries shown in Figure 9-10 has a capacity of 1,300 Ah. Wiring the batteries in series does not increase the capacity. In order to increase capacity, the batteries must be connected in parallel so that both banks are forcing electrons into the circuit at the same time. (To help visualize this concept, picture a car stuck in a snowbank. One person pushing the car may not provide enough energy to get the vehicle out, while a second person giving a hand may easily free it.) The parallel connection of multiple strings of batteries will be discussed further in Chapter 13, "Putting it All Together"

Figure 9-10. This compact battery bank takes up a small amount of room in the utility closet. With a rating of 24 volts at 1,300 amp-hours of capacity, it gives the owners of this home enough energy to ride through several days of grid interruption.

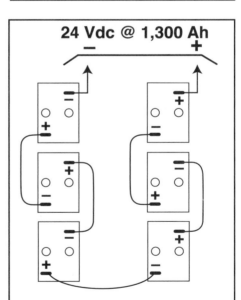

Figure 9-11. The batteries in Figure 10-10 are shown here in a schematic. Each of the six batteries is wired in series, resulting in increased voltage.

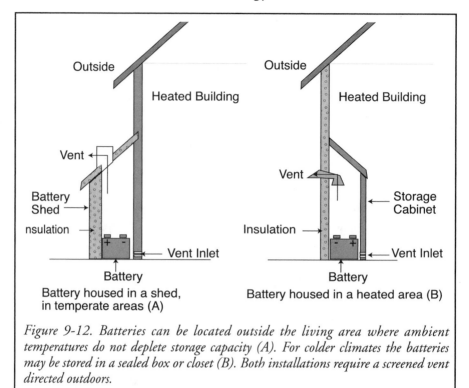

Figure 9-12. Batteries can be located outside the living area where ambient temperatures do not deplete storage capacity (A). For colder climates the batteries may be stored in a sealed box or closet (B). Both installations require a screened vent directed outdoors.

Hydrogen Gas Production

Connecting a battery to a source of voltage will push electrons into the cells, recharging them to a full state of charge. As a liquid-electrolyte battery approaches the fully charged state it will be unable to absorb additional electrons, causing the electrolyte to start bubbling and emit hydrogen gas. "Maintenance-free" gelled-cell batteries are designed to minimize the possibility of hydrogen buildup.

Everyone knows the story of the Hindenburg dirigible: a few sparks and the hydrogen gas exploded into a ball of fire. Preventing this in a renewable energy system is actually fairly easy.

The battery bank should be accessible to allow for servicing and periodic inspection. However, safety is of primary importance. Storing batteries in a locked closet, shed, or cabinet is a wise decision. The storage closet may be vented with a simple pipe that can be made with a dryer vent or plumbing pipe (2"/50 mm) in diameter with screening over the exterior opening. Make sure the pipe slopes slightly upwards from the battery room to the outside as hydrogen is lighter than air and will rise on its way outdoors.

For rooms with long or serpentine vent shafts, a power-venting fan can be installed as shown in Figure 9-13. Pressurizing fans are operated by a voltage-controlled switch contained within the inverter; the voltage indicates the state of charge and hence hydrogen production, activating the fans.

An alternate method of controlling hydrogen gassing is to install a set of reformer caps (Hydrocap Corporation) such as those shown in Figure 9-14. These caps contain a catalyst that converts the hydrogen and oxygen by-products of battery charging into water that simply drips back into the cell. The net effect is eliminated hydrogen gassing and lower battery water usage. Note, however, that hydrogen reformer caps control but do not completely eliminate hydrogen gassing, making some supplementary ventilation necessary.

Figure 9-13. This battery room is power vented. Installing small computer fans in the wall allows the room to be pressurized, forcing hydrogen outside.

Safe Installation of Batteries

Large deep-cycle batteries are heavy and awkward. Some models weigh up to 300 lbs (136 kg) each. It is important that any frames, shelves, and mounting hardware be of sufficient strength to hold them securely. During installation, make sure you have adequate manpower to lift and place the batteries without undue straining. Tipping a battery and spilling liquid electrolyte (acid)

Figure 9-14. HydroCaps replace the standard vent caps on each battery cell. A special catalyzing material converts hydrogen and oxygen byproducts back into water.

Figure 9-15. In temperate climates batteries may be stored in suitable outdoor cabinets, eliminating any concerns regarding hydrogen production. (PSR cabinet courtesy Outback Power Systems Inc.)

all over you is very dangerous. Follow these rules to ensure a proper and safe installation (See Figure 9-19):

• Batteries must be installed in a well-insulated and sealed room, closet, or cabinet, preferably locked. There is an enormous amount of energy stored in the batteries. Children (or curious adults) should not be allowed near them.

• Remove all jewelry, watches, and metal or conductive articles. A spanner wrench or socket set accidentally placed across two battery terminals will immediately weld in place and turn red-hot. This can cause battery damage, possible explosion, and severe burns.

• Hand tools should be wrapped in electrical tape or be sufficiently insulated so as not to come in contact with live battery terminals.

• All benches, shelves, or support structures must

Figure 9-16. This compact battery bank takes up a small amount of room in a utility closet or other out-of-the-way area in the home. (Courtesy Outback Power Systems Inc.)

be strong enough to carry the massive weight of the battery bank. The batteries shown in Figure 9-10 weigh 900 lbs (400 kg), so use extreme caution when handling.

- Wear eye-splash protection and rubber gloves when working with batteries; splashed electrolyte can cause blindness. Also, wear old clothes or coveralls, since electrolyte just loves to eat your $80 jeans.

- Keep a 5 lb (2.3 kg) can of baking soda on hand for any small electrolyte spills. Immediately dusting the spilled electrolyte with baking soda will cause aggressive fizzing, neutralizing the acid and turning it into water. Continue adding soda until the fizzing stops.

Battery and Energy Metering

Electricity is invisible, making it difficult to determine how much energy is actually stored in your battery bank. While it is possible to use nothing more than a hydrometer to measure the stored energy level, this requires a fair bit of tedious fiddling and working with corrosive acid. A better approach is to install a multifunction meter such as the model shown in Figure 9-17.

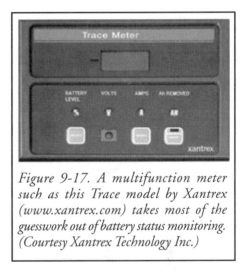

Figure 9-17. A multifunction meter such as this Trace model by Xantrex (www.xantrex.com) takes most of the guesswork out of battery status monitoring. (Courtesy Xantrex Technology Inc.)

These meters operate much like an automobile gas gauge, indicating how full or empty the tank is. Most meters provide dozens of features, the most important being battery voltage measurement, energy capacity in amp-hours and a measurement of the current that is flowing into (charge) or out of (discharge) the battery. These meters can be mounted a long distance from the battery bank, enabling you to read capacity from a convenient location in the house rather than running to the garage or basement.

Meters perform their work through a connection with a device known as a "shunt" (see Figure 9-18), a high-capacity resistor that either bypasses a small amount of the current flowing into or out of the battery or shunts it to the meter.. As the circuit is DC-rated, the direction of current flow is easily recorded. The amount of energy shunted is a calibrated ratio of the total amount of energy flowing in the battery circuit. Increasing the current flow into or out of the battery circuit causes a calibrated current to flow to the meter.

By constantly monitoring the flow into and out of the battery, the meter keeps track of the battery energy level in amp-hours and percentage full. Or it almost keeps track. While there is no doubt that energy meters are very useful and accurate devices, they are not perfect. Batteries are not perfect either. The process of converting electricity

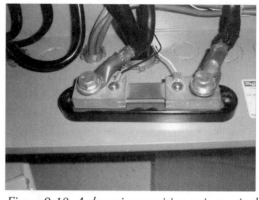

Figure 9-18. A shunt is a precision resistor wired in series with the negative lead of the battery. A small current in proportion to the battery charge or load current is "shunted" to a multifunction meter such as the one shown in Figure 9-17.

into chemical storage within the electrolyte is not 100% efficient. During the charge and discharge cycling of the battery, some of the electrical energy we put in is lost due to conversion inefficiencies. Worse yet, this inefficiency is not the same for adjacent cells, nor does the inefficiency level remain constant over the life of the cell. As a result, the energy meter will require recalibration periodically when the battery is at a known fully charged state (after measurement with a hydrometer).

Electrolyte specific gravity and battery voltage are directly related:

Battery Voltage = Electrolyte specific gravity + 0.84

For this equation and the hydrometer reading to be accurate, however, the batteries should be under light or no load, with the electrolyte near room temperature. It is also important to perform this measurement approximately two hours after charging the batteries to ensure that gas bubbles in the electrolyte do not lower the specific gravity reading.

The only way of absolutely knowing the energy level and general "health" of a battery is to use a hydrometer and measure the specific gravity. A thermometer should be used to correct the specific gravity reading of very cold or overly hot electrolyte according to the manufacturer's ratings. Correlate the corrected specific gravity to the data in Table 9-1 to determine the actual state of charge.

Batteries are charged to a known level a few times a year by performing an equalization charge using grid or renewable energy. During this process the batteries will reach their known fully charged state and the meter may be reset, calibrating the battery level with the meter reading for the next couple of months. Equalization charging will be covered in Chapter 17 – "Living with Renewable Energy".

In addition to the "main" features described above, metering may also contain some or all of the following "secondary" indicators:

- number of days since batteries were fully charged
- total number of amp-hours charged/discharged since installation
- time to recharge battery
- voltage too low for proper operation
- battery charging
- battery fully charged

Summary

In this chapter we have discussed battery selection and safe installation. Chapter 10 – "DC Voltage Regulation" discusses the equipment necessary to ensure proper charging and regulation of the battery bank. Future chapters deal with the specifics of interconnecting cells, safe wiring practices, and maintenance. After all, when the grid goes down, your home should be a shining beacon of bragging rights for all to see.

Figure 9-19. Battery installations should be neat and tidy and performed with the necessary tools on hand to measure electrolyte, specific gravity, and temperature. A clipboard and written record of each cell's "health" will ensure long battery life.

10
DC Voltage Regulation

A charge controller is an important element of a renewable energy electrical system equipped with batteries and is also essential for some models of wind and hydro turbines. Charge controllers are designed to provide protective functions, ensuring that batteries are not under- or overcharged, thereby prolonging battery life. Wind and hydro turbines may require an electrical load at all times. If your system is configured in this way, a diversion load controller will provide a constant load in the event of grid or inverter failure.

Grid-interconnected systems not equipped with batteries or these types of turbines do not require any form of charge controller, as all of the available electrical energy is pumped directly into the grid. If your system is grid-connected *without batteries*, or if you are using PV only, feel free to skip this chapter.

All renewable energy systems sometimes generate more energy than can be reasonably used in the home. Even small configurations will generate excess energy if no one is there to use this energy.

Off-grid systems are equipped with a battery bank with charge controllers forming an integral part of the system design.

As a battery's state of charge increases, its voltage also increases. The renewable energy source produces power at a high voltage in order to provide sufficient "pressure" to cause electrons to flow into the battery. We learned in Chapter 6, "Photovoltaic Electricity Generation," that a typical PV cell will produce 17 volts (V) for a 12 V nominal battery system. The 5 V difference in pressure allows electrons to flow from the higher voltage source to the

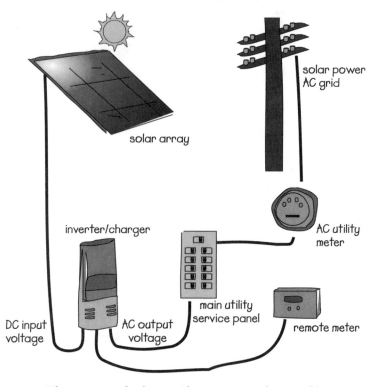

solar power
AC grid

solar array

inverter/charger

AC utility
meter

DC input
voltage

AC output
voltage

main utility
service panel

remote meter

Figure 10-1. The majority of urban, grid-interconnected renewable energy systems are not equipped with battery backup and do not require any DC voltage regulation.

lower one, providing the charging current to the battery bank.

As the battery bank "fills," its voltage rises. Upon nearing the fully charged state, the battery will start outgassing hydrogen and oxygen gas as well as heat. This condition is known as overcharging, and if it continues for prolonged periods damage to the battery will occur. This is one example of where a charge controller is required.

Charge controllers are available in two distinct categories known as *series controllers* and *diversion controllers*. Small off-grid systems and the majority of urban grid-interconnected designs use the series control method. Large, off-grid systems, particularly those utilizing wind and micro-hydro sources, use the diversion control method.

Series Controller

As the name implies, the series controller is wired in series between the PV array and the battery bank as shown in Figure 10-3. In this arrangement, the voltage output from the PV array is fed through the series-connected charge controller prior to being supplied to the battery. The charge controller monitors the battery voltage, and provided it is below a fully charged condition, PV power is allowed to flow into the battery. When the battery voltage level is sufficiently high to fully charge the battery, the charge controller will taper charging current or completely disconnect the PV array from

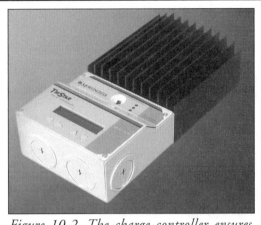

Figure 10-2. The charge controller ensures that your battery bank is properly charged and operating within given operating parameters. (Courtesy Morningstar Corporation)

the battery, slowing or stopping the charging cycle. As the battery is subjected to household electrical loading, the charge controller senses the drop in battery voltage as energy is removed from the cells. The controller reconnects the PV array, restarting the charging process. Think of the series controller as an automatic light switch that turns the flow of current to the batteries on

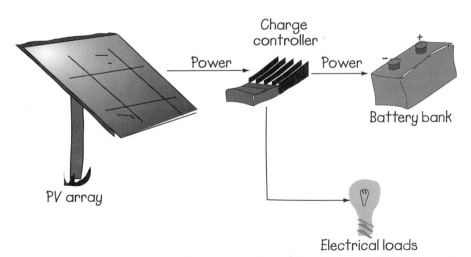

Figure 10-3. The series-connected charge controller is able to connect and disconnect the PV array from the battery. This action limits charging current and voltage to safe levels, prolonging battery life.

and off depending on whether the batteries require additional energy.

During the night, the charge controller performs no function other than to electrically isolate the battery bank from the sleeping PV array. This is prevents energy from flowing *backwards* from the batteries into the array, which can occur at night when the battery voltage is at a higher level than the voltage of the PV array. The large surface area of the PV array may absorb a small amount of energy from the battery and dissipate it as heat. Over the course of a long winter night, this energy loss is measurable and can contribute enough inefficiency to become a concern.

For this reason, most series charge controllers contain a "PV array nighttime reverse-current protection" feature to eliminate the possibility of electricity backflow.

Maximum Power Point Tracking

A word is required about *maximum power point trackers* or MPPT controllers. This technology, which is relatively new to the renewable energy market, is now available in charge controllers manufactured by several companies. All MPPT charge controllers provide the series charge control functions described above. Where these devices surpass standard controllers is in their ability to significantly improve the overall efficiency and power output of a renewable energy source, in essence producing more energy dollars from the same amount of equipment.

Figure 10-4. This charge controller is rated for 60 Amps of continuous load and is complemented with Maximum Power Point Tracking capability. (Courtesy Outback Power Systems)

The MPPT function is a little bit difficult to understand but well worth a few moments to review. The power of a PV module is the product of the voltage and the current. This can be expressed by multiplying the rated current by the rated voltage under load, resulting in the rated power of the module, shown in graphical form in Figure 10-5. The output voltage of a given module is shown on the "x" axis with the output current on the "y" axis. In Chapter 8 we discussed the fact that a PV module generates a very high open circuit voltage when not connected to a load (i.e. when no current is flowing). This point on the graph is position Voc (Voltage Open Circuit), where zero current is flowing and no work is being done.

As a load is connected to the panel, increasing current causes the voltage to slowly drop. A short circuit (directly connecting the "+" and "-" terminals) will result in high current levels, but the voltage will approach zero, thereby producing no useful power and again, no work can be done.

The most efficient point on the curve occurs when the product of the voltage pushing and the current flowing produce the highest value. This occurs at one point, which is known as the Maximum Power Point (MPP). Unfortunately, our battery voltage is unlikely to be located at this point. As we have learned, battery voltage is quite elastic, so there is a high probability that the PV module output will only transit this point occasionally.

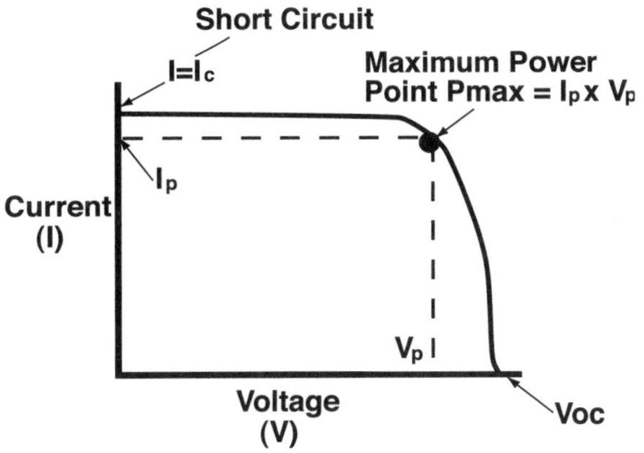

Figure 10-5. The Maximum Power Point (MPP) for a PV module is the point at which current and voltage outputs provide the maximum amount of power (Power = Voltage x Current).

The MPPT controller monitors the PV panel or wind turbine output power in relation to battery or inverter load and determines where the MPP is for the level of light illuminating an array or wind powering a turbine.

Testing shows that MPPT systems increase the wattage of a PV array by an average of 15% to 30%. For a PV array pushing 800 W into the utility grid or battery bank, this will provide an average increase of 120 W with a peak of double this figure. This is approximately equal to one or two modules of "free" power and more dollars from the sale of electrical energy.

For PV-only systems, consider an MPPT controller as an alternative or supplement to an active tracking mount to increase daily energy production.

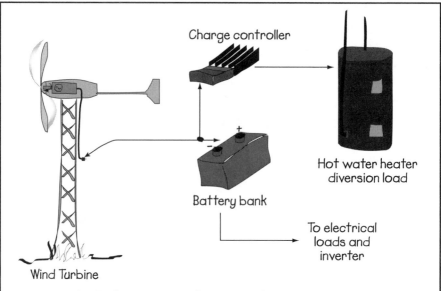

Figure 10-6. The diversion controller strategy shunts or diverts excess energy to an auxiliary air or water heater unit according to battery voltage or state of charge.

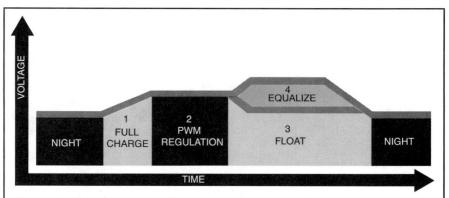

Figure 10-7. A charge controller regulates the amount of energy flowing into the batteries, ensuring a proper charge cycle. Quality charge controllers offer several charging modes. (Courtesy Morningstar Corporation)

Diversion Charge Controller

The diversion controller method (see Figure 10-6) is used primarily for larger off-grid systems or grid-interconnected designs that incorporate wind or micro-hydro turbine generators. In this configuration, the renewable energy source is connected either to a battery bank or to an inverter. All of the energy produced by the turbine flows into the battery or inverter, providing a load connection at all times. This is an important distinction between the two designs of charge controller. Where a PV array can be connected and disconnected at will, many wind and hydro turbines must have an electrical load connected at all times. Removal of the load during operation can cause the turbine to accelerate to a speed where damage can occur. This is similar to a car engine: driving while holding "the pedal to the metal" results in speeding tickets but no engine damage; however, doing the same with the car in neutral will almost certainly destroy the engine, as there is no load to limit its speed.

To ensure a constant load for the turbine without overcharging the battery, the controller will "shunt" or divert excess energy to the diversion load. The diversion load in Figure 10-6 is a water heater, although an air heating or other dump load can be used. With this system, the diversion controller is able to divert all or part of the turbine's energy depending on the battery bank voltage and state of charge, while maintaining a proper load level for the renewable source.

Wind and hydro turbines require a constant load and use the diversion controller in a slightly different manner. In this configuration, the turbine supplies power to both the inverter and the diversion controller simultaneously.

When the turbine output is within its normal operating parameters, the diversion controller is programmed to allow all of the generated energy to flow into the battery. When the battery reaches its fully charged state, the output voltage of the turbine will rise as a result of the lack of load. The controller senses this increase in voltage and shunts the excess energy to the diversion load.

Voltage Regulator Selection

Regardless of which voltage regulator design is selected, it is important to match its electrical rating to your system. The regulator rating is based on the maximum current that is allowed to flow into the batteries (series regulator) or into the diversion load (diversion regulator). It will be necessary to add the peak charging/diversion currents expected in your system. Once this value is determined a derating or safety factor of 25% is normally added. If your charge or diversion loads require a higher capacity rating, multiple controllers may be wired in parallel to divide the electrical load.

peak charging or diversion current x 25% safety factor = regulator rating in Amps

For further details refer to Chapter 13, "Putting it All Together Safely".

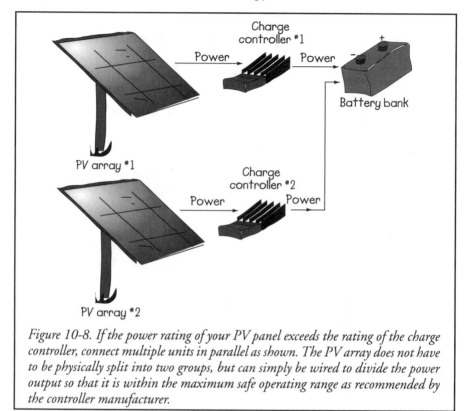

Figure 10-8. If the power rating of your PV panel exceeds the rating of the charge controller, connect multiple units in parallel as shown. The PV array does not have to be physically split into two groups, but can simply be wired to divide the power output so that it is within the maximum safe operating range as recommended by the controller manufacturer.

Charging Strategy (all controller configurations)

When the voltage is below a level that indicates the battery is not fully charged, all of the power from the renewable source is allowed to pour in. As the battery voltage rises, the series charge controller rapidly cycles on and off (several hundred times per second), pulsing the generated power into the battery. In a similar manner the diversion controller rapidly cycles power to the diversion load. The net effect is to direct energy in a controlled manner, maintaining the battery voltage within preset limits.

To visualize this concept, go into a darkened room and flip on an incandescent lamp. The room will be lit at maximum brightness. If you rapidly flip the switch on and off, the bulb will alternate between states of full and zero brightness. The effect on your eyes (notwithstanding the flicker) is that the room is seen in a dimmer light. If the light switch could be controlled so that the switch spent more time on than off, the brightness would increase. Likewise, if the switch spent more time off than on, the brightness would decrease. Engineers term this rapid on-and-off switching *Pulse Width Modulation* or PWM.

To understand how the PWM concept is incorporated into a charge control system, let's take a look at a typical twenty-four-hour day in the life of a battery bank during a grid blackout, assuming that PV arrays are used for daily charging. Figure 10-7 shows the effect of battery voltage over one complete night/day/night cycle. The graph shows time on the horizontal or "x" axis, while battery voltage is shown on the vertical or "y" axis.

Nighttime

Starting at the left-hand side of the graph, the battery voltage is shown at its lowest state at night. During the night lights are used, the fridge operates, and other loads draw energy from the battery bank. As we learned in Chapter 10, "Battery Selection and Design", the battery voltage will decrease as energy is removed from the battery. During this period, the charge controller prevents electrical backflow (discussed earlier) and waits patiently for the sun to rise.

Sunrise – Full or Bulk Charge

As the sun rises and shines on the PV array, electrical energy will start flowing and begin charging the battery. The battery voltage will start to rise according to the battery state of charge, as discussed in Chapter 10.

Until the battery state of charge reaches approximately 80%, all of the PV array power is applied directly to the battery, as shown in stage (1) of Figure 10-7. This charging condition is called "full or bulk charge" mode. It terminates when the battery voltage rises to 14.6 V for a nominal 12 V battery bank (consult manufacturers' data sheets for exact charger settings). For 24 V or 48 V battery banks, multiply the described settings by two or four respectively.

Absorption or Tapering Charge

When the battery reaches an approximately 80%-full state of charge, some of the energy is wasted through "boiling" and out-gassing. Boiling is a slang term used to describe the breakdown of water in the battery electrolyte into its component elements, hydrogen and oxygen. This occurs when there is more energy applied to the battery than it can absorb.

Hydrogen is a gas that is lighter than air and very explosive when mixed with oxygen and an accidental spark, open flame, or cigarette. For this reason, batteries should be stored in locked and ventilated cabinets. It is possible to reduce hydrogen production while providing the optimum charging current to the battery by introducing a PWM tapering current, as shown in stage (2) of Figure 10-7.

During this stage, charging current is automatically lowered in an attempt to maintain the battery voltage at the bulk setpoint level of 14.6 V. Absorption charging continues for a defined period of time, which is typically one to two hours.

The excess energy produced during the absorption charging stage, which can be a considerable amount, is either wasted or diverted depending on the type of charge controller installed. A series controller will waste the excess energy. As this energy is non-polluting and often small in quantity, there is no concern. A diversion controller will shunt or redirect the excess energy to a dump load such as an electrical water or air heater, increasing the system's efficiency. (Remember that all of this wasting and diverting occurs only during times when the battery bank is full and there is nowhere else for the energy to go).

Float Stage

Once the battery is fully charged, the charge controller will reduce the applied voltage to 13.4 V and begin float mode, as shown in stage (3) of Figure 10-7. When the battery reaches this stage, very little energy is flowing into it and nearly 100% of the renewable energy will be wasted or shunted to the diversion load. Float mode will remain in effect until the battery voltage dips below a preset amount (typically 80% of full state of charge) or until the renewable energy source stops producing power at night. The process is then repeated for the next cycle.

Equalization Mode

CAUTION!
Sealed or "maintenance free" batteries do not require equalization charging. Placing them in such a condition can result in fire or cause the battery to explode.

To visualize the equalization process, think of batteries containing electrons as buckets containing water, with one bucket equivalent to one battery cell.

Over the course of a few dozen charge cycles, energy is taken out of and replaced into the battery. This is the same as if I asked you to take four cups (one liter) of water out of each "cell" or bucket, representing a day's electrical load consumption. Assume that our PV array produces enough energy to replace three cups (750 ml) of "energy" into the cells, leaving us with a deficiency of one cup (250 ml). This process is repeated with different amounts of water being added or removed over the course of a month or two, until such time as the batteries are back to a full state of charge. In our example, the buckets are also full.

In reality, the water in each of the buckets would not be at exactly the same level. A certain amount of spillage and uneven amounts of water removal will leave the buckets with varying amounts of water in them. This is the exact scenario that plays out in your battery bank. Over time there is a gradual change in the state of charge in the cells, which can be determined by comparing the specific gravity of one cell with that of adjoining cells.

If the difference is allowed to continue for an extended period, the cell with the lower state of charge can fail prematurely and the amount of available energy stored in the battery bank can be reduced.

To correct this situation, a periodic (typically once per month) controlled overcharge is conducted. Known as an equalization charge, this process is illustrated in stage (4) of Figure 10-7. Chapter 17, "Living with Renewable Energy", discusses how to determine when equalization is required.

Equalization charging is normally conducted early in the morning on a sunny or windy day. The charge controller is set to equalization mode and the normal bulk cycle (step 1) is completed. Upon entering equalization mode the controller raises the battery voltage to 15.5 V (step 4). Equalization mode is maintained for approximately two hours, after which battery voltage is reduced and the charge controller automatically enters float mode (step 3).

If we go back to our bucket example, the effect of equalization is similar to using a garden hose to add water to the buckets and deliberately overfilling them. When you stop adding water (equalization completed) the buckets are topped up right to the rim.

Where does the "extra" electricity go during the equalization mode? During this charging stage, the excess energy applied to the batteries will cause violent bubbling of the electrolyte, producing large amounts of hydrogen and oxygen gas as well as heat and water vapor. It may be necessary to monitor electrolyte levels and battery temperature during this charging stage to ensure that battery parameters are within manufacturer ratings.

If your batteries are equipped with hydrogen reformer caps, ensure that they are removed during equalization mode.

Diversion Loads

Wind and hydro turbines normally require that an electrical load be connected at all times. Energy that is generated by these sources and when batteries are full, must go somewhere. This is where the diversion load comes into play.

A diversion load is any load that is large enough to accept the full power of a renewable source. Although the diversion load may simply waste the excess energy applied to it, a better approach is to put this juice to work.

After all, you did pay for the wind turbine, so why waste the "excess" energy other people have to pay for? Suppose you have a 2 kW wind turbine which operates at maximum output for 5 hours during a grid failure while you and your family are on vacation.

2 kW wind turbine x 5 hours of operation = 10,000 Watt-hour production

Assume for a moment that the house loads are zero (no fridge or lights on). What can you do with this energy? 10 kWh of production is a lot of energy. You could operate a thousand 10 W compact fluorescent lamps for one hour or ask your neighbors over for a really big party at your house.

This free energy can be used in many ways to help offset other systems that cost money to operate. A common solution is to use the excess energy for space and water heating. If all of your waste energy is produced during the swimming season, consider dumping the energy into a hot tub or swimming pool. Just remember that this load must not be considered part of your overall energy budget because the diversion load will only be active when the grid has failed.

The most common diversion loads are air and water heaters such as those shown in Figures 10-9 and 10-10. For PV-based systems, the majority of excess energy is produced during the summer months when air heating is not required. Wind systems tend to pro-

Figure 10-9. Using large amounts of diverted electrical energy is easy if heat enters the equation. An air heating element such as the model shown in this photo (available from www.realgoods.com) will absorb up to 1 kW of excess energy and provide some home heating to boot.

vide maximum power during the late winter and spring periods. Consider the choice carefully. No one wants an air heater cooking when the mercury is in the 90s. My personal preference is a water-heating load such as the one in Figure 10-10. This arrangement requires a bit of installation work, but the excess energy will offset water heating fuel costs when the grid goes out.

In this system, a standard electric storage water heater is purchased and the 240 V heating elements are removed. New elements are installed which have the same voltage rating as your battery bank or your wind or hydro turbine. The elements are then wired to the diversion-type charge controller. The water heater is plumbed so that cold water flows into the electric heater and out to the standard model, as shown in Figure 10-11.

During normal operation, the cold water supply enters the electric water heater. If there is no excess energy, the cold water will absorb some room heat, capturing a small amount of supplementary energy before heading to the regular water heater. As this incoming cold water is below the setpoint temperature, the regular heater will supply the energy necessary to meet demand.

During periods of excess energy production resulting from grid failure, the diversion charge controller supplies the low-voltage electric water heater. This energy may heat the water to the desired setpoint temperature or beyond. Feeding preheated water into the regular water heater reduces or eliminates the need for any further heating, reducing your purchased energy requirements and saving you dollars.

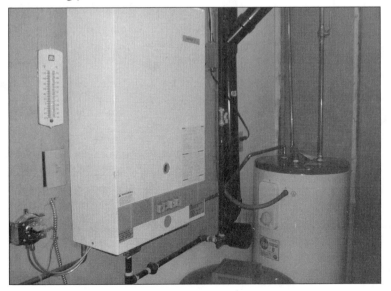

Figure 10-10. This home uses a high-efficiency, in-line gas water heater. The conventional storage water heater is the diversion load. It absorbs waste energy by preheating the cold water fed to the gas water heater.

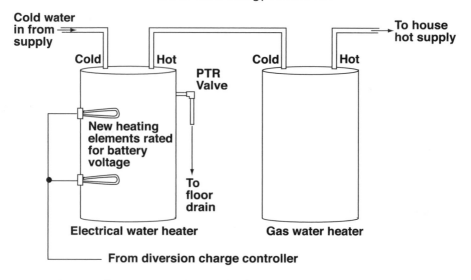

Figure 10-11. A diversion water heating load is connected in series with a standard gas water heater. The elements of the water heater are replaced with heating elements that have the same nominal voltage as the battery bank. They are connected to the diversion charge controller.

It is possible for the temperature of the water in the diversion electric heater to rise above the setpoint temperature, particularly during travel or vacation periods which often occur in the energy-rich summer months. Safety (as well as building codes) dictates that a Pressure and Temperature Relief (PTR) valve be installed on the tank. The outlet pipe should be run to a floor drain or other suitable exhaust to allow very hot water to be safely drained away in the event that the water tank temperature rises to an unsafe level.

For the same reason, your plumber should install a buffering valve. These valves ensure that the hot water supply delivered to the plumbing fixtures is within a comfortable and safe temperature range.

For people considering the addition of a solar water heating system, diversion heating elements may be combined with the required water storage tank.

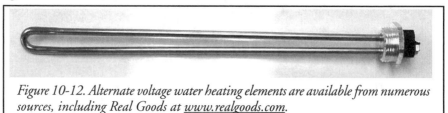

Figure 10-12. Alternate voltage water heating elements are available from numerous sources, including Real Goods at www.realgoods.com.

11
DC to AC Conversion Using Inverters

In earlier chapters, we dealt with renewable energy sources that produce electrical energy in direct current (DC) from a PV array. Wind and hydro turbines generate alternating current (AC), but convert it back to DC for storage in battery banks.

This chapter explains how to convert DC power back to AC power that can be used for electrical appliances and can also be sold to the electrical utility in grid-interconnected systems.

In the "old days," if you wanted to build a renewable energy system you could only do it using 12 Volt (V) direct current appliances similar to those used in early recreational vehicles. Grid-interconnec-

Figure 11-1. The renewable energy world would be lost without utility-grade sine wave inverters and controls such as this. (Courtesy Outback Power Systems)

tion of renewable energy was as far away as the idea of the Internet. Times change, the Internet is here, and so are very high quality inverters for both on- and off-grid applications.

The basic inverter is a device that takes low-voltage DC power (from a battery bank or directly from another energy source such as a PV panel or wind turbine) and converts it to alternating current. It then "steps up" the voltage to match domestically supplied power from your utility. In practice, many inverters offer a whole host of additional features and functions:

- battery charging capability
- transferability of house power between a generator and the inverter
- low- and high-voltage alarms and disconnection (LVD function) of the battery bank
- energy savings—sleep mode turns inverter "off" but it comes back on at a flick of the first house light switch
- automatic start and stop of a backup generator
- maximum Power Point Tracking for grid-interconnected systems
- full safety protection for both homeowners and utility workers

An inverter installed for off-grid operation is shown in schematic form in Figure 11-2. Energy from the renewable source is stored in the battery and directed to the inverter's internal components. A power supply and controller determine the sequence of events required to make the unit function as desired. When the inverter is activated, the controller starts a high-powered

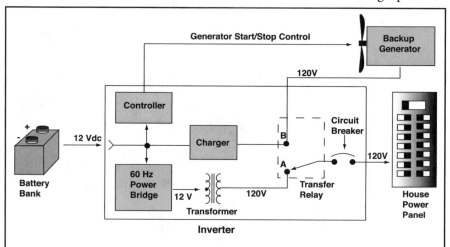

Figure 11-2. The inverter is a marvel of complex technology and is surprisingly easy to operate. A basic inverter for off-grid installation is shown in schematic form.

oscillator or power bridge that generates AC power at the voltage of the renewable source or battery bank, typically 12 V, 24 V, or 48 V.

The low-voltage AC power is set at a frequency of 60 cycles per second or 60 Hertz for the North American market. Inverters in Europe and Asia operate at 50 Hertz. You will recall from Chapter 1.3 that 60 Hertz means that the polarity of the voltage is switched back and forth 60 times per second. Imagine a simple C cell battery inserted into a flashlight first one way then the other 60 times per second, as shown in Figure 11-3. If we plotted the polarity of the battery each time it was inserted into and out of the battery clips shown in "A", we would obtain the plot shown in "B".

The advantage of AC power over direct current is that the voltage can easily be stepped up or down using a transformer. A transformer inside the inverter receives the low-voltage AC power and steps it up to the utility-standard 120 or 240 volts. In the inverter shown in Figure 11-2, the transformer is designed to accept a nominal 24 V AC input from the power bridge and convert it to a 120 V output. This power is then fed through a protective circuit breaker and into the power panel, which distributes it throughout the house. If the system is producing more energy than the house requires, the excess is automatically "exported" to the electrical utility via an electrical energy meter. Likewise, if the system is producing less energy than the house requires, the difference in energy demand is purchased or "imported" from the grid.

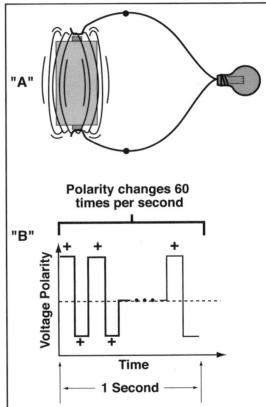

Figure 11-3. A battery flipped back and forth in its socket 60 times per second will supply alternating current to the light bulb load. The waveform at "B" shows the resulting plot of the changing polarity over a one-second interval.

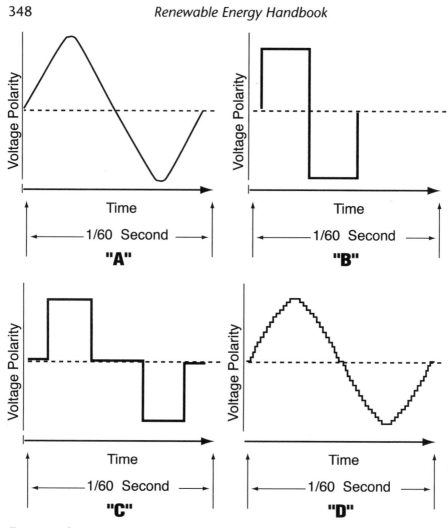

Figure 11-4

Inverter AC Waveforms

The electrical utility generates its AC power using mechanical generators connected to turbines. When a rotary generator creates an AC cycle, the waveform is sinusoidal, as shown in Figure 11-4 A. Compare this waveform with that of the simple inverter shown in Figure 11-3 and again in Figure 11-4 B. Each waveform has been magnified to show what is known as one cycle, one-sixtieth of a second long. Obviously, sixty of these cycles occur in one second.

The main difference between the two waveforms is that the utility-generated power shown in "A" has the voltage ramping up slowly until it reaches a peak level. It then falls back down slowly as the polarity reverses. In contrast, the simple inverter shown in "B" snaps quickly between points of opposite polarity, thereby acquiring the name "square wave." Only the least expensive and oldest inverters operate using the simple square wave.

Improvements in technology led to the "modified square wave" (also known as a "modified sine wave," but that's really pushing it), as shown in waveform "C". These inverters still retain a significant share of the market, but they are only used in off-grid and cottage applications and cannot be used in grid-interconnected systems.

Inverters using a square or modified square wave can operate 98% of all modern electrical appliances with no problem. However, the fast-switching edges of the waveform can produce a buzzing or humming noise in some items such as cheap stereos, ceiling fans, and record players. One primary advantage of this waveform is the ease of producing it. Inverters utilizing this pattern are robust, electrically efficient, and relatively inexpensive. They are not, however, compatible with the electrical utility system.

Many people argue that although these inverters are cheaper they should not be used in the off-grid home because sine wave inverters are better. Two of the homes described in Chapter 4 utilize these inverters without any reported problems. If you have the money, by all means purchase the upgraded sine wave model, but if cost is a concern, these workhorses still have a lot of life in them.

A fairly recent development in inverter technology is the sine wave model that outputs a waveform similar to that shown in "D." This digitally synthesized waveform shape is created

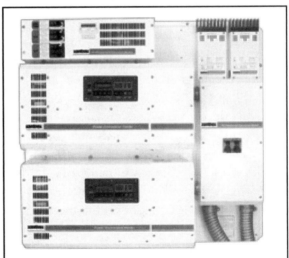

Figure 11-5. Two Xantrex sine wave inverters are "stacked" to provide both 120 V and 240 V AC power at better than utility-quality power. Also shown are two model C60 charge controllers (upper right) and a DC circuit breaker (bottom right). (Courtesy Xantrex Technology Inc.)

using a technology similar to that used to record digital sound on a compact disc MP3 file, and we all know how good those sound. In fact, today's inverter technology offers negligible waveform distortion, with frequency and voltage tracking far better than that of even the most expensive utility power station. In the last few years the number of models and manufacturers has exploded, increasing the range of sizes and features while lowering prices.

Although we are concerned with off-grid homes, we will take a moment to review grid connection of inverters. The grid-connected home with battery storage capacity (known as grid-interactive) truly is an off-grid home when the utility supply fails.

Grid-Interconnection Operation (grid-dependent mode)

A sine wave inverter that generates AC power of utility quality should be able to supply power to the grid in a manner similar to a hydroelectric turbine. Ten years ago this was considered heresy by the electrical utilities (and still is by some). Today, however, the demand for cleaner renewable energy (including political head-knocking by environmental and industry advocates) has weakened the monopoly previously enjoyed by utilities. Of course, some areas of North America are slower than others at getting the point, but the change is already occurring and will inevitably continue. Take California, for example: after continuous problems with rolling blackouts, sky-high energy costs, and never-ending smog, the politicians got the message. Now anyone can install a bunch of PV panels on the roof, connect them all through a grid-interconnected inverter, and watch the electrical meter spin backwards. If that isn't enough, the state will even provide an incentive payment to help make it happen.

The grid-connected system shown in Figure 11-6 is typical of a simple, effective means of allowing renewable sources to supply energy to the utility. In this arrangement, electrical energy flows from the grid, through the utility meter, and into the power panel and electrical loads of the home. This is the normal *import* or purchasing mode that most homes are familiar with. During importation of power, the meter records the power usage on one electronic counter.

When the sun starts to shine on the PV array or the wind or water turbine starts spinning, the inverter synchronizes its own sine wave to that of the utility. In order to *export* power, the inverter creates a waveform that has a higher voltage than that of the utility source, causing current to be pushed onto the grid. During power exportation the utility meter records the generated energy on a second electronic counter. At the end of the billing period, the meter reader will record the power imported and exported to your house.

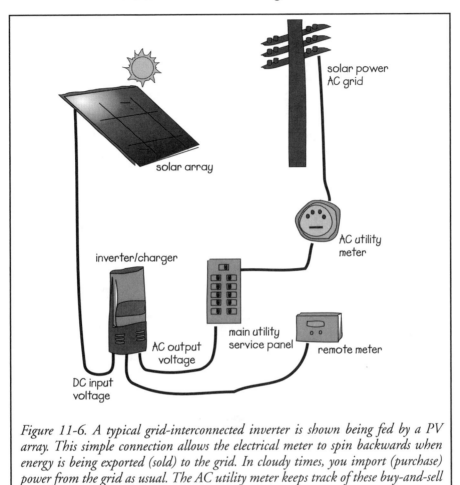

Figure 11-6. A typical grid-interconnected inverter is shown being fed by a PV array. This simple connection allows the electrical meter to spin backwards when energy is being exported (sold) to the grid. In cloudy times, you import (purchase) power from the grid as usual. The AC utility meter keeps track of these buy-and-sell transactions.

The invoice you receive from the utility company will reflect the difference between the two meter readings, providing you with a net energy bill.

For safety reasons, the inverter periodically checks the utility voltage to make sure that it is still present. If it is not, the inverter assumes the grid has failed. It then shuts down and waits until several minutes after grid power has returned to normal before re-applying power. This test eliminates the possibility of a condition known as *islanding*, which occurs if power from the grid is knocked out but the inverter continues to feed electricity into the local area. If this exporting of power to the grid occurs and a utility worker unsuspectingly touches the wires in the islanded area, serious injury can result.

Grid-Interconnection with Battery Backup (grid-interactive mode)

Figure 11-7 illustrates the essential components of a grid-interconnected system with battery backup, also known as a grid-interactive mode of operation. Using this arrangement it is possible to provide battery storage to supply power during periods of grid blackout. The major difference between grid-dependent and grid-interactive modes is the requirement for battery storage and a secondary electrical supply panel to feed essential loads within the house.

When the utility supply is present, electricity is imported and exported in the same manner as in a grid-dependent system. During times of grid interruption, you will recall that the inverter in a grid-dependent system will

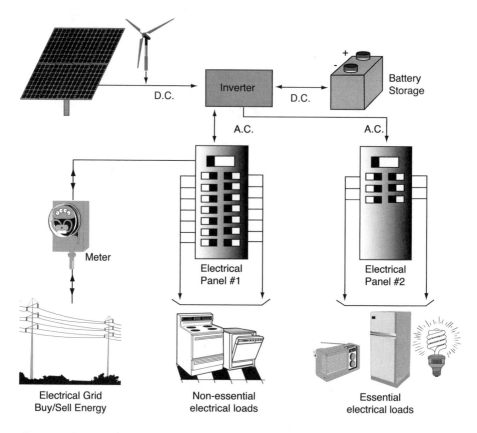

Figure 11-7. A grid-interactive system configuration provides electrical power to essential loads during grid blackout periods. This is accomplished by storing energy in a battery bank and wiring the home with two power panels, one for essential loads the other for non-essential ones.

disconnect itself and enter a standby state waiting for the grid connection to return to normal. With a grid-interactive configuration, the inverter will disconnect from the electrical utility and reconnect itself to an "essential electrical load supply panel." Using energy stored in a battery bank as well as from any renewable sources, the inverter will generate sufficient power to operate these connected loads. It is recommended that all the essential electrical loads be of the highest efficiency and lowest wattage, thereby prolonging battery life and corresponding electrical generation during a blackout.

Inverter Ratings

Important considerations when purchasing an inverter are the electrical ratings of the unit and whether or not you should purchase a "modified square wave" or "sine wave" model. Following is a list of items to look for when shopping for any inverter.

Sine Wave Versus Modified Square Wave Inverter Models

For grid-dependent or grid-interactive systems there is no choice but to use a *Grid Tie Sine Wave Inverter*.

For off-grid or emergency backup power systems that are *not* grid-interconnected you can choose between either sine wave or modified square wave inverter models. Modified square wave inverters tend to be less expensive than their sine wave counterparts and will operate almost all home appliances. The decision in this case will be based on budget as well as on the features you find desirable.

Output Voltage

North American homes are wired to the grid so that loads may be connected at 120 V or 240 V. Normal wall plugs are rated at 120 V, while heavy electrical loads such as electric stoves and furnaces, clothes dryers, and central air conditioning units operate at 240 V.

The voltage rating of most inverters is set to 120 V, allowing the operation of most household appliances within the capacity rating of the renewable energy system. Increasing the size of the renewable energy source beyond the rating capacity of the inverter necessitates the addition of a second inverter. As a general rule of thumb, systems with a load requirement of greater than 4 kW will require two inverters. When a second inverter is added, the electrical connection is in series, making the voltages additive at 240 volts.

In most cases the only 240 V essential loads that can be reliably connected are domestic water pumps. The vast majority of 240 V loads comprise appliances which draw enormous amounts of electrical energy, like central air conditioning, cook stoves, clothes dryers, swimming pools, and hot tubs. Battery capacity is limited and should be used sparingly.

If a 120 V inverter is all that is called for in your application and you require 240 V supply to operate a water pump or other device, an auxiliary step-up transformer may be installed at considerably less cost than a second inverter.

Inverter Continuous Capacity

Inverter capacity refers to the amount of power that the unit can supply continuously. The inverter rating should be at least 25% higher than the maximum power delivered to the connected appliances. Remember that multiple inverters may be wired in series and parallel configurations to increase the capacity as required.

You will recall that power, expressed in watts, is the voltage multiplied by the current. When operating a home, you are likely to use more than one electrical load at a time. In order to determine the inverter continuous capacity required, it is necessary to total all of the essential electrical loads that are likely to be turned on at any given time. For example, assume that a washing machine (500 W), television and stereo (400 W), refrigerator (400 W), and a bunch of CF lamps (100 W) are all on at the same time. The total power requirement of these loads is:

$$500 + 400 + 400 + 100 \, W \quad = \textit{Total Continuous Power}$$
$$= 1,400 \, W$$

Don't forget to add a safety factor to this maximum power requirement by purchasing an inverter that is at least 25% larger than the estimated value. Over time, these electrical loads have a habit of creeping upwards.

Remember to calculate the essential load value when filling out the *Electrical Energy Consumption Worksheet* in Appendix 7. When you are shopping for an inverter, you will find that they tend to be sized in "building block" sizes. There are numerous small models below 1,000 W, although 2,500 and 4,000 W are common sizes.

When trying to read the continuous rating of an inverter, you may be confronted with the term "VA" rather than watts. VA refers to the voltage multiplied by the current that we have understood to mean wattage plus an allowance for power factor. At the risk of trying to split hairs, when an AC motor load is operated, there is an effect known as *power factor* that has to be taken into consideration. In the average home, VA closely approximates watts and may be used interchangeably. It is recommended to reduce the rating of the inverter by as much as 20% if you are operating air conditioning units, pool and spa pumps, or other similar high-wattage *induction* motor loads. If you are operating a shop full of large woodworking tools with many of the motor loads running at the same time, it is best to allow for an inverter derating of approximately 25% or higher to allow for power factor. In practice there is little that you can do other than to be aware of the existence of power factor and reduce simultaneous usage of motor-driven appliances or purchase the next-larger inverter. An additional point is that universal motors that have brushes (sparks may be seen when the motor is running) can be run safely without any derating concerns as they are not affected by power factor issues.

Examples of universal motors include:
- regular and central vacuum cleaners
- food processors and mixers
- drills, routers, shopvacs, radial arm and circular saws
- electric chain saws and hedge trimmers
- electric lawn mowers

If in doubt, contact an inverter dealer or motor repair shop for clarification.

Inverter Surge Capacity

The surge capacity is an indication of how much short-term overload the inverter will be able to handle before it "trips" on this condition. Surge capacity is necessary to allow some large loads to get started, particularly motorized loads requiring starting power two to three times their running power. Although this start period is very brief and lasts a fraction of a second, it should be considered. The main concern is whether or not you have any "unusual" electrical devices such as an arc welder in your home. In addition, it is wise to look at your electrical appliance consumption list and see if there is a likelihood that several large motor loads may start at the same time while operating in battery backup mode. However, in the average home it is highly unlikely that surge capacity will become an issue.

Inverter Temperature Derating

The power protection circuitry for most inverters is temperature compensated, meaning that the maximum load that an inverter can run changes with the ambient temperature. As the temperature of the internal electronics of the power switching bridge increases, the allowable connected load current/power is reduced.

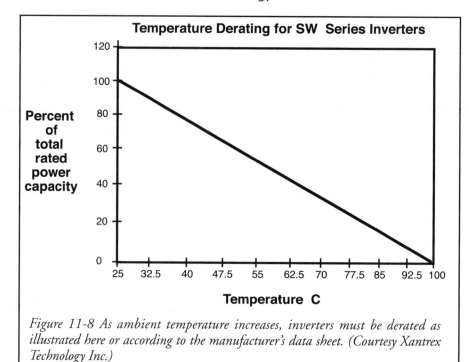

Figure 11-8 As ambient temperature increases, inverters must be derated as illustrated here or according to the manufacturer's data sheet. (Courtesy Xantrex Technology Inc.)

The graph in Figure 11-8 shows the effect of temperature on the capacity of a Xantrex model SW series 4,000 watt inverter to operate connected loads. Notice that the inverter reduces its capacity at temperatures above 77° F (25 °C). The graph also assumes that the inverter is operating at sea level and without any restriction of the airflow around it.

If your grid-interconnected inverter is designed for outdoor installation, it is advisable not to mount it in direct sunlight if at all possible. There have been numerous instances of inverters entering a self-protection mode while basking in the hot Arizona sun.

Battery Charging

You can save money by purchasing an inverter without a battery charger installed rather than a combined inverter/charger unit. If you want to install a battery charger after the inverter is installed, you can purchase a separate battery charger such as the TrueCharge™ series from Xantrex Technology Inc.

Voltage Converters

In a number of cases, it may be necessary to operate small direct current loads at a voltage rating different from the battery. For example, you may wish to operate a 12 Volt car stereo or trouble light from the battery in the event of inverter failure. Another common use for 12 Volt power is to operate a small "micro inverter" which in turn powers small phantom loads such as cell phone and PDA battery chargers.

Most off-grid households will operate the battery voltage at 24 Volts, with larger homes operating at 48 Volts. There is a temptation to use the knowledge gained from series wiring to connect a "jumper" across the battery bank to capture 12 volts. Never do this! Such a connection will cause the battery bank to be drained more quickly from the cells providing the 12 Volt tap circuit. This can lead to premature failure from inadvertent over-discharging of those cells.

A better approach is to purchase a d.c. to d.c. voltage converter that will efficiently drop the higher direct current voltage to the desired lower one. These units will ensure that electrical power drained from the batteries is applied evenly across the entire battery bank. Units are available from companies such as Solar Converters Inc. www.solarconverters.com.

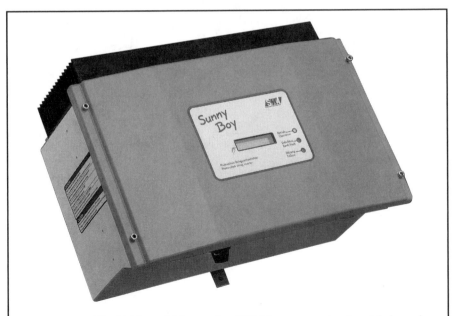

Figure 11-9. The SMA model Sunny Boy 2500 is an example of a grid-dependent inverter supplied without battery charging and other optional features.

Summary

Inverters are so highly advanced and are available in so many sizes that every cabin, cottage, and off-grid home can afford one. This eliminates the need for following in the steps of renewable energy "old timers" who had no choice but to use 12 Volt Recreational Vehicle appliances.

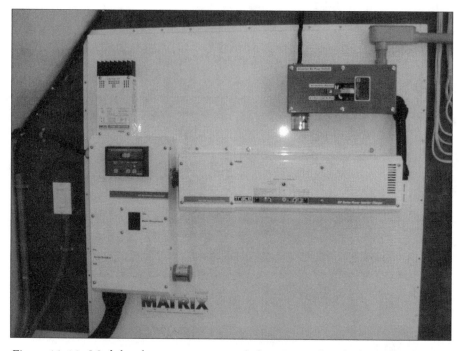

Figure 11-10 Modular, low-cost power panels featuring today's high-quality inverters ensure that off-grid homes enjoy all the comforts of their grid-connected neighbors.

12
Fossil-Fueled Backup Power Sources

If there is an Achilles' heel with off-grid energy systems, it is not related to renewable generation equipment reliability but rather to the need for a backup power source. Not that there is any problem with equipment reliability. In fact, the opposite is true near our home. Over the last dozen years, our neighbors have gone from being more than a little skeptical to recogniz-

ing that while the grid has been down dozens of times as a result of ice storms, blackouts, and general problems, our home has kept on ticking.

No, the problem has more to do with the variability of the weather than with the equipment. During the dark months of November and December, we have what appears to be

Figure 12-1. Fossil-fueled backup generators are the antithesis of what renewable energy stands for. Nevertheless, they are an integral part of any off-grid system. (Courtesy Generac Power System Inc.)

weeks without any sight of the sun or a puff of wind. Once your system meter tells you it's time to charge the batteries, you have to listen.

Bring in the fossil-fueled backup generator, the antithesis of what renewable and clean energy technologies are all about. Life is always full of tradeoffs, and the backup generator is one that can't be ignored.

A well-designed and functional off-grid system will operate at least 90% of the time without the help of a backup generator. But when the weather decides to pull the clouds over the PV panel's eyes, you have no choice but to either shut down the entire system or start the generator. (Remember battery depth of discharge from Chapter 9?)

A backup generator is designed to perform one primary function: charge the battery bank to a full state and then shut down as quickly as possible. Generator power is polluting, noisy, and expensive compared to grid or renewably produced electricity. The less generator running time that is required, the better. To limit running time, the inverter's internal battery charger will "load" the generator to a maximum level during the bulk charging step discussed in chapter 10. If the battery is being charged "normally" the inverter will switch to absorption mode and then float mode to complete the cycle. If equalization is desired the battery will complete the high-voltage equalization charge and return to float mode.

Inverters equipped with automatic controls will then signal the generator to stop, completing the charging cycle. Depending on battery and generator capacity as well as depth of discharge the entire process can take from 5 to 10 hours to complete.

It is common for a renewable energy system to operate for 9 or 10 months of the year without the generator ever switching on. When the cold and dark months arrive, the unit had better start and be ready to go.

Renewable Energy Generators?

Although many folks think of renewable energy sources as wind, water, and sun, it is possible to fuel a generator with liquid renewable energy, a.k.a. biodiesel, and use the generator as the *only* source of electrical power generation. Chapter 16 will discuss some of the ins and outs of biodiesel fuel selection.

Many systems have been developed using this perfectly acceptable method. It has the advantage of easing the entry cost for people who might not be able to afford a roof full of PV panels. The process is quite simple: if an off-grid system is going to have a backup generator in any event, why not simply start with it charging the battery bank? Then, as finances allow, invest in PV, wind, or micro hydro, tying their energy output into the system. Over time, the addition of these renewable sources will cause the generator's running time to drop, which will lower fuel costs, reduce noise, and increase generator life.

Generator Types

A backup generator with a recipro-cating internal combustion engine is more correctly known as a *genset* (a *gen*erator and motor *set*). There are many shapes and sizes and a variety of fuel supply choices. Before we look at models suitable for emergency systems let's review what types not to buy.

Small generators such as the ultra-portable model shown in Figure 12-2 are not suited to full home-emergency or battery-charg-ing applications. As a quick rule of thumb, if you can lift the genera-

Figure 12-2. This small 1,000 W ultra-portable unit from Yamaha is fabulous for operating small power tools or emergency lights, but it is not suited for a house full of appliances or for battery charging. (www.yamaha.com)

tor, it is probably too small. The Honda model EM5000S shown in Figure 12-3 is about the smallest (and least expensive) generator recommended for this application. If you have an old 4,000 W unit kicking around in the garage, by all means put it to use. Keep in mind that performance, fuel economy, and battery charging time will be compromised with undersized models.

Generator Rating

Generators, like inverters, are rated in watts (W) or more correctly volt-amps (VA). The reason for this is that when motor loads are connected to an electrical circuit they behave differently from resistive loads such as lights and heating units. Although the generator may have sufficient nameplate rating capacity, certain loads might not run because of the effects of power factor, necessitating the distinction between VA and W .

Inexpensive gensets will have inexpensive generators and support electronics. This is not a problem for many applications until big motor-driven devices are activated. Anyone familiar with gensets will know that air compressors, well pumps, and furnaces may have a hard time getting started even if the genset has a higher power rating than the connected load. Motor loads have high starting power requirements that can exceed their normal requirements by three or four times.

Similarly, battery charging can be very hard on smaller, less expensive gensets because a battery charger consumes power only from the very peak of the alternating current waveform. Smaller units, typically less than 5,000 W, have difficulty providing power in this mode, even though the charging power applied to the bat-

Figure 12-3. The Honda EM5000S or equivalent-sized models are the smallest units that should be considered for emergency backup power.

tery may be a fraction of the generator's rating.

A gas engine driving a generator is a reciprocating device. You will recall from earlier chapters that rotating generators produce an alternating current (AC) voltage which traces a sine wave pattern. An example of this wave form is shown in Figure 12-4, with voltage represented on the "y" axis and time on the horizontal "x" axis. Starting from the extreme left of the waveform, the voltage starts out with zero amplitude and slowly rises until it reaches a peak of 170, at which time the polarity starts to reverse and the voltage drops back to zero before starting on the negative half of the cycle.

You may be wondering why the genset output voltage is 170 as opposed to the 120 volt level we are accustomed to. Because AC sine wave voltages change with time, we are faced with the issue of determining where on the waveform the voltage measurement should take place. A mathematical formula called the "root mean square" can be applied to such a waveform to calculate the voltage present in the "area under the curve". Perhaps a simpler, less technical way of visualizing the voltage of a sinewave is to refer to it as the average level. If you were to take instantaneous voltage measurements of the sine waveform you would find that it rises from zero and peeks at approximately 170 volts, as discussed above. It is the instantaneous voltage level that provides battery-charging capability.

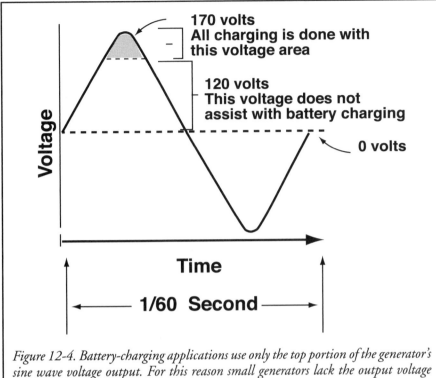

Figure 12-4. Battery-charging applications use only the top portion of the generator's sine wave voltage output. For this reason small generators lack the output voltage required to charge batteries.

The generator's peak output voltage must be high enough for battery charging to occur. The battery-charging unit inside an inverter contains a transformer that is capable of stepping up or down the applied voltages depending on which mode of operation is selected. For example, with a 12 V battery connected, the inverter can step this voltage up to 120 V AC to operate connected loads. To charge a battery, the inverter can reverse this process by stepping the generator's 120 V AC down to a DC voltage level sufficient to "push" current into the battery under charge. The waveform in Figure 12-4 shows that the generator's AC voltage starts at zero and climbs to a peak of 170. There will be a period where the generator's "stepped-down" voltage is less than that of the battery bank. No charging current will flow at this time.

Once the generator's *instantaneous* voltage exceeds 120 V, the inverter's transformer will step it down by the appropriate ratio, convert the AC to DC, and feed the voltage into the battery. The example shown in Figure 12-5 will help to clarify this point. The generator instantaneous voltage has risen to

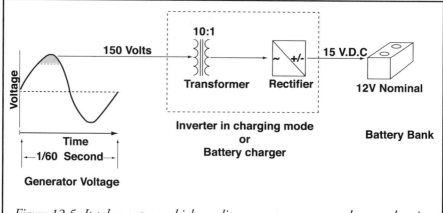

Figure 12-5. It takes a strong, high-quality generator to support battery charging and operate large motor loads such as furnaces and well pumps. Weak or inexpensive generators burn excessive amounts of fuel during the long runtime required to charge a battery.

150 V. The applied voltage reaches the inverter's battery charging transformer where it is stepped down to the ratio determined by the nominal battery bank voltage, which is 10:1 in this example. The transformer output voltage is 15 V AC. This voltage is applied to a rectifier that converts the alternating current to direct current. As the 15 Vdc output is greater than the battery's 12.18 V (or 50% discharged) reading, current will flow into the battery.

The "squashing" of the peak of the sine wave identifies gensets with weak or small generators that have insufficient power to support battery charging. Longer charging time and increased fuel consumption result in unnecessary genset wear and a higher cost for each watt of electricity stored in the battery.

It is highly recommend that you purchase a genset with a rating of 7,500 W (120 V/62 A or 240 V/31 A) or higher. Check with the manufacturer to determine if the unit is equipped with an electronic voltage regulator module within the generator. High-quality generators often have a peak voltage adjustment that ensures rapid battery charging and immediate starting of difficult motor loads. Discuss this issue with the generator sales staff. If they are not familiar with battery charging applications for a particular model of generator, provide them with a copy of the above text and ask them to review the issue with the factory. It is pretty tough to return a generator that is too small for your application.

Generator Type	Inverter Type	Typical Maximum Charging Current (Amps)
Honda 800	DR1512	43
Honda 2200	DR1512	57
Homelite 2500	DR1512	11
Honda 3500	DR1512	39
Westerbeke 12.5 kW	DR1512	65

Table 12-1. This table produced by Xantrex Technology Inc. compares the charging current of several generator models when using a 12 V inverter/charger. The Westerbeke 12.5 kW model can charge a battery bank six times faster than a Homelite 2500. This equates to reduced generator running time, wear, and fuel consumption.

Voltage Selection
The voltage selection of the genset will be determined by the connection method with the home and whether or not essential loads require 240 volts. It will be necessary to review voltage selection with your electrician during the planning phase.

See Chapter 13, "Putting It All Together Safely", for more information on generator voltage selection and connection.

Fuel Type and Economy
Gensets are available in several fuel choices including gasoline, natural gas, propane, and diesel. The less expensive units tend to be equipped with gasoline engines. Larger industrial-grade models are normally fueled using propane or diesel. The choice depends on several factors such as cost of the unit, proximity to a fuel source, and desire for economy and ease of use. From an environmental perspective, look for engines that have an EPA (Environmental Protection Agency) rating, are four stroke, are slow speed (typically 1800 rpm or slower), and preferably burn natural gas. Fuel types that you can choose from are as follows:

• **Gasoline:** Everyone is familiar with the small gasoline engines that are ubiquitous throughout North America. Cheap and easily fueled, they have a very short life span when used in demanding applications. The majority of gasoline engines operate at 3,600 revolutions per minute (RPM), which results in rapid wear and high noise levels. Expect a life span of five years before a major rebuild is required.

- **Natural Gas:** Natural gas engines are offered in two varieties: converted gasoline and full-size industrial. The converted engine is really no better than a gasoline engine, except that it offers the advantage of no fuel handling because it can be directly connected to the gas supply line. Industrial-sized natural gas engines are of a heavier design and operate more slowly, typically at 1800 RPM. This increases engine life and greatly reduces engine noise. However, natural gas is not available in all areas because the fuel is transported via pipelines.

- **Propane:** Propane is similar to natural gas with the exception that this is the fuel of choice for off-grid and rural applications; propane may already be the fuel source for other appliances in your home.

- **Diesel:** The diesel engine has the best track record for longevity. Diesel units are heavy, long-lasting machines that operate at slow speeds. Fuel economy is highest with a diesel engine. Besides offering superior fuel economy, modern diesel engines have excellent cold-weather starting capabilities and are clean burning.

- **Biodiesel:** As the name implies, biodiesel is a clean burning, alternative fuel produced from domestic, renewable resources. Biodiesel contains no petroleum, but can be blended at any level with petroleum diesel to create a biodiesel blend (most often a blend called B20 with a ratio of 80% petroleum diesel to 20% biodiesel). It can be used in diesel engines with no major (or any) modifications. Biodiesel is simple to use, biodegradable, non-toxic, and essentially free of sulfur and aromatics. (See further discussion concerning biodiesel fuel in Chapter 16).

A typical 8-kilowatt propane genset consumes 1.93 gallons (7.3 liters) per hour when operated at 100% capacity. An equivalent genset in a natural gas-fueled model will consume 144 cubic feet per hour (4,077 lph). An equivalent diesel model from China Diesel Imports (www.chinadiesel.com) requires only 0.78 gallons per hour (3 lph). A diesel (or biodiesel) model requires 40% *less* fuel than a propane model for the same amount of appliance operation or battery charging. Over the considerable life span of either model, this translates into a significant savings in operating cost. The fuel economy of a gasoline engine is comparatively poor.

Another argument against gasoline is the requirement to pay "road tax" when you fill up at the local gas pump. Less expensive colored or "off road" gasoline can be purchased but may be difficult to locate. Even with the reduction in road taxes, a high-speed (3,600 RPM) gasoline model will not be as economical (or as quiet) as a diesel or low-speed natural gas or propane model.

Generator Noise and Heat

It's annoying enough to have to run a genset in the first place, but it's even more aggravating to have to listen to it running. The best way to eliminate this problem is not to operate one at all. The next best plan is to locate the unit a reasonable distance from the house and enclose it in a noise-reducing shed or chassis.

The Kohler natural gas genset shown in Figure 12-6 and the equivalent Generac model shown in Figure 12-1 are mounted in a noise-deadening, weatherproof chassis which may be mounted on a cement pad in a similar manner to central air conditioning units.

You can either build a noise-reducing shed using common building materials or purchase a wood-framed toolshed building. The shed should be fabricated with a floating deck floor that does not contact the walls of the building. This construction prevents engine noise and vibration from radiating outside the building. The walls should be packed solidly with rock wool, fiberglass or, best of all, cellulose insulation. The insulation should then be covered with plywood or other finishing material, further deadening sound levels.

All internal combustion engines create an enormous amount of waste heat. A little dryer vent or hole in the wall won't cut it. The unit shown in

Figure 12-6. This Kohler natural gas genset is manufactured with a noise-deadening, waterproof chassis and provided with fully automatic controls. It will automatically start and provide power to the house as soon as the electrical grid fails. Once utility power returns, the unit will reconnect the house to the grid and shut itself down. Units can even be programmed to automatically start and stop periodically, ensuring that they will be ready at the next blackout.

Figure 12-7 is mounted so that the 18" x 18" (0.5m x 0.5 m) radiator and fan assembly blows outside the building pointing away from the main house. This arrangement also requires an air intake, which is provided by air passing under the floating deck of the building.

The exhaust gas leaves the muffler vertically and passes overhead to a second automobile-style muffler before exiting the building, further reducing noise.

Generator Operation

If you want to move the genset away from the house, there are two drawbacks you need to consider. A suitable power feed cable of sufficient capacity will have to be run either overhead or in an underground trench, and long cable runs increase the cost of installation. The second consideration involves starting the genset. If the unit is equipped for manual starting, it will be necessary

Figure 12-7. Installing a genset in a well-insulated shed with a "floating" floor prevents motor noise and vibration from radiating outside the building. Include a large vent fan to remove waste heat, being sure to direct it away from the main house.

to go to the machine shed each time you wish to start and stop the unit. This is no problem on a nice summer day, but it can become a bit trying during winter storms when you need the darn thing the most.

Gensets equipped for automatic operation as discussed above solve the starting problem completely. Alternatively, when the AC supply cable is placed in a trench, a second "control" cable can be run alongside it. This cable is connected to the genset's control unit and the other end is connected to a manual start/stop switch inside the house or to the automatic generator controls contained within an inverter.

Automatic controls do not add an appreciable amount to the cost of the genset and will greatly improve your relationship with the beast.

Other Considerations

A genset should have a long life, particularly if the unit is well maintained. Many underestimate how much television they watch, and I suspect that estimations of generator running time are a close second. Running time is not a problem, but knowing when to perform periodic maintenance on the unit is. For example, the Lister-Petter diesel engine shown in Figure 12-7 requires various servicing functions at 125, 250, 500 and 1,000-hour intervals. Order the unit with a running time meter at a small extra cost.

Oil, air, and fuel filters must be changed periodically. Order a service manual and sufficient spare parts for your model. The dealer will be able to recommend a suggested spare parts list.

Have engine oil on hand as well as the necessary tools to change filters and drain the engine crankcase. Inquire at your local garage about where used engine oil can be dropped off for recycling. Most garages will be happy to oblige, especially if you deal with them for automotive service. Never pour used motor oil into the ground or down a storm sewer. One quart (one liter) of oil can easily destroy 100,000 quarts of ground water, seriously damaging the environment.

To ease winter starting, use synthetic oil rated for operation with your generator model.

Further Notes on Biodiesel Fuel

In recent years, there have been remarkable strides in the "greening" of diesel engines. Anyone who thinks that diesel engines are slow, clunky, and smelly obviously hasn't been introduced to the new "common rail diesel engine" technologies of the past few years. Witness the Mercedes Benz E320 CDI series or ultra-cute Smart™ Car. These cars are just as quiet as their gasoline counterparts and faster in both acceleration and top speed. Cold weather starting problems are also a thing of the past. Not only can you forget about block heaters, but the whole glow plug thing has gone the way of carburetors, muscle cars, and fuzzy dice. (OK, maybe not the fuzzy dice.)

As a further advantage, biodiesel fuel (manufactured using renewable soya, canola, and other grains as well as waste grease and animal fats) may be available in your area. Burning biodiesel is considered green since it has lower-life-cycle carbon dioxide and smog-producing emissions than its fossil-fuel counterpart. There may be a slight cost penalty for its use, but considering the minimal running time of the genset and the symbiotic relationship this fuel offers with the renewable energy system, it is well worth the price. Biodiesel might even improve your relationship with the genset.

You will find more information on biodiesel in Chapter 16.

13
Putting It All Together Safely

We have finally made it to the point where we can stop talking about how all the bits and pieces that make up a renewable energy electrical system work and start putting it all together; *safely*. This chapter deals with interconnecting the various system components in a neat and effective arrangement.

If you are not familiar with electrical wiring, conduit, and general construction work, it's still well worth having a look at this section in order to understand what your electrician is talking about *before* you "throw the switch." This chapter and the relative appendices can also act as a reference should your electrician not be familiar with some of the details of working with direct current (DC), PV modules, and wind turbines.

> ### WARNING!
> Off-grid systems are a fairly rare phenomenon in North America and the electrical inspector in your jurisdiction may not be familiar with them. Be sure to check with your electrical inspection authority *before* you commit to such a system.

A Word or Two about Safety

Obviously, you want the installation work to be done correctly and safely. Owning and operating a renewable energy system is quite enjoyable, and you can almost forget that you have one at times. Although it is pretty cool stuff, it is not a toy and can cause electrocution or fire hazards if not respected. My first discussion with electrical contractors and inspection people left me bewildered. I clearly remember one person saying eleven years ago: "Why would you want one of those systems? You won't be able to run a toaster." Although I still chuckle at this while eating my morning toast, it goes to show that not everyone is up to speed with the technology.

Owning a renewable energy system is no different from owning and operating a standard electrical power station. Size doesn't matter. You can be killed or be seriously injured with battery or inverter power, just as you can with energy from the grid.

Electrical Codes and Regulatory Issues

In North America, electrical installation work is authorized by local electrical safety inspection offices that issue work permits and review the work in accordance with national standards. In the United States, the National Electrical Code (NEC) has been developed over the last century to include almost all aspects of electrical wiring, PV, battery, and wind turbine installation. In Canada, the Canadian Electrical Code (CEC) performs the same function as the NEC.

The NEC and CEC comprise the Part 1 Installation Codes which regulate the interconnection and distribution of electricity to industrial, commercial, and residential buildings. These codes also deal directly with the internal wiring of your home.

Many people believe that because they have their own renewable energy system the code rules do not apply to them. This is wrong. With few exceptions, the installation of PV and wind systems must comply with the requirements of the code. In fact, way back in 1984 Article 690 was added to the NEC to deal specifically with the installation of PV systems.

In addition to the CEC/NEC rules, a Part 2 product standard is required to certify every electrical appliance that operates at 120/240 volts (Vac). Where safety concerns exist, this standard may be extended to lower voltage products such as battery-operated power tools. Many people are familiar with the Canadian Standards Association in Canada and Underwriters Laboratories in the United States. Working in conjunction with the CEC/NEC, these safety agencies are charged with the development of electrical and fire safety standards for household appliances. When a manufacturer develops a new product, the design must undergo extensive safety-related tests by these agencies. Products that meet the requirements are eligible to carry a "certification mark" which tells electrical inspectors that when properly installed they will be safe.

Legitimate manufacturers have their products undergo such testing and are eligible to use the UL, CSA, ETL or other authorized testing laboratory seal of approval. When comparing and purchasing products, look for this seal as a sign of a safe, quality design and be aware that electrical devices without it will not be allowed to connect to the grid.

Figure 13-1. Renewable energy electrical installation is not difficult, as this prewired, integrated panel from Outback Power System shows. Viewing the panel from left to right you see the AC circuit breakers, 2 x 120 Volt inverters "stacked" for 240 Volt output, DC circuit breakers for PV panel, and optional battery bank and MPPT series voltage regulator.

When you or your electrician is ready to begin wiring, it will be necessary to apply for an electrical permit. This permit will authorize you to:

- Perform all electrical wiring according to NEC/CEC codes and any local ordinances in effect at the time of installation.
- Install only electrical equipment that is properly certified. Each device must have a UL, CSA or other approval agency certification "mark."
- Provide the inspector with copies of wiring plans, proof of certification, or other engineering or technical documentation to aid in his or her understanding of the renewable energy system.

Give the inspector written notification that the work is ready for inspection. You must not cover or hide any part of the wiring work, including backfilling of trenches, until the inspector has completed the inspection and provided an authorization certificate.

Your electrical inspector is not working against you. If he or she is asking a lot of questions it is to understand what you are doing. Renewable energy systems are not yet considered mainstream technology, and some inspectors may not be familiar with the specifics. On the other hand, your inspector will know an awful lot about wiring and installation details, and most professionals will be more than happy to assist with guidance and pointers.

Electrical inspectors will review the design and installation work and check for certification marks on the various appliances. Some system components on the market may not have test certification markings, but all electrical code rules require that products *must* have them. It is highly recommended that you check any products before you purchase them to ensure proper compliance. If you require a product and the manufacturer has not had it tested, discuss this with your inspector before you buy. A certified product may be available, albeit at a higher cost. Alternatively, field inspection on site may be allowed in your jurisdiction for an additional fee.

CAUTION!

As code rules are updated on a regular basis and may have subtle differences from one locale to another, use the information in this chapter as a guide but discuss the details with your electrician and inspector before proceeding with installation work.

What Goes Where?

In previous chapters we have dealt with each component as a separate piece of the pie. We are now ready to begin planning the wiring installation to interconnect all of the components.

The wiring overview in Figure 13-2 illustrates how interconnections are made between each component of the off grid system. PV arrays always generate direct current (DC) and may be connected directly to the inverter for conversion to alternating current. Wind turbines may operate using direct or alternating current (AC); although in the case of the latter configuration a unit known as a *rectifier bridge* will convert the voltage to DC for supplying energy to the battery bank and inverter.

By contrast, a *grid-interactive* system which sells excess electricity to the grid and provides emergency backup power during blackout periods is shown in Figure 13-3. The feed from the PV panel or wind turbine, battery bank, energy meter, charge control, diversion load, and inverter circuits will all be

An Off-Grid System Overview

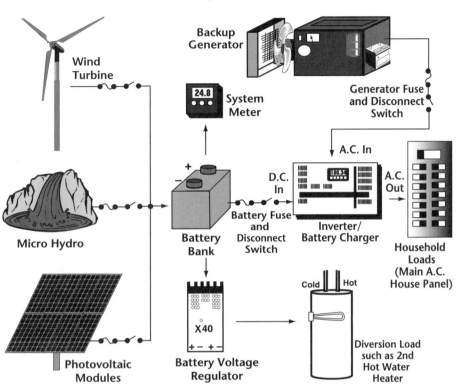

Figure 13-2. This off-grid system overview shows the placement and basic interconnection of each component. Although most systems have only one or two sources, in this case three renewable energy sources (wind, micro hydro, and solar electric PV) charge a battery bank. Power from the battery feeds an inverter which converts direct current to alternating current suitable for supplying household plug loads. A battery voltage regulator and diversion load ensure the battery is not overcharged. System metering provides system status. A backup generator is used to re-charge the battery bank should the renewable energy sources have insufficient output to maintain electrical loads.

completed using direct current in the same manner as an off grid system.

In either system the AC connections between the inverter and house supply panel follow standard household electrical wiring.

The distinction between DC and AC wiring is very profound. Wire, connectors, fuses, and switches are generally not interchangeable. Because the current on the DC side is very high (owing to the lower voltage), wire size tends to be much larger than on the AC side. For this reason we will review each wiring component separately.

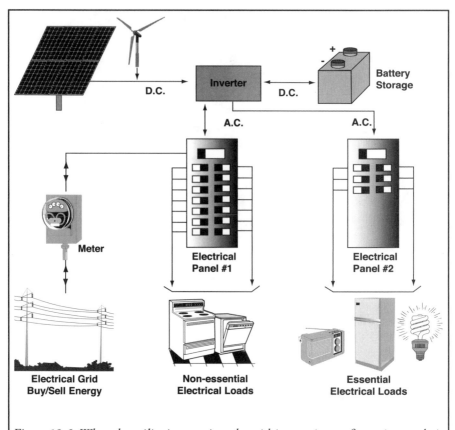

Figure 13-3. When the utility is operating, the grid-interactive configuration works in the same manner as the grid-dependent design. During a blackout, energy is supplied to essential house loads either from the renewable energy source or from a battery bank.

Direct Current (DC) Wiring Overview

Energy stored in a battery bank or supplied by a PV panel is typically supplied at a low DC voltage. When energy is required for our homes, the DC voltage is converted to AC by the inverter and stepped up to either 120 V or 240 V. If we assume that the inverter will supply a maximum house load of 1,500 watts (W) we know that 1,500 W plus an allowance for inefficiencies has to flow out of the battery bank. The wattage, voltage, and current relationship for the household side of the inverter are:

1,500 W load ÷ 120 V house supply = 12.5 amps (A) current flow

And on the low voltage input side of the inverter:

1,500 W load ÷ 12 V battery supply = 125 A current flow

A quick rule of thumb is to remember that the low-voltage side of the inverter current is 10 times greater than the 120 Vac side when dealing with 12 Vdc configurations, 5 times greater for 24 Vdc, or 2.5 times greater for 48 Vdc.

It stands to reason that, since the wattage is the product of the voltage and current, lowering the voltage will cause a corresponding rise in current and vice versa. It requires a wire the size of a Polish sausage to carry 125 amps of current in a 12 V circuit yielding 1,500 W. The same 1,500 W can be supplied through a light-duty extension cord if the voltage is cranked up to 120 V.

Large-gauge copper wire is expensive and difficult to work with, offering plenty of reason to stick with higher DC voltages where possible. This is also the reason that large AC energy consumers in the home such as electric stoves, furnaces, central air conditioning, and dryers are always rated 240 V, allowing the use of smaller wire sizes.

Low-voltage electricity is also difficult to transmit any significant distance. As current flow increases to compensate for lower voltage, more of the power is lost due to wire resistance. The only way to address this problem is to raise the voltage, decrease the transmission distance, or increase the size of the wire. Often it is necessary to do all three. In short, keep DC wiring runs (between PV panels, inverter and battery) as short as possible and increase the voltage where practical. Appendix 10 contains tables that relate the system voltage and load current to wire size and provide a maximum one-way cable length. This chart assumes a 1% electrical voltage loss, which in turn causes a corresponding loss in power. If you can tolerate a higher level of loss, then each of the applicable distances and losses may be doubled, tripled, etc.

For example, in the table for 24 V systems in Appendix 10, assume that a PV array is delivering 40 A of current and requires a cable run of 24 ft (7.3 m). The chart indicates that a #0 size of wire (some trades refer to this as "#1/0", while "#00" is referred to as "#2/0") will be required to maintain 1% electrical loss. Wire of this size is pretty big (about the diameter of a pencil), hard to work with, and fairly costly. Let's consider some alternatives:

1. Do nothing. Use the #0 wire and call it a day. There is nothing really *wrong* with using #0 wire; bigger wire is simply more expensive, harder to work with, and difficult to interconnect.
2. Increase the system voltage to 48V, thereby decreasing the cable size to #6 gauge. (Remember, doubling voltage halves the current to 20 amps.)
3. Allow an increase in voltage loss. Leave the system voltage at 24 V, but allow the system voltage drop to increase to 4%. Use the 24 V data

Figure 13-4. Direct-current wiring circuits such as the battery interconnection cables shown here are expensive and difficult to work with. Keep wiring runs as short as possible and purchase premanufactured cable such as this whenever possible.

table and reduce wire size as desired. For example, using a #6 wire will increase losses to 4%. (#6 wire will carry 40 amps with 1% loss 6 feet. Increasing this length to 12 and 24 feet will increase losses to 2% and 4% respectively).

4. Check the table in Appendix 11 to be sure the selected wire size can carry the required current. Number 6 wire is rated for a maximum of 75 amps. Don't go overboard with losses to increase wire-run distances, as losses will reduce valuable power supplied to the inverter.

5. Run two sets of smaller wire. Suppose that we are connecting a PV array to a battery bank. It is possible to "split" the wiring of the array in two, reducing the current in each set to 20 A. If each set of arrays is connected with #4 wire, we can still maintain our voltage drop at 1% and use smaller wire, which is easier to handle. The two arrays would be connected in parallel back to the inverter or battery bank, combining to produce our 40 A supply.

The low-voltage side of an inverter requires an enormous amount of current, so the larger the cables connected to the inverter the better. Undersized cables result in additional stress on the inverter, lower efficiency, reduced surge power (required to start motor-operated loads) and lower peak output voltage. Don't use cables that are too small and degrade the efficiency that you have worked so hard to achieve.

In addition, keep the cable runs as short as possible. If necessary, rearrange your electrical panel to reduce the distance between the batteries and the inverter. The lower the DC system voltage, the shorter the cable run allowed. If long cables are required, either oversize them substantially or switch to a higher system voltage as discussed above.

Although large cable may seem expensive, spending an additional few dollars to ensure proper performance of your system will be well worth the investment. Xantrex Technology recommends that the positive and negative wires supplying an inverter be

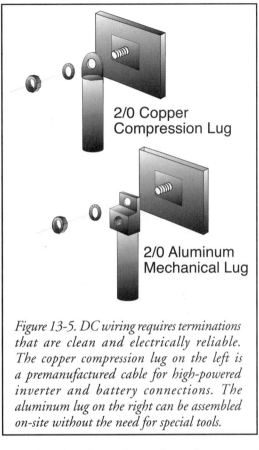

Figure 13-5. DC wiring requires terminations that are clean and electrically reliable. The copper compression lug on the left is a premanufactured cable for high-powered inverter and battery connections. The aluminum lug on the right can be assembled on-site without the need for special tools.

taped together to form a parallel set of leads. This reduces the inductance of the wires, resulting in better inverter performance.

AC/DC Disconnection and Over-Current Protection

For safety reasons, and to comply with local and national electrical codes, it is necessary to provide over-current protection and a disconnection means for all sources of voltage in the ungrounded conductor. This includes the connection between the PV array, wind turbine, and batteries as well as the connection between the batteries and inverter. Most AC sources are already provided with a protection and disconnection device as an integral part of the house or appliance wiring. Generator and inverter units are generally provided with their own internal certified fuse or circuit breaker device.

Standard AC-rated circuit breakers and fuses will not work with DC circuits, and such an installation should never be attempted. Fuses and circuit breakers such as those shown in Figure 13-7 may be used, as they are rated

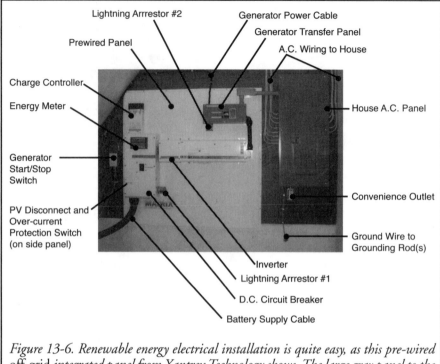

Figure 13-6. Renewable energy electrical installation is quite easy, as this pre-wired off-grid integrated panel from Xantrex Technology shows. The large gray panel to the right is where the standard 120 V house wiring begins.

for breaking a direct current electrical source.

Fuses and circuit breakers are similar to safety valves. When the flow of current through a conductor or appliance exceeds its specified rating in amps, a fuse which is wired in series will "blow," opening the electrical path. A circuit breaker works in the same manner, except that it may be used as a temporary servicing switch and can also be reset after a trip condition.

Each circuit path will have a maximum current based on worst-case conditions. This current level must be calculated by reading the manufacturer's data sheets for the appliance, wire, inverter, PV panel, or other device. The current rating should be given a safety factor of 25%. Therefore, when sizing a cable for a run between a PV array and inverter, the cable-run distance, system voltage, and worst-case current and wire current-carrying capacity have to be considered. In addition, if the cable is contained in conduit that is exposed to the summer sun, the insulation temperature rating must also be considered. For example:

Cable Size Required	Rating in Conduit	Maximum Breaker Size	Wire Rating in Air	Maximum Fuse Size
#2 AWG	115 amps	125 amps	170 amps	175 amps
00 AWG	175 amps	175 amps	265 amps	300 amps
0000 AWG	250 amps	250 amps	360 amps	400 amps

Table 13-1. Battery and inverter cable sizing chart. Note that as a result of heat loss, wires run inside a conduit have a much lower rating than those exposed to air. Battery-to-inverter cable runs should not exceed ten feet in one direction. It is recommended that you use #0000 AWG-size wire for all inverter runs, regardless of length.

1. A PV array outputs a maximum of 30 A under all conditions and the one-way wiring distance is 40 ft (12 m). The system voltage is 24 Vdc.
2. From the chart in Appendix 10, we see that a #00 (#2/0) wire is required to carry this current with a maximum loss of 1%.
3. From the chart in Appendix 11, we see that a #00 wire is capable of carrying a maximum current of 195 A.
4. A safety factor of 25% is added to our maximum PV array current (30 A x 1.25 = 37.5 A).
5. The worst-case current calculated in #4 above is compared to the maximum current rating of the wire calculated in #3. If the worst-case current is less than the desired wire-size capacity, it is acceptable.

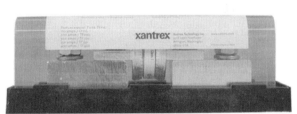

Figure 13-7. A "T" series DC fuse is shown above. On the left is a direct current disconnection and over-current protection circuit breaker and mounting box. (Courtesy Xantrex Technology Inc.)

6. The circuit breaker or fuse size must be equal to or less than the maximum rating of the wire capacity as defined in the NEC/CEC. Note that electrical cables run in conduits or raceways must be derated because of heating effects.

7. The ambient temperature of the environment and other conditions affecting wire insulation temperature ratings are discussed in Chapter 2 of the NEC and Chapter 12 of the CEC.

These notes show you the complexity of wire selection, placement, and type. If you are not familiar with the above issues, purchase a copy of the latest NEC/CEC and review the appropriate standards. Alternatively, you can discuss these items with your electrical inspector at the planning stages.

To purchase the NEC contact:
National Fire Protection Association
1 Batterymarch Park,
Quincy, Massachusetts
USA 02169-7471
www.nfpa.org

To purchase the CEC contact:
Canadian Standards Association
5060 Spectrum Way
Mississauga, Ontario
L4W 5N6
ww.csa.ca

Battery Cables

According to the NEC, battery cables must not be made with arc welding wire or other non-approved wire types. Standard building-grade wire must be used. The CEC has no such restriction.

Wiring Color Codes

Wiring color codes are an important part of keeping

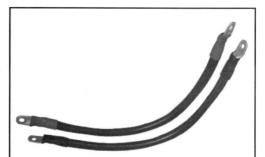

Figure 13-8.Battery cables for use with NEC-approved systems must be made with approved building wire. The CEC does not have this restriction.

the interconnection circuits straight when installing, troubleshooting, or upgrading the system at a later date. The standard color schemes used are discussed below.

Bare copper, green, or green with a yellow stripe
Used to bond exposed bare metal in PV modules, frames, inverter chassis, control cabinets, and circuit breakers to a common ground connection (discussed later). The ground wire does not carry any electrical current except during times of electrical fault.

White, natural gray colored insulation
This cable wire may also be any color at all, other than green, provided the ends of the cable are wrapped with colored tape to clearly identify it as white. This wire carries current and is normally the negative conductor of the battery, PV, wind turbine, or inverter supply. The white wire is also bonded to the system ground connection, as detailed in Figure 13-9 and as discussed later in the text.

Red or other color
Convention requires that the red conductor of a two-wire system be the positive or ungrounded conductor of the electrical system. However, the ungrounded conductor may be any color except green or white.

The majority of DC systems are based on this standard color code. The AC side of the circuit uses a similar approach to color coding except that the ungrounded conductor(s) are generally black and red (for the second wire).

System Grounding
Grounding provides a method of safely dissipating electrical energy in a fault condition. Yes, that third pin you cut from your extension cord really does do something. It provides a path for electrical energy when the insulation system within an inverter or cable covering fails.

Imagine a teakettle for a moment. Two wires from the house supply enter the teakettle, plus a ground wire. During normal operation, the electricity flows from the house electrical panel via the ungrounded "hot" conductor to the kettle. Current flows through the heater element and back to the panel via the "neutral" conductor that is grounded. A separate ground wire connects the metal housing of the kettle (via the pesky third prong) to a large conductive stake driven into the earth just outside the house.

If the insulation or hot wire were to be damaged inside the kettle, it could touch the metal chassis. Because the chassis is bonded to ground (assuming you didn't cut the pin), electrical energy will travel from the chassis, through the ground wire to the conductive stake. This flow of current is unrestricted due to the bypassing of the element, causing overheating of the electrical wires. If it were not for the circuit breaker or fuse limiting this excessive current flow, a fire could start.

Electrical energy has an affinity for a grounded or "zero potential" object and will do whatever it takes to get to there. If there were no ground connection on the defective kettle's chassis, electricity would simply stay there until an opportunity arose to jump to ground. If you were to touch the kettle and simultaneously touch the sink or be standing on a wet surface, the electricity would find its path through your body. This is not a good situation.

In a similar manner, the entire exposed metal and chassis of the system components are bonded to a *common* ground point as illustrated in Figure 13-9. The white or negative wire of the system is also bonded to this point, saving us from adding a second set of fuses and a disconnection device to satisfy NEC/CEC requirements.

The size of the ground wire is determined by the NEC/CEC and is based in part on the size of the main over-current protection device rating, as shown in Table 13-2.

Lightning Protection

If you look carefully at Figure 13-6, you will notice two lightning arrestors attached to the main DC disconnection panel and the generator transfer panel. These devices contain an electronic gizmo known as a Metal Oxide Varistor (MOV), which connects between the DC +/- conductors and ground as well as the AC hot/neutral conductors and ground. A lightning arrestor is about the cheapest piece of insurance you can purchase to protect your power system. Connect one on every cable run that strings across your property. The system shown in Figure 13-6 has a roof-mounted PV panel and a generator located in a remote building. Cables running here and there can attract lightning on its way to ground potential. The arrestor "clamps" this voltage and passes it safely to the grounding conductor.

Interconnecting the Parts

A mechanical layout plan of your desired installation will help you determine the material required and assist you in visualizing the layout of the power station. It is also important to determine what functions you require from the overall design. Some configurations include:
- PV array only;
- PV and wind hybrid system;

Size of Largest Over-Current Device	Minimum Size of Ground Conductor
Up to 60 A	#10 AWG
100 A	#8 AWG
200 A	#6 AWG
300 A	#4 AWG
400 A	#3 AWG

Table 13-2. This table shows the minimum size of the grounding conductor based on the largest over-current device supplying the DC side of the inverter.

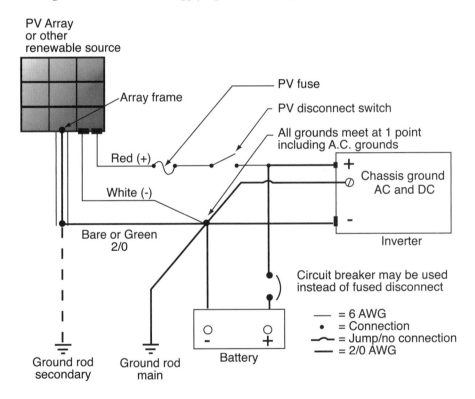

Figure 13-9. This view of a PV array, battery, and inverter DC interconnection shows the placement of over-current protection fuses and disconnect switch in the ungrounded (+) conductor. The negative (-) conductor (white) is bonded to ground and therefore requires no disconnection and over-current protection device. The ground wire (bare or green) connects at one point to a ground rod. A secondary ground rod may be used where the array or generator is a long distance from the main rod. Note the use of an interconnecting cable between both rods.

- Micro hydro only;
- Any of the above with generator backup.

The number of configurations is extensive and it is not possible to cover every installation arrangement within these pages. Fortunately, the installation manuals for the inverter and equipment will assist with reconfiguring for custom-design requirements. By way of example, look at the schematic off-grid design using PV and wind power shown in Figure 13-10. For simplicity, no ground wiring is shown in this view. Refer to Figure 13-9 for grounding requirements.

PV Array

PV arrays may be mounted on a house roof, ground mount, or sun tracking unit. The interconnection of the modules is similar in each case, with one exception. The NEC requires that every roof-mounted PV array be equipped with a device known as a *Ground Fault Interrupter* (GFI), which is shown in Figure 13-11. It automatically disconnects the PV array in the event of an over-current, insulation, or water leakage fault that could cause overheating and a possible fire. A GFI is not required on ground, pole, or tracker mounts.

The first step in wiring your PV array is to determine the battery voltage you will be using. In Chapter 6, "Photovoltaic Electricity Production," we discussed the fact that standard PV modules are manufactured for 12 or 24 V output. Figures 6-8 and 6-15 illustrate the steps required to increase PV array voltage by wiring modules in series as well as in parallel connection for increased current flow. Figure 13-12 shows a group of four 12 V modules interconnected in series to form a 48 V string. Each module is provided with a weatherproof junction box and knockout holes for liquid-tight strain-relief bushings. These bushings press into the hole in the junction box and are held in place by a retaining nut. The flexible cable is then passed between junction boxes and the series or parallel interconnection is completed. The schematic in Figure 13-10 shows two 24 V PV panels wired in series to create 48 VDC.

If a standard voltage regulator such as the Xantrex Model C40 is utilized, the PV array will be interconnected to supply an output voltage which matches the battery bank nominal voltage.

If a Maximum Power Point Tracking (MPPT) regulator such as the Outback MX60 is used, the array output voltage is wired for twice the nominal battery bank voltage.

Be careful to check for proper wire gauge and type to be sure it is suited

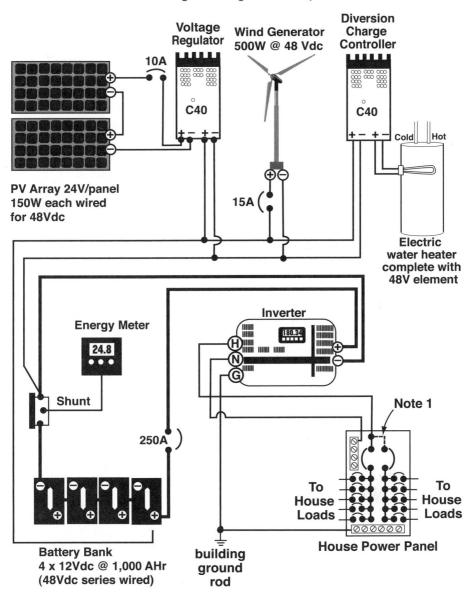

Figure 13-10a. This is one of the most complex versions of off-grid design using PV and wind energy sources, battery backup, series battery charge control, diversion load charge regulation (for the wind turbine), and energy metering.

*Note 1: A jumper wire may be fitted when the house panel is wired **ONLY** for 120 VAC. Using this configuration, only one inverter is required. For 120/240 Volt operation, 2 inverters are required and this jumper wire is not installed.*

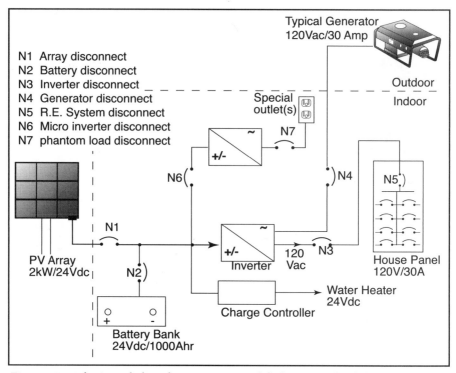

Figure 13-10 b. A single line diagram is a simplified version of a full schematic. Your electrical inspector and/or electrician can help you configure the safety elements of the systems to meet your particular needs.

for outdoor, wet installation. Review the wire choice with your electrical inspector or the NEC/CEC code rulebook.

The output from a grouping of several modules then meets at a combiner box mounted on the rear of a sun tracker unit, as shown in Figure 13-13. The array comprises a total of sixteen 12 V modules which are connected in eight sets of two pairs. Each module is rated 75 W. Two modules are wired in series forming a 24 V set. The wires for the eight sets are then directed to the combiner box where all of the negative wires are connected together at one point. The positive leads are each directed to an individual fuse or circuit breaker (5 A rating per set). The output of the circuit breaker is then fed to the common positive (+) terminal of the series-wired voltage controller, as shown in Figure 13-10. The parallel connection can be completed using terminal strips to connect all of the appropriately sized wires to the main DC supply cable.

Digressing for a moment, let's look at how this works when fully connected:

a) *16 modules x 75 W per module = 1,200 W peak for the array*
b) *2 x 12 V modules wired in series = 24 V*
c) *Current from array = 1,200 W peak ÷ 24 V = 50 A maximum*

If we had simply left the array wired at 12 V and paralleled the 16 modules we would have had to deal with twice the current (100 A). Likewise, if the array were to be wired at 48 V, the current would be 25 A. Power in watts remains the same regardless of how we interconnect the modules. Higher voltage allows the use of a smaller conductor cable, reducing cost and energy losses.

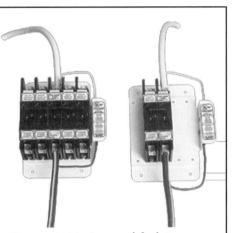

Figure 13-11. A ground fault interrupter provides protection against internal PV module faults that may cause overheating and fire. They are required when the PV array is roof mounted on a dwelling.

Wind and Hydro Turbine Connection

Connection of a wind or hydro turbine is essentially the same as for the PV array. The installation manual for each model will provide a connection point or suggest a wiring interconnection method. For example, Bergey Windpower recommends strapping a weatherproof junction box near the top of the tower, drilled to allow the feed/connection wires from the turbine to enter using a liquidtight fitting to provide strain relief for the wires and keep the weather out.

Figure 13-12. A PV array detail showing four 12 V modules interconnected in series, forming a 48 V string. The output of each of the four strings feeds into the combiner box shown at the right.

A feed cable is then run down the tower to a second servicing junction box. The feed cable running the length of the tower (and finally to the house) should be installed in either a PVC plastic conduit or metal conduit secured to a tower leg with non-corrosive metal clamps. An equally acceptable alternative is a flexible cable/conduit combination known as teck cable. This material is more expensive than traditional conduit but installs in seconds and is just as strong. Use ultraviolet light-resistant cable ties to secure the teck lead wire to the tower leg.

Renewable Source to Battery Feed Cable

The electricity from the PV array (or wind turbine) must be routed to the battery room. Cables may be fed in conduit down the side of the house or wired overhead or underground if the generator is located some distance away. Underground cable connections tend to be the most common because of simplicity and lower installation cost. The NEC/CEC has provisions for both direct burial cable and cable protected by conduit. In either case, a trench is dug, typically 18" (0.5 m) deep, between the source and the inverter/battery room or house. Where the cable exits the ground to enter the house or travel up the array/turbine support leg, it is necessary to use a length of conduit to protect the wire from damage. The cable is placed in the trench and covered

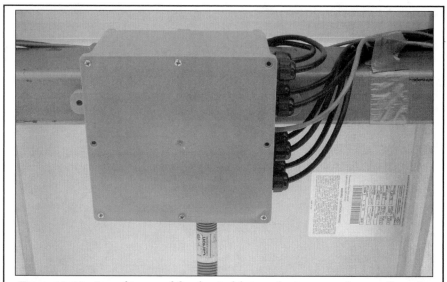

Figure 13-13. A combiner and fuse box is fabricated using a weatherproof junction box and liquid-tight strain relief. The feed wires from the array enter the box and are parallel connected, with each positive lead passing through a series-connected fuse or circuit breaker.

with 6" (15 cm) of soil. At this point, an "underground wire" marking tape is placed into the trench. The intent of the tape is to warn anyone digging in the area that a buried cable lies below. The trench is then filled in.

Over-Current Protection and Disconnect Devices

When the energy source is connected to the inverter (grid-dependent) or batteries (grid-interactive), the NEC/CEC requires that an over-current protection and disconnection device be installed. This may be done in one of two ways: you can purchase individual fused disconnect switches for each source, as shown in Figure 13-14; or you can have auxiliary circuit breakers added to the main battery/inverter disconnect box. The latter was chosen for the system outlined in Figure 13-1.

Every manufacturer of inverter-based systems provides some form of

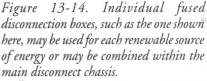

Figure 13-14. Individual fused disconnection boxes, such as the one shown here, may be used for each renewable source of energy or may be combined within the main disconnect chassis.

integrated panel as shown in Figure 13-1. Although initial cost will be higher for an integrated wiring panel, final cost will be the same or lower compared to purchasing the parts "a la carte" and wiring them together. In addition, any problems with incompatibility or "finger pointing" are eliminated when the complete system is provided by one manufacturer.

Battery Wiring

As you learned in Chapter 9, "Battery Selection and Design", wiring batteries in series increases the voltage (Figure 13-15A), and connecting them in parallel increases the capacity (Figure 13-15B). Battery manufacturers offer many voltages and capacity ratings to suit your specific requirements. The most important considerations for installation are voltage, capacity in amp-hours (Ah), and distance from the inverter. You will have to determine your essential load power requirements and the length of time you expect these items to

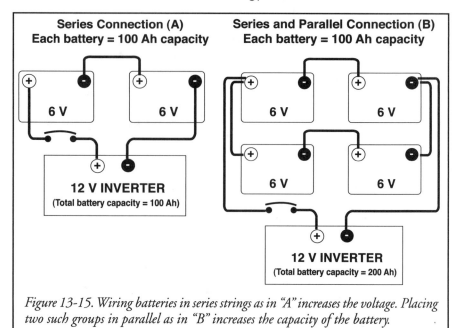

Figure 13-15. Wiring batteries in series strings as in "A" increases the voltage. Placing two such groups in parallel as in "B" increases the capacity of the battery.

operate during a blackout before selecting a battery configuration. Refer to Appendix 7, "Electrical Energy Consumption Worksheet", to calculate the backup energy requirement and resulting battery capacity.

As with integrated wiring panels, some manufacturers are offering inverter/battery combination packages to take the guesswork out of these calculations. Beacon Power offers its grid-interactive *Smart Power M5* and Outback Power Systems provides the *PS1-GVFX3648* system for outdoor installation.

As the size of the battery bank physically increases (providing power for long-duration blackouts) it becomes more difficult to use short wire lengths to feed the inverter. Plan the battery room layout using graph paper and cutouts of the selected battery model to determine the best physical layout, paying close attention to the cable connections and the length of the runs. Be sure to allow for battery disconnection and over-current protection within the wiring layout. Refer to Chapter 9 for further details regarding series/parallel battery theory.

Battery capacity and physical size can become so large that the length from one end of the battery bank to the inverter may exceed the manufacturer's recommended cable length for inverter operation. To circumvent this problem, interconnection wires should be bumped up to the largest size possible (#4/0) with the positive and negative leads taped together along their lengths to reduce cable inductance. Sometimes there is only so much you can do!

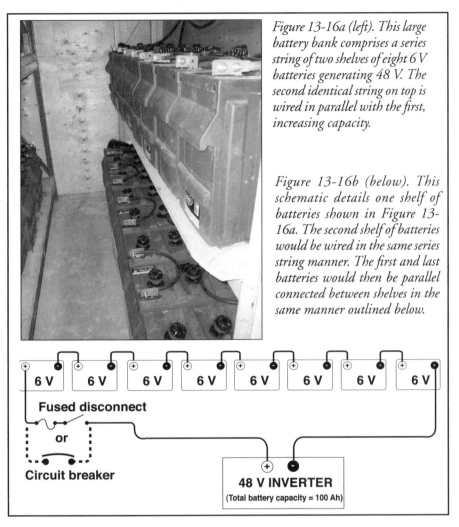

Figure 13-16a (left). This large battery bank comprises a series string of two shelves of eight 6 V batteries generating 48 V. The second identical string on top is wired in parallel with the first, increasing capacity.

Figure 13-16b (below). This schematic details one shelf of batteries shown in Figure 13-16a. The second shelf of batteries would be wired in the same series string manner. The first and last batteries would then be parallel connected between shelves in the same manner outlined below.

Battery Voltage Regulation

Battery voltage is regulated using a charge controller, which was discussed in Chapter 10, "DC Voltage Regulation". The charge controller may be connected as either a series regulator or a diversion regulator. The series regulator is the simplest arrangement, simply turning the PV input to the batteries on and off based on their state of charge. If your system includes a wind turbine, it's almost always necessary to maintain a load on the generator, requiring a shunt or diversion load arrangement.

Figure 13-10 illustrates a system designed with a series controller and a diversion charge controller with an electric water heater operating as the diversion load.

Series Regulator

Series regulation is by far the simplest and least expensive system to install. The downside with series regulation is that excess energy is wasted when the unit is regulating battery voltage. Morningstar Corporation indicates that series regulation adds less than 5% to the cost of a mid- to large-sized PV array and battery system. Maximum power point tracking options will increase this expense somewhat.

In Figure 13-10, a PV array has been configured by series wiring two 24 V panels to provide a nominal output of 48 volts DC. As you learned in Chapter 6, the open circuit voltage (PV panel in full sun without a connected load) will develop approximately 68 volts for a nominal 48 volt array. If this array were connected directly to a 48 volt battery bank, the output voltage of the PV array would match that of the battery.

The PV array will generate maximum power at only one point on its current/voltage rating. It is certain that the battery voltage will not coincide with the maximum power point of the PV array. The series controller connected to the PV array may contain a maximum power point tracking feature which allows the array voltage to track the maximum power point. The output

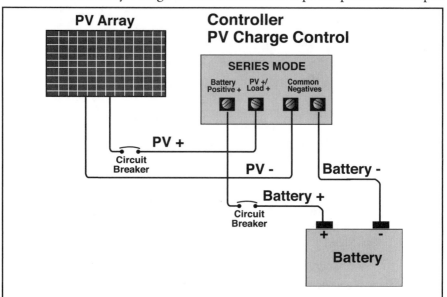

Figure 13-17 shows the connection details for a Xantrex Technology C-40 model charge controller which is not equipped with maximum power point tracking. The connection details can't get much simpler than this. Note that circuit breakers (or fused disconnect switches) must be used in each leg of an ungrounded wire (positive lead) fed by a source of voltage.

of the controller "converts" this input power to a level that maximizes battery charging capability. In addition, the controller will automatically taper the charging current as illustrated in Figure 10 -7. When using an MPPT function, the array voltage is wired higher than the battery bank nominal voltage, as indicated in the controller's manual.

Shunt or Diversion Regulator

Figure 13-18 shows a Xantrex model C40 charge controller configured for diversion load control. This configuration is required when the battery bank is fully charged. When these conditions are accompanied by strong wind, the wind generator will produce more energy than is required by the system. As discussed previously, wind and micro hydro turbines often require an electrical load to be connected at all times.

When the batteries are full their voltage will rise, telling the PV array series controller to stop charging (Figure 10-7). This releases the electrical load on the PV array. If the electrical load were released from the wind generator, the blades could possibly enter an over speed mode, destroying the unit. To prevent this from happening, the diversion charge controller "shunts"

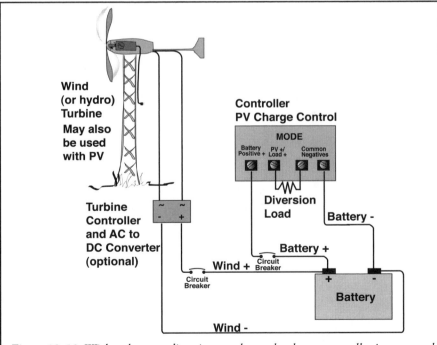

Figure 13-18. With a shunt or diversion regulator, the charge controller is connected to a diversion load such as a hot water heater or electric air-heating element.

excess energy from the turbine to the diversion load (in this case a water heater), maintaining the required load.

Review Chapter 10 for details on plumbing and electric heating element swapping. The standard 120/240 V elements of an electric water heater used as a diversion load **MUST** be replaced with new ones rated at the nominal wind turbine/battery bank voltage. Ensure that internal thermostats and over-temperature protection devices within the water heater are bypassed, either by removing them from the wiring circuit or installing a suitable heavy-gauge jumper wire. You do not want an internal thermostat opening, thereby disconnecting the load from the turbine.

Figure 13-19. Absorbing large amounts of diverted electrical energy is easy if heat enters the equation. An air heating element such as the model shown here will absorb up to 1 kW of excess energy and provide some home heating to boot.

Air-heating elements such as the model shown in Figure 13-19 can be used as a diversion load. However, air heating tends to be wasteful, as the majority of the heat may be generated outside the heating season. If an air-heating load is used and the byproduct heat is not required, consider mounting the heating element in a rainproof outdoor chassis.

The electrical rating of the diversion load should be 25% greater than the capacity of the devices feeding it. For example, the wind turbine shown in Figure 13-10 is rated 500 W and the PV array does not contribute to the diversion energy because it is equipped with its own series charge controller. Therefore, the diversion load rating should be 500 W plus a 25% safety factor, requiring an electric heating element of 625 W capacity. The rating of the diversion controller must also be calculated to ensure that it is able to supply the maximum power to the diversion load. The charge controller rating is calculated by dividing the diversion heater rating (in watts) by the nominal system voltage:

625 W diversion load ÷ 48 V system voltage = 13 amp diversion controller rating

This is well within the Xantrex C40 load rating capacity of 40 amps.

Directing the excess heat to a hot tub or spa is another good place to "dump" excess energy. Use caution when connecting the diversion elements in this application. Both Underwriters Laboratories and Canadian Standards Association safety standards require the use of safety current collectors and GFI protection in spa systems[1].

The Inverter
DC Input Connection
The inverter is the heart and brains of the renewable energy system. Inverters are also pretty darned heavy, and if your system is designed to operate 240 V house loads it may be necessary to have two inverters "stacked" together to generate this voltage. Figure 13-21 shows a prewired power panel from Xantrex Technology Inc. that contains:

- two 4,000 W sine wave inverters
- two series-wired charge controllers
- dual 250 A over-current disconnect units
- AC wiring chassis for generator and house panel connection

Purchasing your system prewired like the ones shown in Figures 13-1 and 13-21 is a wise decision. As the panels are assembled at the factory and approved to applicable safety standards, there is no problem satisfying your electrical inspector. In addition, costly errors in wiring are eliminated.

If you wish to complete your own wiring a la carte style, your electrical inspector will require a simplified wiring schematic such as the one shown in Figure 13-10. Note that this schematic drawing is in no way a full interpretation of the NEC/CEC code rules, but it does provide you with a basis for discussion with your inspector.

AC Output Connection
The AC side of the inverter will be well known to any electrician. Single-phase alternating current voltage at 120 volts potential is supplied by one black "hot" wire and a white "neutral" return wire. A safety ground is also required as discussed above. If the configuration requires a 240 volt supply, a second black "hot" wire is provided. The first and second hot wires are commonly

---

[1]   UL Standard UL 1795 and CSA standard C22.2 #218.1 require the use of current collectors to prevent shock hazard in the event of heater failure. 120 V spa-heating elements may be purchased from your local pool and spa supply store. Consult with the applicable standards before connecting the diversion controller in this situation.

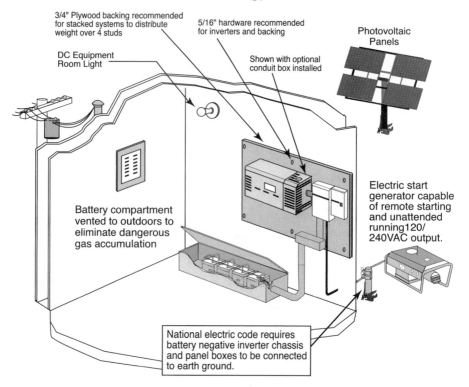

3/4" Plywood backing recommended
for stacked systems to distribute
weight over 4 studs

5/16" hardware recommended
for inverters and backing

Photovoltaic
Panels

DC Equipment
Room Light

Shown with optional
conduit box installed

Electric start
generator capable
of remote starting
and unattended
running120/
240VAC output.

Battery compartment
vented to outdoors to
eliminate dangerous
gas accumulation

National electric code requires
battery negative inverter chassis
and panel boxes to be connected
to earth ground.

Figure 13-20. This drawing details many of the issues related to the DC connection of the batteries, inverter, charge controller, and disconnect and over-current protection system. Consult with the NEC/CEC and your electrical inspector to determine the details specific to your installation. (Drawing based on Xantrex Technology Inc. installation designs)

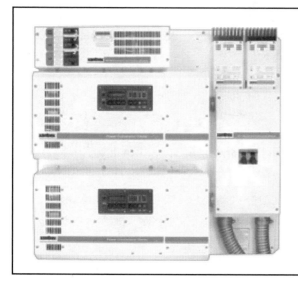

Figure 13-21. This prewired electrical system is compact, neat, properly installed and, most importantly, certified, which reduces problems with sometimes wary electrical inspectors. (Courtesy Xantrex Technology Inc.)

referred to as line one and line two respectively. Systems that require a 240 volt supply will require two inverters connected in a "synchronized-stacked" arrangement providing additive (120 + 120 V = 240 V) supply.

Residential home wiring panels are designed exclusively for 240 volt supply. Where a home requires only 120 volt feeds, a jumper wire may be added to the panel, supplying a single 120 volt supply to both "legs" of the electrical panel. In this configuration, 240 volt loads may not be directly connected to the house supply panel. Note that it is possible to use a step-up transformer to convert 120 volt to 240 volt supply for the operation of specific appliances such as well pumps. It is generally less expensive to operate a single large 240 volt load from such a transformer than to purchase a second inverter and run it in "stacked" operation. However, if the load capacity of your home's electrical system is greater than a single large inverter can supply, the required second transformer will automatically provide the required 240 output when connected with the first inverter.

The Generator

Generators supplied in North America are available with 120/240 V split phase or 120 V only output configurations. If your renewable energy system is wired to provide 120 V only, having your supplier prewire the generator for 120 V will make installation simpler. The same is true for 240 V systems.

The requirement of having all ground and neutral points bonded at one location (see Figure 13-22) will necessitate the removal of the ground-to-neutral connection inside the generator wiring chassis.

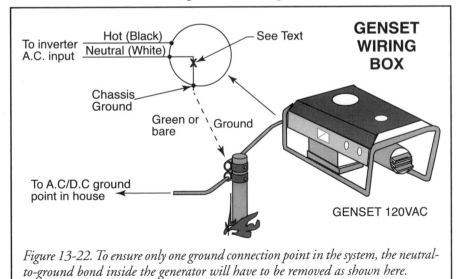

Figure 13-22. To ensure only one ground connection point in the system, the neutral-to-ground bond inside the generator will have to be removed as shown here.

Energy Meters

Earlier we discussed how an electrical energy meter records the number of electrons flowing into or out of the battery (current exported or imported). A device known as a shunt (detailed in Figures 13-10 and 13-23) converts this current flow into a signal that can be measured by the meter, allowing the calculation of energy consumed and generated. The shunt is usually mounted inside the breaker chassis box, ready for wiring in series with the battery bank negative terminal.

If you are purchasing a prewired system it is well worth a few extra dollars to have a shunt and energy meter prewired. An example of the energy meter is shown in Figure 9-17. These handy little devices record energy consumed, produced, and, most importantly, battery state of charge, which will help prevent damaging over-discharge conditions.

If you are considering wiring your own system refer to Figure 13-23, which details the connection of the shunt. Refer to the specific energy-meter wiring-connection diagrams for the balance of the system connection. Note that where the small-gauge signal wire (typically #22 to #28 AWG) cable exits a hole in the breaker chassis, an anti-rubbing bushing must be added to prevent damage.

As with prewired panels it is wise to have the shunt installed by the manufacturer, as the large wire gauge and tight chassis conditions make installation after the fact almost impossible.

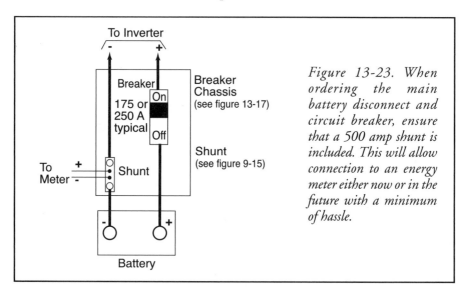

Figure 13-23. When ordering the main battery disconnect and circuit breaker, ensure that a 500 amp shunt is included. This will allow connection to an energy meter either now or in the future with a minimum of hassle.

14
Pools, Hot Tubs, and Saunas Powered by Sol

Since the ancient Romans and Turks built their public bath houses, people have enjoyed the benefits and restorative powers of water and heat on the human body. Although times change, people are still flocking to pool, hot tub and sauna dealers in record numbers to purchase these wares. In many suburban homes, the property isn't considered complete unless one of these appliances graces the grounds.

Unfortunately, these much sought-after products are energy and water sink holes that are difficult to manage and typically aren't even considered for the off-grid home. As with all other construction techniques and building codes in North America, many of these products are not manufactured with energy and resource efficiency in mind. As a result, poorly insulated and designed hot tubs draw more energy than most off-grid homes will produce in a week.

If you are going to consider using these items, it will be necessary to understand how they consume energy and to develop ways to keep this consumption to a minimum.

Swimming Pools and Hot Tubs (Spas)
Swimming pools consume energy in a number of different ways. The primary consideration with any pool, whether above or below ground, indoors or out, is the requirement to keep the water clean and fresh looking. A pool

that is cloudy or discolored will not only be a turn off at your next summer block party; this condition is also an indication that things have gone very wrong with your water. Poor water quality can damage circulation equipment because of chemical imbalance, and it can also sting the eyes of swimmers or even cause unpleasantries such as "beaver itch" and other equally distasteful or dangerous conditions.

Water purification in a pool or spa is achieved by the judicious balancing of water chemistry, which includes pH level, alkalinity, calcium hardness, and total dissolved solids. Purification involves the use of sanitizers such as chlorine, bromine, and ozone as well as the continuous, or nearly continuous, circulation of the pool or spa water.

Water is very heavy, and it requires a considerable amount of energy to move the large volumes of water from the pool through the plumbing and filter media and back to the pool. For homes connected to the utility grid, this is not a big deal; simply pay for more electricity each month. A typical pool will have a circulation pump of between ¾ and 1 horsepower (approximately 900 and 1,200 Watts respectively), which you will be obligated to operate approximately 10 hours per day. On-grid this will amount to approximately $1.00 per day of increased electrical cost to perform the required filtration.

Off-grid, pool water circulation is more problematic as energy requirements can exceed sensible power generating system capabilities.

Pool circulation time may exceed this basic level as a result excessive sunshine causing algae growth, large numbers of swimmers, or airborne debris and dust blown into the pool. In addition, highly restrictive flow passages due to small diameter circulation pipes, sharp plumbing corners, excessive lengths of pipe runs between equipment and pool, as well as small surface area filters all work together to lower pumping efficiency and flow, requiring even more energy.

If you live in the northern United States or Canada, you might already be aware of the painfully short swimming pool season. The tomato growing season appears to be a few weeks at best and the pool season isn't much different, unless you have a very high tolerance for cold water. As a result of the abysmal swimming season, many people resort to heating their pools with fossil fuels or electricity, pumping out on average 20 tons (18 tonnes) of greenhouse gases and other air pollutants per season. As discussed in Chapter 5.4, the Canadian Solar Industry Association states that "the average Canadian pool requires more energy and costs more money to heat than the average home." Given the level of energy required to heat a home, it's obvious that using electricity produced off-grid to heat a swimming pool is simply not going to work.

A pumping system that utilizes a 1,000 Watt pump operating for 10 hours per day will require 10 kWh of energy just to circulate the pool water. Do you remember the Hubickis' home we visited in Chapter 4.5? Mike Hubicki estimates that his entire house requires this amount of energy to operate per day. The cost of the renewable-energy equipment necessary to produce this amount of power: $42,000! Many off-gridders assume that they can simply circulate the pool water for an hour or two in an effort to save energy. This is simply not the case, as it results in a murky, unpleasant pond rather than the desired crystal-clear pool.

Ironically, a hot tub or spa requires a similar amount of energy even though the volume of water is greatly reduced. The reason: hot water. Although water circulation in a spa can be achieved with a smaller and more efficient circulation pump, enormous amounts of energy are required to heat and circulate the water in the spa vessel. As with a pool's circulation system, electric spa heating while operating an off-grid home or cottage is simply not practical.

Saunas (and Steam Rooms)

Although there is no denying the pleasure of a hot sauna or steam bath after a day's skiing or excessive Christmas shopping, the energy required to heat these units electrically off-grid is simply not feasible.

A common theme running through each of the above technologies is the requirement for heat. In each case, the assumption is that electricity produced using an off-grid system is too valuable to use for these luxuries, unless money is no object. Fortunately, there are ways to solve these problems, given a little bit of ingenuity.

14.1 Conventional Swimming Pools Off-Grid
(including a discussion on spa circulation systems)

Regardless of whether you choose to install a simple above-ground pool or an elaborate in-ground model, site location is very important. A home should be designed to incorporate passive solar technology and so to should your swimming pool. Figure 14.1-1 shows the ideal layout for a pool, following the same rules used to develop a passive solar house, namely to ensure that the long axis of the pool is oriented east/west and that late spring, summer, and early fall sunlight can fall directly on the pool surface. Although the sun's ultraviolet light can contribute to the oxidation of pool sanitizers, resulting in increased chemical consumption, the added energy warming the pool will greatly offset this loss.

A swimming pool that absorbs the sun's energy during the day will quickly lose it at night unless there is a means of insulating the pool against this energy loss. A pool cover such as the model shown in Figure 14.1-2 will contribute greatly to retaining heat energy in the water, with many models providing security against children and animals falling in. Additionally, a cover that is used judiciously will help reduce water circulation time, as sanitizers will not

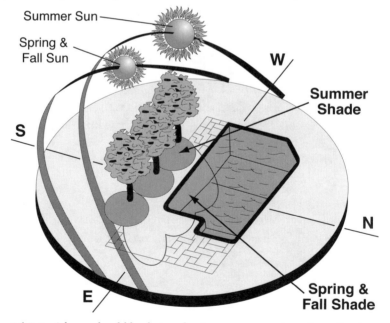

Figure 14.1-1. A home should be designed to incorporate passive solar technology, and so too should your swimming pool. Ensure the long axis of the pool is oriented east/west and that late spring, summer, and early fall sunlight can fall directly on the pool surface.

Figure 14.1-2. According to the Department of Energy, the most effective means of reducing pool heating costs is to cover the pool when it is not in use, with potential savings of between 50 and 70%. (Courtesy Cover-Pools Inc. www.coverpools.com).

oxidize as quickly and dust, leaves, and other debris will not enter the pool, further contaminating the water.

The cover manufactured by Cover-Pools Inc. of Salt Lake City, Utah can be automatically activated when the pool is not in use, possibly eliminating the requirement for a safety fence (depending on local codes) and helping to ensure the cover is properly utilized.

Rick Clark, president of Cover-Pools, Inc., likes to refer consumers to the Web site for the U.S. Department of Energy (DOE), Office of Energy Efficiency and Renewable Energy (www.eere.energy.gov/consumerinfo/factsheets/pool_covers.html). According to the DOE, the most effective means of reducing pool heating costs is to cover the pool when it is not in use, with potential savings of between 50 and 70%. "An automatic solid-vinyl safety cover adds to the convenience of energy conservation. Since an automatic cover is easy to use, the pool owner is more likely to cover the pool every day during the swimming season," says Mr. Clark. Although the automatic pool cover offers excellent convenience, it remains a premium product. If you cannot afford such a model, consider purchasing a manually applied "solar blanket" available through your local swimming pool dealer. Just remember: regardless of which type or model of cover you purchase, if you don't install it, it won't work.

Figure 14.1-3 outlines a typical swimming pool (or spa) filtration system, shown with an optional solar thermal absorber attached. For readers interested in solar thermal pool heating, refer to Chapter 5.4 for further information.

All swimming pools and spas have some form of plumbing system which

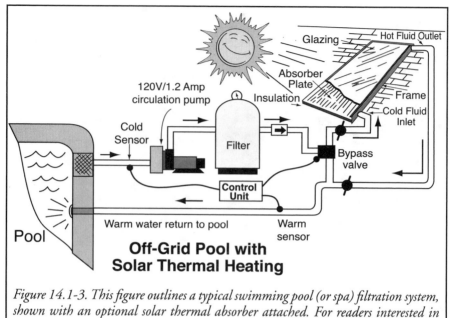

120V/1.2 Amp
circulation pump

Glazing

Hot Fluid Outlet

Absorber
Plate

Insulation

Frame

Cold Fluid
Inlet

Cold
Sensor

Filter

Bypass
valve

Control
Unit

Warm water return to pool

Warm
sensor

Pool

**Off-Grid Pool with
Solar Thermal Heating**

Figure 14.1-3. This figure outlines a typical swimming pool (or spa) filtration system, shown with an optional solar thermal absorber attached. For readers interested in solar thermal pool heating, refer to Chapter 5.4 for further information.

draws water from the surface of the pool (the skimmer intake) and secondary floor drain(s). Water is drawn into a pre-filter strainer basket, usually mounted on the intake of the pump, and then routed to a centrifugal pump where the water is pressurized. Water then flows into a cartridge or sand filter which removes fine particulate matter from the water before returning the water to the pool. Pumping time and energy are increased if the flow of water is impeded by long pipe runs, sharp angles and turns, or dirty strainer baskets and filters. Before installing your system, try to ensure that these conditions can be overcome through efficient site planning. Also consider installing the smallest pool or spa that will meet your needs rather than basing your decision on what your ego dictates. A small pool has a correspondingly small volume of water and will require less filtration time for a given amount of electrical pumping energy.

When developing a new pool, consider the following steps to ensure that plumbing efficiency has been increased to the highest level possible:

• Install the smallest pool that will suit your needs.
• Use oversized plumbing pipe and avoid sharp angles when changing pipe direction. Two 45° elbows spaced slightly apart are better than a single 90° elbow.
• Consider the use of oversized, flexible "spa hose" to further minimize

plumbing friction.

- Ensure that the pump and plumbing lines are nominally horizontal (a slight downward angle to allow water to drain back to the pool is ideal). Having the pump and plumbing lines at different heights increases pumping "head pressure" and reduces water flow.
- Order a filter that has an internal surface area that is at least twice the size required for your pool size. A larger filter surface area reduces "backpressure" on the pump, increasing the flow through the filter and in turn reducing pumping time. You may have to explain to your skeptical pool dealer that you are off-grid and that every drop of energy efficiency counts.
- Ensure that a solar blanket or other pool cover is installed whenever the pool is not in use.
- Maintain a strict pool cleaning and chemical maintenance schedule. Properly maintained water will require less filtration time.
- Encourage pool and spa users to have a shower before entering the pool. This will reduce deposits such as dead skin and deodorants entering the water, increasing filtration time.
- Install a battery-operated timer (one without a phantom load) to operate the filtration pump during the daylight hours in order to capture excess solar energy production. The timer may also be used to limit water circulation during the peak solar hours of the day and thus prevent pool overheating.

As noted above, standard swimming pools often use very large centrifugal pumps rated ¾ to 1 horsepower or larger. Provided the restrictions in the plumbing have been minimized, smaller, higher efficiency pumps can be employed. An example of a high-efficiency pump is shown in Figure 14.1-4. The "Circ-Master" is manufactured by Aqua-Flo Inc. of Chino, California and is designed to replace standard high-power water pumps where a smaller model may suffice. This model is rated 1/15th horsepower and

Figure 14.1-4. Provided restrictions in the pool plumbing have been minimized, smaller, high-efficiency pumps can be used, resulting in energy savings of up to 10 times.

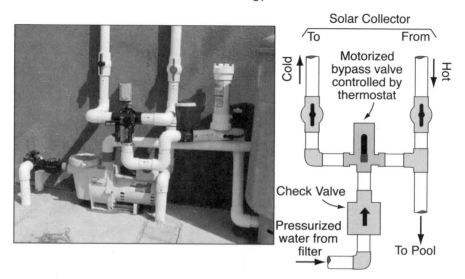

Figure 14.1-5. It is a very simple task to plumb a swimming pool to accept a solar thermal heating system. In this figure the plumbing photograph is illustrated in a schematic view shown on the right. Pressurized water from the filter enters a one-way check valve. An optional motorized bypass valve controlled by a thermostat directs water to the cold side of a solar thermal collector panel, with the return water directed back to the pool.

draws 1.3 amps when connected to a standard 120 Volt utility service connection or inverter. The total power required to operate this pump is 156 Watts and, with carefully reduced filtration time based on the above rules, the pool may require as little as 900 watt-hours (0.9 kWh) of energy per day, with energy comsumption reduced by 10 times compared to a standard pool.

Plumbing a pool for solar thermal heating is accomplished by adding a check valve and optional motorized bypass valve as shown in Figure 14.1-5. The schematic view indicates that pressurized water exiting the filter flows into a one-way check valve. (This prevents a thermo-siphon backflow condition that occurs at night when the circulation pump is turned off.) Pressurized water then flows to an optional thermostat-controlled, motorized bypass valve which directs water to return directly to the pool when it is sufficiently warm or through the solar collector panels when heat is required.

Pools installed with a small number of solar collectors may not require a motorized bypass valve, as these smaller installations are not likely to overheat the pool. It is also possible to control the pool's water temperature by limiting the pump operation through the peak daylight hours using a programmable timer.

14.2 An Alternative Way Forward: The Eco-Pool Concept

Overview

The attraction of a swimming pool cannot be denied. For many children, visiting summer camp or going to a cottage and swimming away the hot summer days is an image that is burned into their collective memory, and recalled in later years as those youthful summers are re-lived.

Figure 14.2-1. The Hubicki family eco-pool forms a lovely vista from the living room patio doors. This natural pool is built with recycled and local materials and is maintained by nature's rains and a pond ecosystem.

Figure 14.2-2. The eco-pool water is oxygenated as it spills over a waterfall and nourishes the water plants of the pond.

Figure 14.2-3. This view details the "negative edge" of the pool. Whenever the pump is activated the pool "overfills," causing a lovely waterfall from pool to pond.

Figure 14.2-4. The water plants making up the clarifying pond are clearly visible in this view.

Figure 14.2-5. Long after the swimming season has ended, the eco-pool and pond still provide a beautiful setting.

Figure 14.2-6. In this view, Mike is shown with the settling filter lid open. During operation, water is forced to swirl around the inside the container, causing heavy sand or other particles from the pond to settle before the water is returned to the pool.

If it is your desire is to recreate the natural setting of past summers, then why start with a concrete and vinyl circular swimming pool when what you really want is a lake or "swimming hole"? In an effort to resolve the dichotomy of a modern pool and a natural setting, people working in the area of permaculture have developed the concept of the eco-pool.

Tina and Mike Hubicki are one couple who wished for the pleasures of a swimming pool but didn't want to stray from their approach of using natural materials and sustainable methods in their quest. "In 2004 we built our eco-pool based on work done in Germany and other locations around the world," explains Mike. "The pool is constructed in a 14 ft x 28 ft (4.3 m x 8.6 m) rectangle which feeds a 15 ft x 30 ft (4.6 m x 9.2 m) natural pond via a 12 in (0.3 m) high waterfall created by a "negative edge" between the pool and the pond."

"The basins for the pool and pond were constructed using recycled tires for earth retention and steps around the inside of the pool," Mike continues. "Old carpet was used as a base to protect the rubber pond liner. Water is circulated using an energy-efficient pump (discussed in Section 14.1 above) and water sterilization is completed without the need for chemicals by using a copper-silver ionizer unit. Total power requirements for the pool and pond system are approximately 1.5 kWh per day. Apart from the initial filling, the pool and pond surface form a catchment basin for rainwater, ensuring sufficient water level."

The eco-pool has minimal chemical requirements for water sterilization owing to the unique natural filtration system. I asked Mike to explain how the pool works as an integrated system. "When the pump is activated, water flows into the pool, filling it beyond its maximum level. This causes water to spill over the negative edge waterfall, aerating the water with oxygen. Oxygenated water spills into the stilling pond, where a natural selection of water plants and microorganisms purifies the water. Pond water is then directed to a settling filter which allows sand and other particulate matter to separate. The water is then pressurized by the pump and returned to the pool. An electrically operated silver-copper water ionizer (available at most pool dealerships) provides final water purification to ensure that the water stays crystal clear."

Eco-Pool Construction

The best way to describe the construction of the eco-pool is to show you by walking step-by-step through the construction process.

"Throughout the planning, financing, approvals, design, and construction stages of our home and eco-pool we found it difficult, if not impossible, to find relevant information to assist us in our decision-making process," explains Mike. "Once our home and its integrated systems were completed, we decided to use them as a teaching prototype. We formed H2OMES (www.h2omes.net), our home-based teaching and consulting service, to promote the ideals we believe in. Readers interested in developing their own eco-pool are invited to contact Tina or Mike through their Web site.

Figure 14.2-7. No, the high hoe operator is not sleeping; he is studying the plans to ensure that the eco-pool elevation and site layout are within tolerance. Although it is possible to excavate by hand, a backhoe or "bobcat" unit will save a lot of wear on your back.

Figure 14.2-8. It is essential to have all the excavation level and set to the correct depth. As you will see in the following pictures, the entire pool structure sits on columns resting on grade. Careful attention to this detail will ensure that the finish decking materials line up with the entrance doorway and guarantees straight and true construction.

Figure 14.2-9. Once the excavation has been roughed in, the tires are placed around the perimeter of the pool and set back to form a series of steps, easing the entrance into and exit from the pool. Each tire is tamped with earth to make a solid, secure structure.

Figure 14.2-10. Hand-dug trenches hold the plumbing lines which run to and from the pump, pool, and sediment wells.

Figure 14.2-11. The pool floor drain is moved into position and a plastic bag is placed over the mouth of the inlet to prevent dirt from entering.

Figure 14.2-12. Gate valves are placed in the plumbing lines to facilitate servicing the pump and filter without having to drain the pool.

Figure 14.2-13. This view details the connections of the pool plumbing to the sediment wells.

Figure 14.2-14. A series of cement patio stones is placed over the tires that will form the negative edge waterfall seen at the top of the picture.

Figure 14.2-15. Once the tires and plumbing lines are roughed into position, a series of decking joists is temporarily placed in position to ensure correct alignment between the desired decking surface height and the pool. Note that the decking is a "floating" design that is not directly attached to the house structure.

Figure 14.2-16. Mike has used a piece of sewer pipe as the vessel for the hot tub planned for the future. The hot tub will share the pool's hydraulic system and decking.

Figure 14.2-17. This view shows the roughed-in clarifying pond, with one edge located directly under the negative edge waterfall of the pool. The pond has been deliberately designed to have a freeform, natural look.

Figure 14.2-18. Cement patio blocks form the foundation for the negative edge waterfall. It is imperative that this surface be level or the waterfall will not be uniform.

Figure 14.2-19. Deck support beams are made by laying pressure-treated 2" x 10" (51 mm x 255 mm) boards on their sides, providing a wide footing for the deck finish boards.

Figure 14.2-20. Standard concrete "cottage blocks" are installed along the finish grade, which in turn supports 2" x 10" pressure-treated joists. It is very important to ensure level orientation between the joists and tire support boards to ensure a quality deck finish.

Figure 14.2-21. The boys are smiling because they have completed the most difficult and heavy part of the construction process.

Figure 14.2-22. At this point, the entire construction team enjoys a coffee break.

Figure 14.2-23. What a great way to recycle pink and lime green carpets! Placing carpets on top of the rough excavation protects the rubber pond liner against cuts and abrasions, increasing its operating life.

Figure 14.2-24. With a mighty heave, the crew (recruits of wary friends) carefully unrolls the pool liner into the eco-pool excavation.

Figure 14.2-25. The liner is pulled into position and aligned parallel to the pool sides.

Figure 14.2-26. The alignment continues, ensuring that folds and pressure points are eliminated.

Figure 14.2-27. Careful attention is paid to the corners as well as the "steps" formed by the tires. The pool liner must rest on a solid surface so as not to stretch when water is added.

Figure 14.2-28. Corner folding is perhaps the trickiest part of the installation.

Figure 14.2-29. Sufficient liner material must overlap the pool edges to be held in position by the deck support boards.

Figure 14.2-30. Once the liner is in position, the water delivery truck is called in.

Figure 14.2-31 Mike always did want to become a firefighter.

Figure 14.2-32. As water is added to the pool, the crew continues to check the liner and make final adjustments.

Figure 14.2-33. The liner will have a tendency to "pull away" from the walls of the pool and slide towards the bottom. Here the crew continues to press the liner into its proper position to ensure that it stays in place.

Figure 14.2-34. Once the pool is filled, the overlap of liner material is gently stretched into position and fixed under the deck support boards placed on the tires. The weight of the deck material will hold the liner in position.

Figure 14.2-35. The decking material is now nailed into position starting at the house end of the deck.

Figure 14.2-36. Joists in the vicinity of the patio doors run perpendicular to the main deck and necessitate decking material to be installed at right angles to the rest of the deck.

Figure 14.2-37. The pond liner is fitted into position. Less precision is required with this area as the "lines" of the pond should be more fluid. Note the use of stream stones to hold the liner in position and define the boundaries of the pond.

Figure 14.2-38. Stone dust walkways have been integrated into the perimeter of the pond and intersect with the main deck. This makes a perfect transition from the formal deck to the natural pond.

Figure 14.2-39. This view details the pool negative edgewaterfall-to-pond transition area. The pool liner folds into the pond liner, ensuring that 100% of the circulation water stays within the confines of the hydraulic system.

Figure 14.2-40. Additional stream stones and perennial plants edge the completed deck and hide the support structure underneath.

Figure 14.2-41. The hot tub (concrete sewer pipe) is planked with vertical cedar boards and held together with a cable band, giving the impression of a large coopered barrel.

Figure 14.2-42. The inside of the hot tub is lined with Styrofoam®-brand closed-cell insulation cut to size and held in position with compatible "insulation adhesive."

Figure 14.2-43. The insulation is clamped into position and allowed to set. Later the inside of the hot tub will be lined with waterproof and rot-resistant red cedar and plumbed into a new solar thermal heating system.

Figure 14.2-44. This is the boundary between the waterfall and clarifying pond.

Figure 14.2-45. Once the pool is filled, the black liner and pool water surface play tricks on the eye. The negative edge waterfall is a delight to see and hear.

Figure 14.2-46. Water plants coupled with the random placement of the rounded river rocks form a natural pond which starts the water purification process.

Figure 14.2-47. Viewed from the pond surface, the eco-pool almost disappears.

14.3 Hot Tubs and Solar Power: A Marriage Made in Heaven

We discussed in Section 14.1 how the circulation system of a typical spa operates. What we did not discuss is the enormous amount of energy that is required to heat a spa to its normal operating temperature of 104°F (40°C). You may remember from Chapter 1.3 that thermal energy is often measured in British Thermal Units (BTU), from which we can calculate the amount of electrical energy required to heat a hot tub or spa. (This calculation can also be completed using the metric system of Joules or Calories.)

A small three-person spa may have a water capacity of approximately 260 gallons (≈1,000 liters) which will have a mass of 454 pounds (206 kg). If we assume the initial temperature of the water used to fill the spa to be approximately 50°F (10°C), then the amount of heat energy required can be calculated by multiplying the mass of the water by the differential between its cold and desired water temperatures (an additional amount for efficiency factor and heating losses is also required, but this "efficiency factor" will depend on the boiler or heater system):

454 pounds of water x 54°F heat rise   = *BTU of energy required*
= *24,500 BTU*

From Appendix 1, we can find the amount of energy contained in various energy sources. For example, each kWh of electricity has sufficient energy to provide 3,413 BTU of heat:

24,500 BTU to heat spa ÷ 3,413 BTU / kWh of electricity = kWh of electricity required
= *7.2 kWh of electricity*

This amount of energy is slightly more than that required to run the entire Kemp household (discussed in Chapter 4.4), and approximately $31,000 worth of off-grid electrical equipment would be required. Furthermore, this calculation neither includes the energy required to circulate the spa water nor factors in the heating losses of the system. (The day-to-day average operating energy would be considerably less than the initial heat-up energy, but it would nonetheless be a considerable amount.) The point of this exercise is simply to say that heating a spa with off-grid-generated electricity is simply not in your best financial interests.

There are several alternatives to using electricity to heat a hot tub unit. The simplest and most common is to purchase a propane pool/spa heater (Figure 14-3.2). Based on the energy content of propane, approximately 1 quart (1 liter) will provide sufficient heat (23,000 BTU) for the spa discussed

Figure 14.3-1. If you are going to consider an electrically heated hot tub for off-grid use, consider a Softub models such as the one pictured here. These units are manufactured with pool-grade vinyl and are heavily insulated with a closed-cell foam material. The hot tub captures the waste heat of the electric circulating pump and transfers this energy into the water. Although the Softub units require approximately the same amount of energy as other well-insulated units, they tend to be smaller (requiring less water and heating energy) as well as being more energy efficient. (Courtesy Softub Canada).

above, with less required to maintain the water temperature on a day-to-day basis. Of course propane is not a renewable fuel, so let's examine other options that use renewable energy heating methods.

• Wood-Heated Hot Tubs

Wood-fired hot tubs are manufactured both by Snorkel and by Madawaska Millworks and are designed for use where electrical energy is either at a premium or nonexistent, making them perfect for the cottage or for occasional use. Many are designed for fill-use-drain operation where water is pumped from a lake, heated, and drained from the unit after a few days use. Another big advantage of

Figure 14.3-2. There are several alternatives to using electricity to heat a hot tub unit. The simplest and most common is to purchase a propane pool/spa heater.

wood-firing is that the water heats very rapidly.

Wood is an environmentally sound energy source for hot tubs as the burner units consume enormous amounts of air, resulting in clean and low-particulate smoke output. One person who regularly uses one on winter trips to his cabin draws lake water from a hole drilled in the ice, gets the fire

Figure 14.3-3. The wood-heated hot tub may remind people of Gilligan's Island and the ever-present headhunters; however, these units are very enjoyable and extremely fast heating, which is a blessing for weekend or cottage use. (Courtesy Snorkel Stove Company)

Figure 14.3-4. Wood-heated hot tubs are perfect for the cottage or for occasional use. Many are designed for fill-use-drain operation where water is pumped from a lake, heated, and drained from the unit after a few days of use, possibly eliminating the need for chemical sanitizers. (Courtesy www.madawaskamillworks.com).

rolling, and hops in, all within just under two hours. When the fun is over, he pulls the plug and the water drains directly on the ground—a system that integrates very well with nature.

Chemical sterilizers can be used with these models, although the lack of a circulation pump and filtration system makes water purification very tricky. If you decide to add a filtration system, ensure that the plumbing and electrical components are installed in accordance with electrical standards, that a ground fault circuit interrupter (GFCI) is connected to the electrical mains connection (yes, even off-grid homes require these safety devices), and that your electrical inspector approves the installation.

• Solar-Heated Hot Tubs

Robert (Bob) Owens is a self-described solar-tinkerer who decided that if a solar thermal panel or two could provide enough hot water for the family home in Bradenton, Florida, he could probably heat his spa with the same sort of system.

Bob installed two flat plate solar thermal panels on the roof of his house and plumbed them into the standard spa circulation lines. A plywood box/seat was built to contain the temperature differential control and hot water circulation pump (Figure 14.3-7). Referring back to Chapter 5.4, you will recall that a typical solar thermal system operates on the basis of an electronic control module that measures the outlet or "hot side" of a solar thermal panel and compares this temperature to the coldest point in the circulation system, typically at the pump intake. When the temperature differential is greater than a preset number of degrees (typically 10°F or 6°C) and the spa temperature is lower than a preset upper limit (typically between 100°F and 104°F/38°C and 40°C for spas), the circulation pump activates, forcing spa water into the solar collector panel, extracting heat, and returning warm water to the spa.

Figure 14.3-5. Robert (Bob) Owens of Florida decided that there was no need to pay to heat his spa when a little bit of ingenuity and solar thermal technology could do it for him. Here Bob toasts the successful marriage of sun and water. Note the water "spout" to Bob's right, which returns hot water to the spa from the solar thermal system. (Courtesy Robert Owens).

Temperature measurement requires the use of devices known as "thermistors" which change their electrical resistance as a function of change in temperature. (Figure 14.3-8). Bob did not want to drill holes to mount the thermistors directly in the water path, so he opted to glue them to the side of the spa surface after carefully removing a

Figure 14.3-6. Using two flat plate solar collectors such as these, Bob is able to heat his large spa from a cold fill in less than three days. Requiring just a few running hours per week, the system maintains the desired water temperature of 100°F (38°C). (Courtesy EnerWorks Inc.)

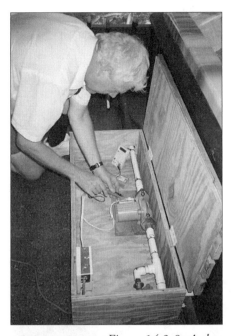

Figure 14.3-7. Unfortunately, commercial spa and hot tub manufacturers don't see the beauty or environmental advantage in running their units from solar energy. Bob Owens decided to take things into his own hands and make the necessary modifications himself. (Courtesy Robert Owens)

Figure 14.3-9. One thermistor is mounted on the "cold" point, which corresponds to the water temperature in the spa. The other sensor measures the temperature of the "hot side" or output of the solar thermal collector. (Courtesy Robert Owens)

Figure 14.3-8. A thermistor device such as this model measures water temperature and sends an electric signal to a control module. When the spa temperature is below the desired setpoint and the solar thermal panels are a few degrees warmer than the spa water, the control module activates the pump.

Figure 14.3-10. Regardless of what fuel you use to heat your hot tub, there is no point in wasting it. A solid thermal cover such as the model pictured will save at least 75% of your heating energy and considerably more in winter.

patch of insulation (Figure 14.3-9). Once installed, the insulation was reapplied and the system was ready to run.

Regardless of what fuel you use to heat your hot tub, there is no point in wasting it. A solid thermal cover such as the model shown in Figure 14.3-10 will save at least 75% of your heating energy and considerably more in winter. Spa covers also provide a degree of safety, preventing small children from entering the spa while unattended. Many models come with locking hardware that makes unauthorized removal of the cover virtually impossible.

Figure 14.3-11. Unlike a hot tub, a sauna does not have any way of buffering or storing heat from one use to the next. (Courtesy Madawaska Mill Works).

Figure 14.3-12. Since saunas are usually heated electrically, these models are not usually appropriate for off-grid living. This coopered barrel design can be supplied with an electric heater (which could be powered with a biodiesel or high-concentration ethanol gasoline generator to maximize environmental sustainability) or with a traditional wood stove heater operating from renewable energy sources. (Courtesy Madawaska Mill Works – www.ma dawaskamillworks.com).

15
Communications beyond the "End of the Line"

Living free of the electricity grid is extremely gratifying, but at the same time communicating with the outside world can be a challenge. This is because the utility poles that normally bring electricity into the home usually supply other services, including cable television and telephone. Fortunately there is no need to worry. Just because the telephone poles don't quite make it to your home it doesn't mean that you have to remain disconnected from the rest of society. Although your first reaction might be to call your local phone company to run the lines into your home, there are other options for you to consider.

If the phone lines don't reach your house and you want the convenience of telephone, Internet or email service, the number of solutions on the market is growing by leaps and bounds. But do they really work as well as the slick ads and Web sites suggest? The technological forest of communications options requires a bit of thinning to find out what really works, what doesn't, and if these "solutions" are within the budget of real people.

The following discussion provides an overview of various methods of communicating over the last few miles (or more) of the rural "wire chasm," where the phone company may not extend service and reviews the following alternate communication technologies:

- Cellular phone service
- Whole home (fixed) cell service
- Point-to-point phone service extender
- Fixed-point, broadband Code Division Multiple Access (CDMA) (digital) wireless service (provided by the phone company)
- Radiotelephone service
- Satellite phone service
- Wide-area wireless high-speed Internet service
- Satellite Internet service
- Voice over Internet protocol (VoIP) service

Cellular Phone Service

With the spread of cellular service across North America, this should be the second option to consider after standard wireline telephone service. Most cellular providers try to "optimize" their investment in infrastructure, so they tend to concentrate service in high density urban areas and along major highway corridors. Providing cell service to a few dozen rural folks is clearly not a priority. However, cellular coverage is beginning to blanket the continent, and with the use of high-gain directional antennas, service coverage may be extended well beyond what the carrier's "coverage maps" actually indicate.

Figure 15-1. With the spread of cellular phone service across North America, rural homeowners should consider this their second option after standard wireline telephone service.

When cellular (cell) service was first introduced, it was based on analog technology which required high-power and heavy transmitters to supply the desired 3 Watts of radio signal. These early units were typically installed in automobiles and came equipped with a fixed antenna on the roof or window. Later models were introduced as a "bag phone" which was also equipped with a small antenna, battery, and charging device.

As cell technology improved and became ubiquitous, service providers began the transition to low-power digital handheld phones with limited transmitting range. As most users live in high-density urban areas and desire small, portable units with long battery life, limited radio signal strength is not a concern. Additionally, high-frequency digital systems also allow higher call density per cell transceiver tower as well as expanded features such as voicemail, email, and data transmission. Unfortunately, low-density rural would-be subscribers are often left out of the cell game.

Figure 15-2. As cell technology improved and became ubiquitous, service providers began the transition to low-power digital handheld phones, with limited transmitting range. High-frequency digital systems allow higher call density per cell transceiver tower as well as expanded features such as voicemail, email, and data transmission. Unfortunately, low-density rural would-be subscribers are often left out of the cell game.

The greater the power output of your phone, the more likely it is that you'll be able to link up with a cellular antenna tower in your area. Your first task will be to find which cellular provider offers the best service in your area and then try and track down the highest wattage phone, amplifier, and high-gain antenna you can. You may wish to inquire about older analog transceivers. Low-frequency high-power analog units have greater range than their higher frequency, low-power digital counterparts. Keep in mind that the U.S. Federal Communications Commission (FCC) is no longer requiring cell service providers to upgrade or even maintain analog service as the drive to a digital world continues.

Cellular service comprises a multiplicity of transceiver and tower units sprinkled across the countryside to provide service to a series of coverage areas or "cells." As you travel along a highway, you may notice your phone service becoming weaker and suddenly strong again. This is the process of one cell automatically "handing over" the call to the next cell.

In a rural location, you may find some or all of your calls are long distance. This is because you may be quite a distance from a cell tower and the serving areas do not match those of the standard wireline service providers. This may also occur if your home is within range of more than one cell tower, in which case some calls are billed as local calls while others are charged a higher "roaming" rate.

This occurs due to the system scanning for the nearest cell tower that has both the capacity and signal strength to handle your call. Each time you call you may get bumped from one cell tower to another, as if the calls have originated from different locations.

Cell phones are normally equipped with omni-directional antennas that broadcast their signals equally and in all directions. A directional antenna, commonly called a "yagi" (Figure 15-5), can be mounted on any convenient elevated tower or on the side of a building. It will direct the cell phone transmission directly at the desired service provider cell tower, eliminating "tower bouncing." In addition to providing a directional signal, yagis also increase signal strength or gain by focusing the majority of the transmitted signal in one direction, thus providing improved clarity and transmission range. It is also possible to ask your service provider to select your local calling area to prevent unnecessary roaming charges.

Whole Home (Fixed) Cell Service

As an alternative to using portable cellular phone devices such as those shown in Figures 15-1 and 15-2, consider installing a whole home or fixed cell system such as the model shown in Figure 15-3. The Phonecell® SX5e can be thought of as a wireless version of a regular household phone system. The off-grid home is wired for telephone service in the normal manner, but rather than being connected to a landline system, the wires are routed to the Phonecell® terminal.

The terminal simulates the normal wireline phone service operation, allowing it to connect with standard house phones (including cordless models) as well as fax machines. The device also provides Internet and email connectivity using the digital cellular phone system. The internal 2 Watt radio

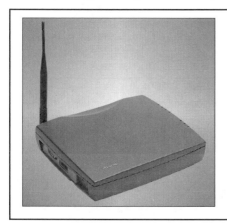

Figure 15-3. As an alternative to using portable cellular phone devices, consider installing a whole home or fixed cell system such as the Phonecell® SX5e. It can be thought of as a wireless version of a regular household phone system. (Courtesy Telular Corporation)

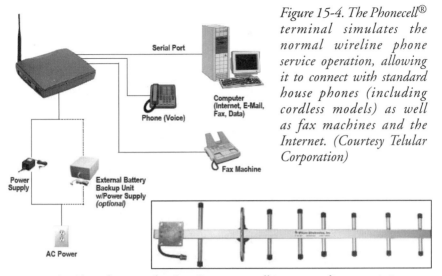

*Figure 15-4. The Phonecell®
terminal simulates the
normal wireline phone
service operation, allowing
it to connect with standard
house phones (including
cordless models) as well
as fax machines and the
Internet. (Courtesy Telular
Corporation)*

*Figure 15-5 (inset). A directional or "yagi" antenna will increase radio transmission range
by focusing the signal directly at the distant cell tower system. Directional antennas must
be elevated and may be mounted on any convenient tower or to the side of a building
structure. (Courtesy Wilson Electronics Inc.)*

transmitter and optional external directional antenna ensure maximum signal
strength.

Costs for fixed cell terminals are in the range of $500 plus external anten-
nas and installation (if required). Operating costs are the same as for regular
cell phones, with many service providers offering "family plans" or bundles
where a group of cell phones share a pool of minutes at a reduced rate.

Point-to-Point Phone Service Extender

If you live in an area not covered by cellular service and you have access to a
phone line within 20 miles (32 kilometers) of your home, you may want to
consider a point-to-point phone service extender. This system will involve
positioning a device at the end of the distant phone line that converts the
telephone calls into a radio signal. A similar device located at your home
will convert the radio signal back to a phone signal that is compatible with
regular phone, fax, and computer modem connections. You will in fact be
using two radio frequencies (duplex): one to send information (voice or data)
and a second to receive it.

This type of system will require a broadcasting license, as radio channels are considered public property. A yearly fee is collected by the Federal Communications Commission (FCC) in the United States and Industry Canada in Canada for the use of this radio "airspace."

Point-to-point service extenders have advantages and disadvantages over other communications technologies:

Advantages

- These systems mimic a real phone line, so you will be able to use your fax machine and computer modem to log onto the Internet and use your email.
- Local calls are not billed by time in the same manner as they are with cell phones.
- You maintain the same access to long distance service providers that you would normally have with a regular landline telephone.

Disadvantages

- The purchase price of these systems can be significant.

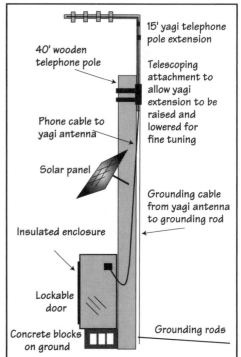

Figure 15-6. If you live in an area not covered by cellular service and have access to a phone line within 20 miles (32 kilometers) of your home, you may want to consider a point-to-point phone service extender. This system requires positioning a device at the end of a distant phone line that converts telephone calls into a radio signal. A similar device located at your home converts the radio signal back to a phone signal that is compatible with regular phone, fax, and computer modem connections.

Systems typically cost between $2,000 and $5,000+ to install. If you are running a business from your home and can amortize this over time, it may not be a concern, but it can hurt the wallet of mere mortals.

- The "bandwidth" on some (typically low-frequency) systems will be limited, which means that Internet service will be slower than with regular landline dialup service. If you want to download large files, these systems can be prohibitively slow.

- As you are using radio frequencies, connection and audio quality may be subject to the whims of the weather.
- You will need a location to install the transmission box and antenna at the distant end of the phone line. Your local phone company may allow you to install it on one of their poles, but they may not be able to assist in all instances. If the end-of-the-line telephone poles are on private property, you may be able persuade the landowners to allow you to use their existing pole and grid-electricity from an existing service to operate the radio transceiver unit. Alternatively, you may use a photovoltaic panel and battery bank to operate the unit.
- Depending on system frequency you may be subject to radio license fees.

Installation Issues

Point-to-point phone service extender systems will require an antenna similar to the model shown in Figure 15-5. All makes of antenna should clear the tree line and be able to "see" the distant antenna located at the wireline end of the system. The higher the frequency, the more important it becomes to have line of sight between the two points, as signals become attenuated (diminished) by radio signal absorption from trees or other features of the landscape. Low-frequency units are affected less by local landscape issues and may be the only system that will work in a given area.

Figure 15-7. If the end-of-the-line telephone poles are on private property, you may be able persuade the landowners to allow you to use their existing utility pole and grid-supplied power to operate the radio transceiver unit. A bottle of wine at Christmas will more than cover the electricity charge. Alternatively, you may use a photovoltaic panel and battery bank as seen in this photograph.

Figure 15-8. Phone line extenders require a radio transceiver box, battery, charger, and antenna at the end-of-the-line phone service. Your local phone company may allow you to install the transceiver unit on one of their poles, with the electrical utility providing power to operate the unit.

Insulated enclosure for transmit unit and solar power/ batteries for Optaphone

Inside dimensions are 28" wide, 20" tall, and 9" deep.
Back panel should allow mounting of voltage regulator.
Unit has a locking handle with door to prevent vandalism.
Unit must be mouse/animal proof.

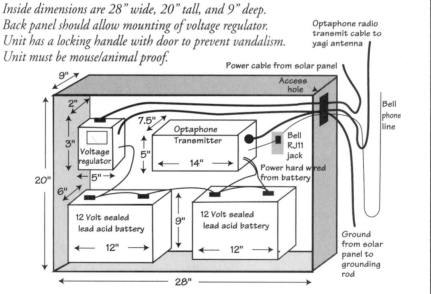

Figure 15-9. Mechanical view of the insulated enclosure for the radio transceiver unit and PV-charged battery pack.

At the home end, the transceiver base station is powered just like any AC or DC appliance. (This is an excellent example of a necessary phantom load.) As telephone service is normally desired on a 24/7 basis, it is wise to connect the base station directly to the battery bank through a dedicated DC circuit at the appropriate voltage. If the transceiver is not designed for direct connection to the house battery bank, a DC to DC voltage converter (See Voltage Converters at the end of Chapter 11) may be installed to power the unit directly via its DC voltage input connection. Alternatively, a small phantom load inverter or even the regular house inverter can be used to power the unit.

At the remote location you will have to deal with how to provide power to the unit. If the unit is located at a neighbor's home, power should not be a problem, provided a power line extension from their existing service is allowed.

If you decide to locate the unit in a remote utility box, chances are that you can purchase power from your local electrical utility, since where there are phone lines there *may* be electricity as well. You will have to contact the electric utility and arrange to pay the expense for this new service, requiring a meter and service outlet. A drawback to this system is that phone service will also be affected by power outages and radio service interruptions.

An alternative to powering the remote phone transceiver with grid power is to make your own electricity, just as you are doing at your home. The communication company that sells you the phone system should be able to help you with this. In the installation shown in Figures 15-6 through 15-10, power is provided by one small photovoltaic panel sized to accommodate the system energy requirements. In this example, PV-supplied electricity charges two gel-cell sealed lead acid batteries. The batteries are surrounded with closed-cell insulation to allow the batteries to

Figure 15-10. An alternative to powering the remote phone transceiver with grid power is to make your own electricity, just as you are doing at your home. A small photovoltaic panel and suitable battery bank will easily do the job. Your local radio system provider will be able to provide the necessary design (and possibly components) for this project.

remain as warm as possible.

This installation is based on the system being able to store enough DC power in the batteries to run the phone system for four hours per day for two weeks with no sun, especially during the cold winter months (when the batteries will be cold and not perform as well). In the six years this system has been in operation, sometimes well in excess of four hours per day, there has always been enough power.

The utility box also contains a series type solar charger (Chapter 10) similar to those used in a home, as well as the service box where the phone company supplies their connection. The directional yagi antenna is grounded to two 8 ft grounding rods, providing necessary lightning protection. The three lines that come into the box (the phone line, the power line from the solar panel, and the antenna feed), all form a "U" or drip line before they enter the box. This allows water running down the lines during a rainstorm to "drip" off rather than run into the electrical equipment. Keeping water out of this utility box is absolutely essential.

• High-Frequency Phone Service Extender Systems

High-frequency (2.5 GHz. or 2,500 MHz) phone service extender systems are similar to the low-frequency models described above, with the exception that radio signals are more easily absorbed than with low-frequency (approximately 400 MHz) models.

In the 2.5 GHz range, obstacles as small as raindrops can interfere with the signal, so it is imperative to have good "line of sight" between the two antennas. This means that if you're at the top of one of the antennas you should be able to see the top of the other antenna with nothing impeding the view.

An advantage of these newer, high-frequency systems is that they provide greater "bandwidth," which will allow faster Internet service. Remember that the system will only be able to provide Internet speeds as fast as that provided by the phone company, and rural service is often much slower than it is in an urban area. The further you are from a major telephone trunk line, the slower Internet access will be.

Fixed-Point Broadband CDMA (Digital) Wireless Service (Provided by the Phone Company)

Your local phone company may actually offer a service similar to the phone extender technology described above but capable of servicing multiple customers. The advantage to the phone company is that it will save the cost of stringing copper phone cable along "the last mile" junction between high-speed phone and data trunk lines the rural home. As the cost of CDMA/wireless technology continues to drop and the cost of installing utility poles goes up, you may find your phone company receptive to installing a system for you.

Radiotelephone Service

Yet another technology to consider is phone company-supplied radiotelephone service, a glorified walkie-talkie system that allows connection between remote sites and a powerful radio tower operated by the phone company. These systems fall into two categories: operator assisted or automatic.

Manual or operator-assisted systems require the remote location to press a "call" button, which alerts a mobile operator to acknowledge your request. As in the old days, you provide the operator with the number you wish to call and the connection is completed. Automatic systems replace the operator with a touch-tone keypad, allowing direct access to the desired number.

In either system, the radio frequency is shared by a number of users in a county-wide party line system. Although radio reception quality ranges from "OK" to "Would you repeat that please?", for many people this may be the only economic means of communication.

Satellite Phone Service

If your dream home is in the wilds of Alaska one further communication system to consider is the satellite phone. You may already be aware of these units installed in the seat backs of planes or used by journalists covering news stories from remote reaches of the globe.

These remarkable phone units are capable of operating anywhere in the world and provide email and Internet connection as well. Another amazing feature of these systems is their ability to drain your bank account. One model was recently priced

Figure 15-11. If your dream home is in the wilds of Alaska or beyond the range of cellular or extender technology, one further communication system to consider is the satellite phone. These remarkable phone units are capable of operating anywhere in the world and provide Internet connection as well. (Courtesy Globalstar USA, LLC.)

at U.S. $700. The airtime fees ranged from $2 per minute for a 50-minute block to $6,000 per year for the 10-hour "economy" package.

I'm sure Bill Gates has a couple of these babies lying around the yacht!

Wide Area Wireless High-Speed Internet Service

With the demand for high-speed Internet access booming, many technology companies are scrambling to find ways to bridge the "last mile." This term was coined to reflect the difficulty of offering high-speed, or broadband, Internet data service down traditional telephone wires, many of which were designed over 100 years ago simply to carry the sound of a human voice.

Over the years, the major telephone companies have been upgrading their line circuits from copper wires to glass optical fibers and digital laser technology. Pumping light down an optical fiber is not only less expensive than using traditional copper, it is faster and less subject to signal degradation. Although the optical fiber network stretches across vast miles of land and sea, where it doesn't go is what counts to most of us: the last mile.

The last mile refers to the distance from your home to the "local central office" (often just a box along the highway), which contains the termination point between the optical fiber and the copper wire lines serving your house. The cost to upgrade the last mile is prohibitive.

Cable TV companies have already bridged the last mile, offering Internet service that dovetails nicely with their primary business of delivering television programming, which requires data bandwidth similar to that required for Internet data. Unfortunately for rural off-grid dwellers, they don't get cable service either.

Figure 15-12. With the demand for high-speed Internet access booming, many technology companies are scrambling to find ways to bridge the "last mile" between high-speed data trunks and your home. If you are lucky enough to be within the broadcast area of high-speed wireless service, you may have solved your phone and Internet problems in one shot. (Courtesy Storm Internet, www.storm.ca)

Recognizing the last mile dilemma, companies have developed wide-area wireless high-speed Internet that is designed to bridge the last mile—and then some. If you're lucky enough to be in the "and then some" category, you're in luck.

Wide-area wireless service requires the service provider to install a communication antenna and radio on a high point in the local area, usually a municipal water tower. Anyone within visual line-of-sight of the tower (and within a given signal propagation distance) may erect a small, low-cost semi-parabolic dish such as the model shown in Figure 15-12.

The service provides exceptionally high-speed service (generally up to 3 megabits per second in both directions) for about the same cost as internet over cable technology.

As the wave propagation is circular, signal strength may allow reliable connection within a 6 mile (10 km) radius or more from the transmitting tower.

Anyone who pays attention to the news and Internet reports will no doubt be aware of a relatively new phenomenon known as Voice over IP. This technology allows telephone calls using an Internet connection and will be discussed later in this chapter, as it applies to both wide-area wireless and satellite Internet services.

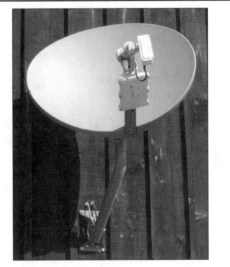

Satellite Internet Service

Many rural homeowners receive outstanding television reception and channel selection through their satellite dish and receiver. Some satellite television companies offer Internet "down" service from their satellites while requiring the client to use a phone line and regular modem to "upload" data to an Internet Service Provider (ISP). Obviously, this is not a workable solution unless you have access to wireline phone service and will therefore not be applicable to most off-grid homeowners.

The alternative is to use a specialized high-speed satellite

Figure 15-13. Many rural homeowners receive outstanding television reception and channel selection through their satellite dish and receiver. Across North America, the cousin to satellite TV reception is a specialized high-speed satellite bi-directional service provider to stay connected with the world.

bi-directional service provider. Powerful radio and computing equipment on board geosynchronous satellites enables Internet service providers to offer high-speed bidirectional Internet connection. Most systems provide 0.5 megabits per second download speed with uploads operating at between 50 and 100 kilobits per second. This is about 15 times faster than dialup for the download direction and between 2 and 4 times faster during uploads.

The electrical requirements for the dish and support electronics are quite reasonable, typically in the order of 15 Watts, or about the same as a compact fluorescent lamp. This level of energy consumption fits well within the energy budget of most off-grid homes. As with any computer and monitor, ensure that the power supply feeding the dish is connected to a power bar or room switch that will turn all the equipment off (no phantom loads) when your work is complete.

Voice Over Internet Protocol (VoIP) Service

The concept of Voice over Internet Protocol (VoIP) is easy enough to understand. It's unravelling the hype from the facts that takes a bit of work. A microphone digitizes a person's voice, converting it into a series of digital data in much the same manner as compact discs or MP3 players convert music into data. Using a reasonably fast Internet connection, the voice data is transmitted to the receiving PC and converted back into sound so that it can be played over a sound card or dedicated Internet phone or headset.

A system that relies on communications between two computers as in the above example is known as peer-to-peer communication. Currently, the hands-down winner in the race to replace the telephone with VoIP technology belongs to Skype

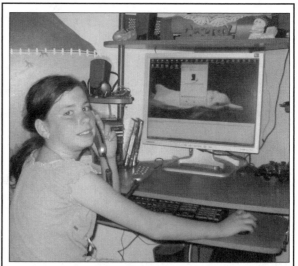

Figure 15-14. Deseray O'Rielly makes Voice over Internet Protocol (VoIP) look like child's play. A system that relies on communication between two computers is known as peer-to-peer communication, with Skype (www.skype.com) the hands-down winner in the race to replace the telephone.

(www.skype.com). This European-based company offers a free software package by the same name which allows free PC to PC calls to anywhere in the world. Skype is able to provide this software without charge because there is no required infrastructure other than the existing Internet connection, a PC, and either a headset and microphone or other speaker/microphone package.

But it doesn't stop there. Peer-to-peer calling is fine if both users are on-

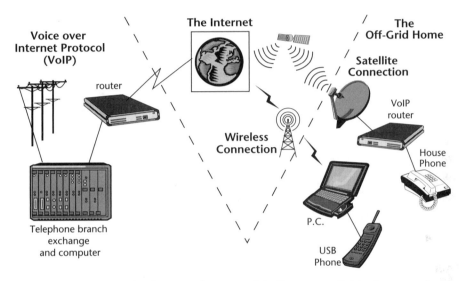

Figure 15-15. VoIP technology uses the power of the Internet to digitize human voice in the same manner as MP3 music is recorded. The voice data is then routed in real time to another computer user (peer-to-peer) or though a service provider to any telephone on the planet.

line and running the appropriate software. But what if you are in Vancouver and you want to call Mom who is holidaying in Florida? Skype is currently running another program known as SkypeOut that allows the calling PC to route the call through the traditional wireline phone system. Because Skype is "renting" the phone system to complete the call, users are charged a flat rate of approximately two cents per minute for calls to North America and Western Europe. Other areas are being added to the system, albeit at differing and higher rates.

At the time of writing (June 2005) Skype is beta testing a product known as SkypeIn, which provides a phone number that people can call from any standard telephone. As with regular Skype, the long distance carrier is the

Internet, allowing the company to offer this subscription service for a very reasonable fee of under $40 per year. Skype is certainly not overcharging for its services, but not everyone wants to be tied to the computer when making a phone call. Additionally, Skype doesn't currently work with cell phones or handheld devices, which may limit widespread adoption of the technology.

And Skype is not alone. Other firms such as Vonage and Net2Phone believe that Skype will not become mainstream because of the peer-to-peer nature of the software and the need to operate on a PC.

Figure 15-16. A similar service to Skype is iConnect, which uses a special USB telephone or Plantronics-style headset and microphone. (Courtesy iConnect.com)

Vonage offers a $20 VoIP box (Figure 15-17) which connects directly to the Internet via a cable/DSL modem and router. This tiny unit connects to the house phone line and provides an invisible connection to the outside world. Unlike Skype, the Vonage unit allows calls to be made with any touch tone-phone and to any phone on Earth. The party you are calling doesn't need to use the Vonage service; nor do they require an Internet connection. If both parties use the Vonage service, there is no charge to either party.

Vonage has virtually duplicated the features of regular landline and cellular calling by providing a host of features such as "911" emergency service, call waiting, call forwarding, etc. And with low minute rates for long distance (non-Vonage system calls), why not sign up immediately?

Figure 15-17. Vonage offers a VoIP box which connects directly to the Internet via a cable/ Digital Subscriber Line (DSL) modem and router. This tiny unit then connects to the house phone line and provides an invisible connection to the outside world.

Does VoIP Technology Work?

Because people have always assumed telephone service to be an essential part of life, it may seem strange to think of it as a byproduct of the Internet. But technology has a way of shaking things up, and VoIP technology is no different.

To test the suitability of VoIP technology in an off-grid home, I decided to recruit some serious technical help from a computer guru friend, Ken O'Rielly. Ken works as a software/hardware developer and runs two Web sites that offer product reviews and critiques of technology equipment manufacturers. Many of the world's technology companies such as Samsung and Intel go to Ken for his review advice (see www.envynews.com and www.gadgetlife.org).

"It is pretty clear that testing VoIP technology for the off-grid home is going to be slightly different than for a typical urban home," remarked Ken, as he started thinking about the project. "Firstly, an off-grid home is not likely to have access to DSL or cable service. Secondly, if an off-grid home has a phone line extender or cell service, there is not much point in using VoIP technology since an extender would have insufficient speed and cell service would be more expensive than the VoIP technology it would interact with."

Ken decided that since some off-grid residents might have access to cable, DSL, or high-speed wireless service it would make sense to test these technologies to provide a benchmark for sound quality and ease of operation.

"When using a peer-to-peer connection, regardless of which broadband technology was used the end result was that Skype and other VoIP technologies were no different than speaking with someone on a regular telephone," Ken explained. "I tried the technology with people all over the United States and Canada and found no flaws or problems with the technology. The USB phone that was supplied by Radio Shack for testing was a nice touch, but I found the software (iConnect) didn't work well with the self-contained keypad. There were some compatibility issues with the computer motherboards I tested (or more correctly, with the USB chipsets used to communicate with the telephone). Windows was unable to recognize the device and considerable tinkering was required to get it going." Ken also noted that Skype works on both PCs and Macs.

When Ken was able to get the USB phone working (and he is sure the problems are very short term), it did provide a sense of familiarity for non-technical people. It also provided a very clear voice signal for both users.

The story changed considerably when we switched the Internet link to one user on cable (urban) and the second user on satellite high-speed (off-grid). "This is a more realistic connection for people living off-grid," Ken said. "The reason people are off-grid in the first place is because of the distance from the power lines, and cable suppliers aren't likely to extend their lines for a few bucks per month."

Ken connected to Cam Mather, who lives off-grid about 15 miles (24 km) from the nearest power line. Cam uses a satellite service from Xplornet (www.xplornet.com) which provides him with 500 kbps download and 100 kbps upload speed, using the large-format 39" (98 cm) dish.

"While speaking with Cam, roughly 75% of the conversation I heard was perfect. The 25% that was unacceptable was either garbled or choppy and I had to constantly ask Cam to repeat himself," said Ken. "At Cam's end, I was coming in perfectly clearly and he had no problems at all. This makes sense because of the download speed of the satellite. One problem that did occur with the satellite system was the inclusion of an approximately three-second delay after one of us spoke before the other would hear it. We opted to play "trucker" and use the old CB radio slang, saying "over" each time one of us finished talking."

Figure 15-18. It looks like off-grid folks are going to have to wait a little while longer before VoIP technology comes to the (off-grid) back country.

Cam was unable to get SkypeOut operating at all, so we opted to test the Vonage technology and phone adaptor shown in Figure 15-17. The purpose of this test was to see if this North American technology would operate from Cam's satellite internet system while dialing out to a regular landline (see Figure 15-15 for details of Vonage connection to landline service). As Ken had experienced with the peer-to-peer technology, I was sure that Cam and R2D2 were calling me from space. Cam could hear me fine, but one-way telephone conversations aren't worth a dime to me.

Cam discussed this matter with Xplornet customer service, and after he had explained that he had the "upgraded" dish and enhanced data package the response finally was: "Unfortunately we don't support the Vonage VoIP because of the latency and upload speeds."

It looks like off-grid folks are going to have to wait a little while longer before VoIP technology comes to the back country.

Conclusion

If you are looking to move away from utility lines, make sure you have a phone strategy and ensure that it will work before you take the plunge. More and more people work out of their homes, or are dependent on phone and Internet service for many of their daily activities such as banking and paying bills. While the dream to "drop out" to Walden Pond remains an admirable one, most of us still require some interaction with the industrial economy, and for most of us the phone system is an integral part of that link.

16
Biofuels

16.1 Biofuels Introduction

Humans have very short memories. If you were of driving age in 1973 you will no doubt recall the long lineups and short supply of gasoline at your local filling station. Even if gasoline was available, OPEC raised the price of oil from $4.90 a barrel to $8.25 a barrel in that year. None of this had anything to do with diminishing world oil supplies, but came about as a result of the American support of Israel and the misguided energy policies of the Nixon administration.

Figure 16.1-1. Misguided energy policies of the Nixon administration coupled with the Arab oil embargo over American support of Israel led to the severe oil supply shortages in the early 1970s. (Courtesy U.S. National Archives and Record Administration)

The demand for energy in the United States and Canada is insatiable, and it appears no one has learned from the mistakes of the 70s. Look around: Sport Utility Vehicles (SUVs) and minivans abound, with the result that fuel economy is nearing an all-time low. At the same time, the square footage of the average North American home has more than doubled and energy efficiency codes have been relaxed, requiring additional energy for heating and air conditioning.

Over the same period domestic oil supplies in the United States have dropped to an insignificant amount of total demand, requiring vast quantities of imported oil to make up the deficit. Current world consumption is around eighty million barrels of oil per day and rising quickly because developing countries are demanding their share of the energy pie. Even though the majority of the world's oil is supplied by regions that are politically unstable and often unfriendly to the West, energy dollars are being exported as fast as the U.S. Treasury Department can print them. According to an October 25, 2003 report in The Economist, OPEC has drained the staggering sum of $7 trillion from American consumers over the past three decades. This massive amount does not even include industry subsidies, cheap access to government land for oil extraction, or military security required to get the sticky stuff safely into North America. If this money had been pumped into solar energy and fuel cell and hydrogen technologies, the internal combustion engine would be as common as a woolly mammoth.

Many North Americans are astounded at the high cost of gasoline in Europe. It is interesting to note that the price of oil (and hence gasoline) is a world price. European countries decided long ago that taxing the daylights

Figure 16.1-2. The United States has put itself in the very difficult position of having to rely almost exclusively on imported oil to keep its economy running. Unlike European governments, decades of North American administrations have not had the backbone to improve the Corporate Average Fuel Economy (CAFE) laws or to tax gasoline in order to reduce consumption.

out of transportation fuel would improve energy efficiency and reduce both the size of motor vehicles and the number of needless miles driven. As prices exceed and are beginning to nudge $50 per barrel, up from $11 in 1998, many people will begin to question the wisdom of $80 fill-ups every couple

Figure 16.1-3. Oil prices in the United States are virtually untaxed and don't include the costs of military muscle in the Middle East or damage to the environment. The after-inflation price of gasoline has actually dropped since the oil crisis of thirty years ago, to the point where a quart of bottled water costs more than a quart of oil.

of days and the need for cars as large as my parents' first house.

On the other hand, gasoline prices in the United States are virtually untaxed and don't include the costs of military muscle in the Middle East or damage to the environment. The after-inflation price of gasoline has actually dropped since the oil crisis of a couple of generations ago, to the point where a quart of bottled water costs more than a quart of oil.

Reducing consumption by increasing efficiency is a fairly easy fix (provided you're prepared to do it), but no current government has the guts to impose the regulations necessary to see it through. If you must keep your beloved Hummer or for that matter your fuel-efficient Toyota running, gasoline and diesel fuel will continue to power The American Dream for some time to come.

An Alternative to Fossil Fuel

While the world waits breathlessly (and probably endlessly) for hydrogen-powered fuel cells and advanced electric vehicles, developments in biofuels are continuing at a rapid rate. Almost everyone has heard of ethanol and some enlightened souls may be familiar with biodiesel. While these technologies are not nearly as sexy as hydrogen and fuel cells, they are available now and offer numerous advantages over imported oil.

Modern internal combustion engines, including those used in transportation vehicles, do not have to run on gasoline. In fact, early automotive pioneers did not have access to refined gasoline and used peanut oil and alcohol for fuel. Coincident with the development of the internal combustion engine was the discovery of large amounts of crude oil in the United States. With ready access to this low-cost energy source, the days of the gasoline-powered buggy were upon us.

Fossil fuels in the form of coal, oil, and natural gas originated eons ago as plant matter and marine plankton growing in the ancient world. All living plants use the sun's energy in a process known as photosynthesis to convert atmospheric carbon dioxide into carbon which is stored in the plant structure. The modern fireplace or woodstove can burn firewood and convert the stored carbon back to heat and atmospheric carbon dioxide. When a plant dies and becomes trapped in mud, decomposition is stalled and under conditions of high heat and pressure fossilization takes place, forming coal. Marine plankton follows a similar fate, converting into oil and natural gas.

With world oil and natural gas supplies reaching their peak outputs and demand continuing to rise unabated, alternative and clean energy sources are required immediately. Some fuels such as waste oil-based biodiesel can even be produced at home.

The following sections will describe the two most common biofuels and help you move towards a cleaner, low-carbon future.

16.2 Ethanol-Blended Fuels

Fossilization is not the only means of extracting energy from plant life. Fermentation of grapes and apples has been fuelling alcohol-induced binges for as long as man can remember. The naturally occurring sugars in the fruit produce wine and cider with a maximum alcohol content of approximately 12%. Applying heat to these beverages, in a process known as distillation, allows extraction of the alcohol at up to 100% concentration.

The primary source of plant sugars can vary, as automotive-grade ethanol can be produced from grains such as corn, wheat, and barley. Recent advances in enzymatic processes even allow the conversion of plant waste in the form of straw and agricultural residue into sugars which can in turn be fermented into ethanol.

Conventional grain-derived and cellulose-based ethanols are the same product and can be easily integrated into the existing gasoline supply chain. Ethanol may be used as a blending agent or as the main fuel source. Gasoline blends with up to 10% ethanol can be used in any vehicle manufactured after 1977. High-level ethanol concentrations of between 60% and 85% can be used in special "flex-fuel vehicles." At the time of writing Ford, General Motors, and DaimlerChrysler warranties allow up to 10% ethanol blends in their standard North American vehicles and 85% concentrations in their flex-fuel products.

Adding ethanol to gasoline increases octane, reducing engine knock and providing cleaner and more complete combustion, which is good for the environment. Compared with gasoline, ethanol reduces local greenhouse gas emissions: a 10% ethanol blend with gasoline (known as E10) will reduce GHG emissions by 4% for grain-produced ethanol and 8% for cellulose-based feedstocks. At concentrations of E85, GHG emissions are reduced by up to 80%.

Greenhouse gas emission reduction was abundantly demonstrated during a recent 6,000-mile (10,000 km) driving tour conducted by Iogen Corporation. An SUV fuelled with 85% cellulose ethanol produced the same GHG emissions as a super-efficient hybrid car.

In addition to making environmental improvements, ethanol is produced from domestic renewable agricultural resources, thereby reducing our dependence on imported oil. Even at low ethanol/gasoline concentration levels, ethanol production is a major source of economic diversity for rural farming economies.

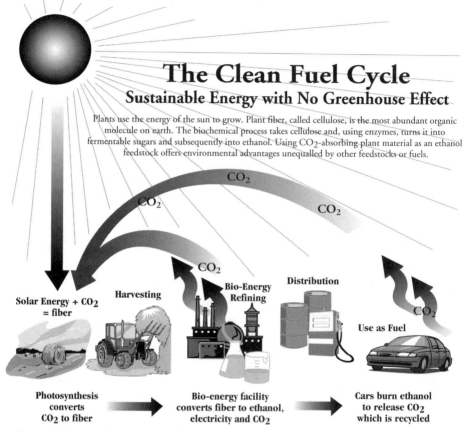

The Clean Fuel Cycle
Sustainable Energy with No Greenhouse Effect

Plants use the energy of the sun to grow. Plant fiber, called cellulose, is the most abundant organic molecule on earth. The biochemical process takes cellulose and, using enzymes, turns it into fermentable sugars and subsequently into ethanol. Using CO_2-absorbing plant material as an ethanol feedstock offers environmental advantages unequalled by other feedstocks or fuels.

Figure 16.2-1. Plant grains and fibre can be converted to sugar, fermented into ethyl alcohol (ethanol), and used as a blending ingredient with gasoline or as the main fuel. High concentrations of ethanol can reduce greenhouse gas emissions by up to 80% relative to gasoline. (Courtesy Iogen Corporation)

Figure 16.2-2. The majority of ethanol is derived from corn. The production of ethanol fuels strengthens agricultural regions and creates new and more stable markets for the farming community.

In the United States, 77 ethanol plants produce over 3.3 billion gallons (12.5 billion liters) of ethanol per year. Canadian production currently stands at 62.6 million U.S. gallons (237 million liters) per year. Of the 77 plants in the United States, 62 use corn as the feedstock. The remainder use a variety of seed corn, corn and barley, corn and beverage waste, cheese whey, brewery waste, corn and wheat starch, sugars, corn and milo, and potato waste. In the United States there are 55 proposed new plants and 11 currently under construction. Canada currently has 6 plants with 1 under construction and 8 new proposals on the drawing board.

There is no question that using domestically grown feedstocks from renewable and clean-burning grains is better than importing fossil fuel from the Middle East. However, there are long-term concerns about diverting food stocks into fuel feedstocks to continue our love affair with the automobile, thereby contributing to urban sprawl and other societal problems. Additionally, the "energy-budget" required for growing, fertilizing, harvesting, and producing food-based ethanol requires more energy "input" than is derived from the ethanol fuel created, leading to problems with supply sustainability and life-cycle GHG reduction.

A reduction in total human population and automotive usage is a key long-term goal, but it is fraught with difficulty in the short term. Until urban planners can wrest the car keys from suburbanites, society might as well use all the clean "transition" fuels at its disposal.

Cellulose ethanol eliminates the diversion of food crop to fuel feedstocks and is an advanced new

Figure 16.2-3. North American ethanol production currently stands at 3.36 billion gallons (12.72 billion liters) and is expected to increase by 28% before the end of the decade. This is just a drop in the bucket compared with fossil-fuel gasoline usage, but with good demand management (through CAFE and properly managed taxation) it could go a long way toward reducing reliance on imported oil. (Courtesy Iogen Corporation)

transportation fuel which has some advantages over grain-based ethanol:

- Unlike grain-based ethanol production, the manufacturing process does not consume fossil fuels for distillation, further reducing greenhouse gas emissions.
- Cellulose ethanol is derived from non-food renewable sources such as straw and corn stover.
- There is a potential for large-scale production since cellulose ethanol is made from agricultural residues which are produced in large quantities and would otherwise be destroyed by burning.

Figure 16.2-4. Adding ethanol to gasoline increases octane, reducing engine knock and providing cleaner and more complete combustion. In addition to performance and environmental improvements, ethanol is produced from domestic renewable agricultural resources, reducing dependence on imported oil and providing a major source of economic diversity for rural farming economies.

Cellulose-based ethanol has, in the past, been very expensive as a result of the inefficient processes required to produce it. The Iogen Corporation, working in conjunction with the Government of Canada and Shell Corporation, has recently launched a cost-effective method for producing what the company refers to as EcoEthanol^{TM}. Once the first phase of production, using straw and corn stover, is under way, it won't be long before other sources can be used, such as wood-processing byproducts or other feed stocks like hay or fast-growing switch grass.

Ethanol Considerations

Aside from the net-energy loss of converting food crops to fuel, there are very few downsides when comparing ethanol fuel to gasoline. Because ethanol contains oxygen it permits cleaner and more complete combustion, which helps keep fuel injection systems deposit free. Be aware, though, that ethanol acts as a solvent and can loosen residues in your car's fuel system, necessitating more frequent fuel filter changes.

Gasoline containing 10% ethanol has approximately 3% less energy than regular gasoline. However, this loss of energy is partially offset by the increased combustion efficiency resulting from the higher octane and oxygen concentration in the ethanol blend. Road tests show that fuel economy may be reduced by approximately 2% when using E10 concentrations. To put this in perspective, driving a passenger vehicle 12 mph (20 kph) over the 60 mph (100 kph) speed limit increases fuel consumption by an average of 20%, by contrast, the 2% reduction in fuel consumption using E10 is insignificant.

Lastly, there may be a slight cost increase for ethanol-blended gasoline in your area. However, compared to the long-term alternatives of the fossil-fuel market, injecting a few pennies per gallon back into the domestic rural economy and helping the environment at the same time is surely the right thing to do.

Figure 16.2-5. Iogen Corporation founder Patrick Foody Sr. stands in a field of straw, which may become the twenty-first century equivalent of a Saudi Arabian oil field. (Courtesy Iogen Corporation)

Figure 16.2-6. Loading straw into an ethanol plant is a lot cleaner and much more sustainable than extracting oil from the North Sea. Waste grasses, straw, and wood products are "carbon-neutral" fuels that are available domestically and in endless supply. Think of these sources as "stored sunshine." (Courtesy Iogen Corporation)

16.3
Biodiesel as a Source of Green Fuel and Heat

Under conditions of extreme heat and pressure, marine plankton are transformed over millennia into crude oil and natural gas. Refining crude oil through a cracking and distillation process produces gasoline, kerosene, diesel fuel, and other hydrocarbon compounds.

Internal combustion engines ignite fuel using one of two methods. A spark ignition engine produces power through the combustion of the gasoline and air mixture contained within the cylinders. An electric spark which jumps across the gap of an electrode ignites this volatile mixture.

The compression ignition or diesel engine, as it is more commonly known, uses heat developed during the compression cycle. (Have you ever noticed how hot a bicycle air pump becomes after a few strokes?) With a compression ratio of 18:1 or higher, sufficient heat is developed to cause diesel fuel sprayed into the cylinder to self-ignite. The higher energy content of diesel fuel as compared with gasoline contributes to improved fuel economy and power, desirable features for fleet and other high-mileage vehicles.

Biodiesel, like its cousin ethanol, is a domestic, clean-burning renewable fuel source for diesel engines and oil-based heating appliances. It is derived from virgin or recycled vegetable oils and animal fat residues from rendering and fish-processing facilities. Biodiesel is produced by chemically reacting these vegetable oils or animal fats with alcohol and a catalyst to produce compounds known as methyl esters (biodiesel) and the byproduct glycerin.

Biodiesel that is destined for use as transportation fuel must meet the requirements of the American Society of Testing and Materials (ASTM) Standard D6751 for pure or "neat" fuel graded B100. Fossil-fuel diesel (or petrodiesel) must meet its own similar requirements within the ASTM standard. Biodiesel and petrodiesel can be blended at any desired rate, with a blend of 5% biodiesel and 95% petrodiesel denoted as B5, for example. The testing and certification of any transportation fuel is a requirement of automotive and engine manufacturers implemented to minimize the risk of damage and related warranty costs.

From an agricultural viewpoint, biodiesel offers many of the same advantages as ethanol in terms of supporting the farming community and contributing to life-cycle energy reduction. The production of biodiesel is an inherently low-energy process, and when produced from waste oils, greases, and animal fats, biodiesel offers superior life-cycle GHG, environmental sustainability and energy advantages compared to food-grade feedstock-produced biodiesel and ethanol.

In addition, the process of making biodiesel is relatively simple and low cost, which may lead to extensive cooperative and rural ownership of processing and production facilities, further increasing farming income and risk diversity. Even the byproduct of the biodiesel manufacturing process, glycerin, has a ready market in the food, cosmetic, and other industries.

By way of example, Milligan Bio-Tech Inc. located in rural Saskatchewan, Canada is a community-owned agri-business venture. The canola seed growers in the region produce a food-grade, crystal clear canola oil for the food industry. Vagaries in the weather may cause the oil seed to produce off-color oil that is rejected by the market. Crop failures are all too common in farming communities, resulting in financial hardship or worse. Progressive thinking in the community led to the creation of Milligan as a way of diversifying crop development and reducing the economic risk inherent in the canola business.

Milligan realized that simply selling the basic seed and food oil would not provide sufficient diversification for the business. Assessing its options, Milligan determined that a mobile seed crushing/oil extraction plant would allow the seed husk to be left at the farm site, providing oil seed meal that could be used as a feed supplement for livestock.

Virgin oils could then be graded and sold as food oil or placed into the biofuel and lubricant processing stream. As discussed earlier, vegetable oil produces biodiesel and the byproduct glycerin. Biodiesel may be sold directly as a transportation fuel or refined into penetrating or lubrication oil. It can also be used as a carrier in diesel fuel additives which reduce emissions and

Figure 16.3-1. Biodiesel fuel is a product of vegetable oils or waste animal fats which produce a renewable, clean-burning fuel source for diesel engines and oil-based heating appliances.

increase lubricity, prolonging engine life. Milligan has even developed a novel use for the glycerin byproduct, creating an ecologically friendly, biodegradable, and long-lasting dust suppressant for gravel roads which can replace salt-based calcium chloride products currently in use.

According to Milligan principal Zenneth Faye, rural communities cannot afford to be dependent on one income stream without taking unnecessary financial risks. Diversification is the key to community growth and sustainability.

It is conceivable that local farming cooperatives will produce high-quality "batch" biodiesel fuel at a production cost lower than that of petrodiesel either by importing oilseed from other regions or by using locally grown crops, as is the case with Milligan. This fuel stock can be produced after the fall harvest when farmers have extra time available, offering a low labor cost. Biodiesel fuel can be winter stored and used at high concentrations (possibly >B50) during the busy and energy-intensive spring/summer/fall growing season.

If the aggregate fuel consumption of the cooperative (or individual farm) is high enough, trading carbon credits acquired as a result of consuming biodiesel can provide further cash income for the farm.

Home brewing of biodiesel has also become somewhat of a national pastime, under the guise of its being a "sustainable, environmentally friendly fuel source." Of course the real reason people make biodiesel is because they can and because it is approximately one-third the cost of petrodiesel. If homemade biodiesel were more expensive than petrodiesel, or more complex to produce, I am sure no one would consider making it; I know of no home brewers of ethanol, for example.

Notwithstanding the hype of home brewing, it is a practice that will continue as long as the cost differential between biodiesel and petrodiesel remains. Section 16.4 will deal with home brewing procedures, safety, and environmental issues in greater detail.

Biodiesel Performance

The modern diesel engine is a far cry from the smoky, anemic model of the 1970s. As a direct result of Mr. Nixon's oil crisis, consumers lined up to purchase Volkswagen Rabbit and Mercedes diesel cars, enticed by fuel economy claims. In addition to being disillusioned by the lack of power and acceleration, consumers found that when the mercury dipped below 32°F (0°C) and a diesel engine wasn't plugged in overnight a bus ride was a sure bet the next morning. Is it any wonder that people thumbed their noses at diesel-powered cars?

Figure 16.3-2. As the price of petrodiesel continues to rise and biodiesel production costs fall, rural communities will produce their own democratic energy to fuel their part of the economy. (Courtesy Lyle Estill/Piedmont Biofuels)

Fast forward to today. Gone are the smelly, smoky, lumbering diesels of old. Witness the new Mercedes E320 family of "common rail, turbo-diesel" engines that offer no "dieseling" noise, smoke, or vibration, achieve superb mileage, and have better acceleration than the same model car equipped with a gasoline engine.

Biodiesel offers some distinct advantages as an automotive fuel:

- It can be substituted (according to vehicle manufacturer blending limits) for diesel fuel in all modern automobiles. B100 may cause failure of fuel system components such as hoses, o-rings, and gaskets that are made with natural rubber, but most manufacturers stopped using natural rubber in favour of synthetic materials in the early 1990s. According to the U.S. Department of Energy, B20 blends minimize these problems. If in doubt, check with your vehicle manufacturer to ensure compliance with warranty and reliability issues.

- Performance is not compromised using biodiesel. According to a 3 ½-year test conducted by the U.S. Department of Energy in 1998, using low blends of canola-based biodiesel provides a small increase in fuel economy. Numerous lab and field trials have shown that biodiesel offers the same horsepower, torque, and haulage rates as petrodiesel.

Figure 16.3-3. The simplicity of a biodiesel production facility is shown in this aerial view of a continuous production plant. Soybeans are delivered to the processing plant for conversion to food-grade oil. Alcohol and a catalyst are added to the oil to produce biodiesel and glycerine. The biodiesel is stored in the tank farm. The glycerine byproduct is sold while the water and alcohol are recycled, creating an environmentally friendly processing cycle. (Courtesy West Central Soy)

- Lubricity (the capacity to reduce engine wear from friction) is considerably higher with biodiesel. Even at very low concentration levels, lubricity is markedly improved. Reductions in the sulphur levels of petrodiesel to meet new, stringent emissions regulations have, at the same time, reduced lubricity levels in petrodiesel dramatically. Biodiesel blending is currently being considered by the petrodiesel industry as a means of circumventing this problem.

- Because of biodiesel's higher cetane rating, engine noise and ignition knocking (the broken motor sound when older diesels are idling) are reduced.

Cold Weather Properties

Automotive petrodiesel is supplied as low-sulphur No. 1 or No. 2 grade. Home heating fuel is similarly graded, although it contains higher levels of sulphur. (It may come as a surprise to many, but home heating oil is essentially the same stuff that is burned in road vehicles). No. 2 grade is the preferred fuel as

Figure 16.3-4. Over 200 public and private fleets in the United States and Canada currently use biodiesel, and the number is increasing rapidly. Environmental stewardship regarding climate change, urban smog, and air quality is creating mass-market acceptance of biodiesel fuel. This bus, equipped with a bike rack, allows commuters to use a combination of public transit and peddle power for their commute. (Courtesy Kingston Transit)

it has a higher energy content per unit volume, giving better economy whether powering a truck or heating a house. Unfortunately, when temperatures drop No. 2 is subject to an increase in viscosity known as "gelling."

The cold weather performance of biodiesel is similar to that of petrodiesel No. 2, although its pour point temperature (the temperature at which it gels) is slightly better than No. 2 at -11°F (-24°C). Since No.1 has a pour point of approximately -40°F (-40°C), fuel suppliers blend No.1 and No.2. diesel in ratios appropriate for the geographical location and time of year. To prevent gelling of No. 2 petrodiesel or biodiesel blends, it is recommended that the fuel be stored in an underground or indoor tank if the fuel is subjected to extremely cold weather conditions. Alternatively, your fuel supplier may be able to blend the appropriate mix of No.1 and No. 2 and biodiesel fuels for outdoor storage.

World Energy Alternatives, LLC conducted a number of tests on cold weather performance in the winter of 2002-2003. Working with distributors throughout the United States, the company delivered B20 blended fuel to Malmstrom Air Force Base in Montana, Warren AFB in Wyoming, Peterson AFB in Colorado Springs, and the Winter X Games in Aspen. World Energy biodiesel was introduced to Toronto Hydro in 2001, the first Canadian commercial fleet to use the fuel to reduce urban smog. Biodiesel is also used to fuel the generators and snowcats at Buttermilk Mountain ski resort in Aspen, Colorado. World Energy indicates that there hasn't been a single complaint or operational problem with their biodiesel fuel, even in these cold-weather locations.

Figure 16.3-5. World Energy Biodiesel is delivered to Aspen ski area to fuel the generators and snowcats at Buttermilk Mountain. Numerous ski and eco-sensitive areas are economically and responsibly using biodiesel to meet greenhouse gas emission targets and promote sustainable fuel options. (Courtesy World Energy Alternatives, LLC)

Rev Up Your Furnace

According to the U.S. Department of Energy, heating fuel oil consumption in the United States is currently hovering around 7 billion gallons (26.5 billion liters) per annum, with the vast majority supplied by imported oil. Blending a mixture of 20% biodiesel with No.2 would reduce fossil-fuel consumption and replace that amount of petrodiesel with domestic, renewable, clean-burning biodiesel. If this strategy were adopted across the United States, biodiesel consumption would increase by 1.4 billion gallons per year (5.3 billion liters per year), reducing carbon monoxide, hydrocarbon, and particulate emissions by approximately 20% while supporting domestic agricultural economies.

Are you ready to start using biodiesel? The downside is that finding biodiesel can be challenging. In the Northeastern U.S., suppliers are starting to fill this niche market, albeit slowly. Many distributors purchase biodiesel from large producers such as World Energy Alternatives, LLC (www.worldenergy.com) or enter a purchasing (and often producing) cooperative such as Piedmont Biofuels of Chatham County, North Carolina (www.biofuels.coop). According to Piedmont partner Lyle Estill, production limitations and price can be a deterrent to many people. Nevertheless, Piedmont cannot keep up with demand even though it charges $3.50 per gallon ($0.92 per liter) for B100. Of course, if you purchase a B20 blend the price difference compared to petrodiesel is quite small while the environmental improvements are still considerable.

Figure 16.3-6. Biodiesel comes to the masses. The author (seen grinning enthusiastically, rear) was the first Canadian to fill up with commercially produced biodiesel at this pumping station in Toronto, Canada, in March 2003. Although North Americans are just beginning to use biodiesel, Europeans have been filling up on it for some time now.

Figure 16.3-7. Blending a mixture of B20 (20% biodiesel to 80% petrodiesel) would reduce carbon monoxide, hydrocarbon, and particulate emissions by approximately 20%. In addition, biodiesel would assist rural domestic suppliers and help keep North American money at home. (Courtesy West Central Soy)

For information regarding supply of biodiesel consult the National Biodiesel Board in the United States at www.biodiesel.org (1-800-841-5849) or, in Canada, the Biodiesel Association of Canada at www.biodiesel-canada.org.

Biodiesel acts as a weak solvent which will soften natural rubber hoses, seals, and gaskets over time. If your furnace was manufactured after 1990 it is likely that these items are made with artificial rubbers such as neoprene which are not damaged by the solvent action. If you are uncertain, contact a furnace service company for advice.

The solvent action of biodiesel will also loosen deposits of "crud and sludge" in the tank and fuel lines. If there is a large amount of this buildup present, it will dislodge and eventually plug the fuel filter. For this reason, studies recommend that you switch to the desired biodiesel blend level (i.e. B20) immediately, rather than building up the concentration over time. This allows the deposits to be released quickly, requiring only one or two fuel filter changes rather than many changes over a period of time.

Biodiesel Considerations

Before you start filling up the furnace fuel tank, take a few moments to review some of the important issues relating to biodiesel B20 as a heating source:

Advantages of Biodiesel
- contains less than 15 ppm sulphur (an ultra-low sulphur fuel)
- 100% biodegradable
- less toxic than table salt
- high flash point—over 260°F (121°C)
- energy content comparable to No.2 diesel fuel
- reduces furnace, heater, or boiler nozzle cleaning
- works in almost any oil-fired appliance that does not have natural rubber seals and pipes

Disadvantages of Biodiesel
- poor availability
- more expensive than regular petrodiesel heating oil
- gels in very cold weather (similar to No.2 heating oil) and must be blended with No.1 and No. 2 heating oil or stored indoors or underground. (Large suppliers will provide the biodiesel pre-blended to suit your climatic conditions.)

- acts as a solvent, removing "crud and sludge" inside fuel storage tanks and lines. Keep spare fuel filters handy and change them a couple of times when first making the conversion to biodiesel-blended fuel.

Summary

Biodiesel fuel will not solve all of North America's transportation and heating supply issues single-handedly. Other energy options will also be required. But since biodiesel is a direct replacement of or blending agent with regular petrodiesel or heating oil, no infrastructure or societal changes are necessary in order to start using this fuel (unlike the "hydrogen economy" concept that would require trillions of dollars to build new distribution infrastructure). Biodiesel provides a clean, renewable, domestic and economically diverse energy source which will help pave the way to a more sustainable future.

Figure 16.3-8. Biodiesel can be used in virtually any oil-burning furnace or boiler provided that fuel lines and gaskets are not made of natural rubber. Because biodiesel will gel (like No.2 heating oil), storage tanks should be installed indoors or underground. Alternatively, your fuel supplier can provide winter-blended heating oil containing both No.1 and No.2 heating oil with biodiesel blended to B20 to achieve the desired low-temperature storage performance.

Figure 16.3-9. Many homeowners want to reduce their ecological footprint and are requesting biodiesel heating oil delivery from major suppliers or local cooperatives, as shown here.

16.4 Brew Your Own – An Introduction to "Biodiesel" Production

People are making their own beer and wine, so why not biodiesel? Many of the diehard back-to-the-landers have been producing a "biofuel" similar to biodiesel for years[1] under the guise of "environmental sustainability." Of course if this environmentally sustainable fuel were more expensive than petrodiesel, almost everyone would drop the masquerade and continue filling up at the petro pumps. However, there is a certain allure to going to the local chip truck or greasy spoon to collect waste cooking oil and turn it into heating and transportation fuel. So if you're going to produce homemade biodiesel in order to save money, let's take a look at some of the quality, safety, and cost issues that will directly impact the value and sensibility of this practice.

Figure 16.4-1. Steve Anderson has made so many 27-gallon (100-liter) batches of biodiesel, he has actually lost count. In fact, Steve says it's so easy that a child can make it. Here, Steve and his young helper are demonstrating the fine art of making biodiesel at an environmental fair.

---

1. "Biodiesel" by definition is "a fuel comprised of mono-alkyl esters of long chain fatty acids derived from vegetable oils and animal fats, designated B100 (pure biodiesel) and complying with ASTM fuel standard D 6751." Homemade fuels will not meet the requirements of the ASTM standard and for this reason alone they are not to be considered the same as commercial biodiesel. In addition, many people extract fuel partway along the step-by-step refining process. To distinguish semi-complete biodiesel from the finished product, the term "raw biodiesel" is used to describe the unrefined product as it "moves along" the chemical assembly line. Further discussion will explain why this is such an important issue. All references to "homemade washed and pH corrected biodiesel" will be referred to as "biodiesel" even though it does not have ASTM certification.

Figure 16.4-2. Used French-fry oil is a disposal problem for the owner of the fry truck and a cheap source of energy for the prospective biodiesel maker. Get friendly with your oil supplier and try to get the best quality feedstock possible. Higher quality used oil results in proportionally higher quality biodiesel and is less expensive and time consuming to make.

Before you consider cutting up the gas credit card and switching to homemade "bio" you must consider several factors related to the processing of biodiesel fuel:

- organic chemistry of biodiesel production
- use of low- or high-blend concentrations of biodiesel/petrodiesel
- cost of biodiesel reactor equipment
- quality of fuel
- cost and supply of raw materials
- personal safety issues
- winter driving and cold-flow issues

Understanding the Organic Chemistry of Biodiesel Production

People who wish to use Waste Vegetable Oils (WVO) for transportation and heating fuels fall into two broad categories: those who want to use Straight Vegetable Oil (SVO) and those who wish to produce a petrodiesel-compatible biodiesel fuel.

I will not hide my bias and waste any time on the subject of SVO use in diesel engines and neither should you, unless you are simply playing or doing it for the "tinker" factor. Yes, some diesel vehicles can be modified to operate on SVO with the addition of a second fuel tank, fuel heat exchanger, and switching valve, but why would anyone bother? SVO **will not work** on newer CDI (Common Rail Direct Injection) diesels, so why risk damaging the fuel system and spend several hundred to a thousand dollars to modify the vehicle for this supposed "technology"? If this process had serious potential, you can be sure the automotive or fuel companies would have picked up on it by now. In addition, numerous tests have shown that due to the higher viscosity of SVO, poor combustion, increased emissions, and environmental damage exceed those of regular petrodiesel. If you are serious about producing a quality fuel, biodiesel production is the only way to go.

The production of biodiesel or refined Fatty Acid Methyl Esters (FAME) is a well-known organic chemical process. Although there are several methods of reacting waste oils, by far the simplest and most common is known as the "base catalyzed transesterification of waste vegetable oil with methanol" using the batch process.

Vegetable oil contains a molecule comprising three hydrocarbon chains (esters) that are attached to a molecule of glycerin, forming a triglyceride. WVO is approximately 20% glycerin content by volume.

A white crystalline material known

Figure 16.4-3. The production of biodiesel or refined Fatty Acid Methyl Esters (FAME) is a well-known organic chemical process. Although there are several methods of reacting waste oils, by far the simplest and most common is known as the "base catalyzed transesterification of waste vegetable oil with methanol" using the batch process.

as sodium hydroxide is used as a catalyst to "break" the triglyceride chain during the formation of biodiesel. It is sold as a common household drain cleaner known as lye or caustic soda and is a strong chemical base, the opposite of acid. During the frying process, heat and the addition of water and

Figure 16.4-4. Waste Vegetable Oil (WVO) is often dumped into recycling containers such as this. Never use oil from a bulk recycling container, as the quality, water, and debris content of the oil cannot be controlled. Many restaurants will be only too happy to supply you with oil direct from the fryer.

food particles make the WVO more acidic than virgin oils. A chemical analysis called titration measures this acidity to determine "how used" the waste oils are. The outcome of the titration test indicates the amount of sodium hydroxide required to neutralize the acidity and start or catalyze the transesterification process.

The correct amount of sodium hydroxide is premixed with methyl alcohol or methanol to form a nasty solution known as sodium methoxide, an extremely corrosive base and volatile solution.

The transesterification process begins when the WVO is prewarmed to a temperature not higher than 131°F (55°C). The sodium methoxide solution is added to the WVO and is used to remove the glycerin molecule from the triglyceride and replace it with a molecule of alcohol. As the reaction proceeds, glycerin, which is denser than the esters, sinks to the bottom of the reaction tank. The fluid floating on top of the glycerin is the reacted methyl esters FAME or raw biodiesel. After the reaction and settling time (usually 24 hours),

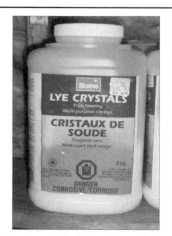

Figure 16.4-5. A white crystalline material known as sodium hydroxide is used as a catalyst to "break" the triglyceride chain during the formation of biodiesel. Sodium hydroxide is sold as a common household drain cleaner known as lye or caustic soda. It is a strong base, the chemical opposite of acid.

Figure 16.4-6. The correct amount of sodium hydroxide is premixed with methyl alcohol or methanol to form a nasty solution known as sodium methoxide, an extremely corrosive base and volatile solution.

the glycerin is drawn off from the bottom of the tank and the raw biodiesel is ready for post-reaction washing, pH adjustment, filtering, and quality inspection.

There is considerable debate among home brewers about the need to wash and chemically adjust the pH of raw biodiesel in order to finish it and create a fuel as close to the ASTM standard as possible. In reality there is no debate at all. Raw biodiesel **must** be washed and treated to remove residual (corrosive) lye and soaps. Modern automotive engines were not designed to run with soap in the fuel system, and given enough time corrosive lye will break down the materials making up the fuel system components. It would be a shame to save a couple of hundred dollars in fuel charges and then prematurely replace a $1,200 fuel injection system.

Washing the FAME is actually a fairly easy process, although it does take a bit of time to do a complete job. Depending on the quality of the WVO and the care taken in ensuring a complete reaction, one to three wash cycles may be required.

A common technique, bubble washing, uses tiny air bubbles

Figure 16.4-7. The transesterification process begins when the WVO is prewarmed to a temperature not higher than 131°F (55°C). The sodium methoxide solution is added to the WVO and is used to remove the glycerin molecule from the triglyceride and replace it with a molecule of alcohol. As the reaction proceeds, glycerin, which is denser than the esters, sinks to the bottom of the reaction tank. The fluid floating on top of the glycerin is the reacted methyl esters (FAME) or raw biodiesel.

formed by passing compressed air through an aquarium air stone. The raw biodiesel is placed in a washing vessel or left in the reaction tank after the glycerin phase has been drained off. Water is added to the tank at a ratio of 1/3 water and 2/3 biodiesel along with a small amount of acetic acid (common white vinegar). The air stone is placed in the tank and the compressor is switched on, causing air bubbles to mix the water and FAME, transferring residues in the biodiesel to the water. After a 24-hour period, the wash water

Figure 16.4-8. Washed raw biodiesel is now ready for testing. Although home testing does not comply with ASTM requirements, many home brewers have achieved satisfactory results.

Figure 16.4-9. Steve explains to a group of curious onlookers that he has personally used thousand of gallons of homemade biodiesel fuel for home and shop heating as well as transportation.

is drained off and the acidity and clarity of the raw biodiesel is checked. This process is repeated as necessary.

Washed biodiesel is now ready for drying and testing. Although home testing does not comply with ASTM requirements, many home brewers have achieved satisfactory results. The home test process may include:

- A visual inspection ensures that the product is not cloudy and that there are no unreacted materials present in the biodiesel.
- pH or acid/base tests are completed to ensure that all unreacted sodium hydroxide has been neutralized by the acetic acid/water wash step.
- A specific gravity or density test is completed to ensure that the viscosity or thickness of the fuel is within limits.
- Viscosity (fuel thickness) is tested with a pipette and compared to that of commercial biodiesel. The biodiesel must have a low viscosity similar to that of commercial biodiesel to ensure fuel system compatibility and atomization into the engine cylinders.
- Gel and cloud point tests are performed to ensure that the fuel meets your requirements for cold-weather flow conditions.
- Final filtering is performed using a 5-micron filter with water extraction feature or agglomerator before supplying fuel to the vehicle.

Assuming that all tests pass your personal requirements, the biodiesel fuel is ready for use.

Use of Low- or High-Blend Concentrations of Biodiesel/ Petrodiesel

North American vehicle manufacturers are a cautious bunch. A recent press release from DaimlerChrysler's Jeep Division announced that diesel vehicles will be shipped from the factory fueled with a 5% (B5) blend of biodiesel fuel. Other vehicle manufacturers are following suit, with some already recommending heady levels of biodiesel of up to B20. Calls to the technical departments of VW of America and Mercedes Benz yielded similar, conservative replies when I asked for their positions on biodiesel. Mercedes indicated that biodiesel fuels were not compatible with their vehicles and that the fuel should not be added even in low concentrations. VW indicated that "they were pleased to extend vehicle warranties for up to B5 on all US-market diesel vehicles." These are interesting comments given that biodiesel has been used in Europe, at up to B100 concentrations, for a number of years now.

Although using B5 blended biodiesel is considered a landmark decision by the major automotive manufacturers, hard-core B100 advocates look askance at this effort. Despite the conservative approach taken by the major automotive manufacturers, thousands of people claim to be operating their cars, trucks, and tractors on B100 without any problems. Why the discrepancy?

To find out, I contacted Mr. Steve Howell, a partner in MARC-IV Consulting. Steve has a long history of working in the commercial biodiesel industry and is the chairman of the ASTM D6751 biodiesel fuel standard. "You really need to have ASTM-certified fuel if you want your engine and fuel system components to operate correctly and over the long term," Steve explains. "Sure, there are lots of people who use substandard fuel, but they are generally operating older equipment that is not equipped with computerized engine management systems. Diesel engines are designed to last hundreds of thousands of miles, and home brew and substandard fuels are going to reduce the life of these vehicles. Time after time after time, at every OEM (Original Equipment Engine Manufacturer), we have found long-term problems from using substandard fuels." Steve goes on to explain that the mono and diglycerides present in homemade biodiesel accelerate the instability of the fuel, causing more filter clogging and fuel injector coking problems and leading to expensive engine overhauls.

Steve goes on to discuss the fact that education and field experience will help change the conservative stance taken by the automotive industry. "The engine manufacturers can be queasy about supporting even a 5% or 10% blend (of biodiesel). Most of their reticence is due simply to lack of knowledge rather than demonstrated issues or potential problems with the fuel. It is really more a problem of being conservative and preventing potential liability issues from developing."

"I have been working with almost all of the major fleet manufacturers to find out what the true impacts of B20 blended into the commercial fuel stocks of North America really are," Steve continues. "The results will be posted on the NBB (National Biodiesel Board www.biodiesel.org) Web site by July 2005. The results won't surprise anyone, as the field results and fuel handling guidelines are the same ones we have been preaching time and time again: quality certified fuels and proper handling practices. I am fairly certain that over the next year or two, we will have across-the-board support for B20."

"There is a huge chasm between B20 and B100. The European fuel standards are designed around B100, while the US specifications are designed for a maximum blending profile of B20. The Europeans have a great deal more long-term experience using B100 than we have in North America. As a result, the fuel specifications in Europe are even tighter than they are here in the US," Steve explains. "Assuming we are still talking about certified biodiesel, the real issues going from B20 to B100 are cold flow issues, higher viscosity (which places additional stress on the fuel injection system, particularly during cold weather), and fuel stability. It is possible to produce and use fuel faster to overcome the problem of stability, or to impede microbial growth through the use of additives. Another problem with B100 is its higher viscosity, leading to atomization and fuel spray pattern problems in the engine. Since biodiesel has a higher vapor pressure than petro-diesel it can spray farther into the engine cylinder, leading to uncombusted fuel and increased atmospheric emissions as well as fuel leakage into the engine oil. Lastly, there are material compatibility issues with seals and other elastomers in the fuel system. Although it is generally true that older vehicles are more susceptible to compatibility issues, many current engine manufacturers have no idea if their fuel system materials are compatible with B100."

Steve maintains that the increasing complexity of today's automobiles is leading to other engine and fuel system issues. "The Europeans are also beginning to see problems with the new common rail diesel systems that are equipped with highly advanced engine management systems. These newer systems monitor engine performance in real time and make adjustments to the fuel spray pattern. With the lower BTU content and higher viscosity of biodiesel compared to petrodiesel, we have seen situations where the engine simply won't run or just sputters along."

Steve concludes our discussion by explaining that many people have used and do use B100 with great success, but caution is warranted. "Using B100 requires a fair amount of technical knowledge and tender loving care. Understand that OEMs in Europe are reversing their trend from supporting B100 and taking the more cautious approach seen in North America."

To summarize all of this information, B100 may or may not work in your application. However, if the fuel is ASTM certified or of the highest quality you can produce, start by using B20 blends and slowly increase the concentration of biodiesel blend. It may require a bit of a leap of faith to get everything right, but remember the old adage that "anything worth doing is always worth doing well."

Cost of Biodiesel Reactor Equipment

The transformation of waste oils and greases into biodiesel is a relatively easy affair which does not require vast amounts of processing equipment or the budget of Shell Oil. Industrial chemical reactors such as the model shown in Figure 16.4-3 are expensive devices, for sure; however, homemade reactors or even simple commercial units are available for the home brewer. How much do they cost? A simple reactor using a water tank and standard plumbing pieces may be manufactured for under $300. The commercial unit such as the model shown in Figure 16.4-10 is available from Biodiesel Solutions for approximately $3,000. Add the cost of scales, pH meter, safety gear, storage tanks, and a secure and ventilated shop and it can be easy to make the sky the limit.

Whatever your budget, just don't skip on safety or quality issues, as that is never an economically sound decision.

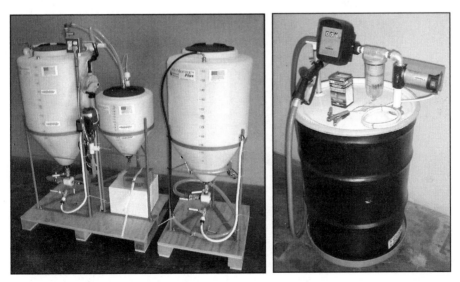

Figure 16.4-10. The Fuel Meister™ by Biodiesel Solutions is an example of a commercially available biodiesel production, washing, and dispensing system that is capable of producing over 40 gallons (151 liters) of fuel with as little as 24 hours of processing time and 1 hour of operator time. (Courtesy Biodiesel Solutions)

Quality of Fuel

Quality control is another concern: ensuring that batch-to-batch processing is consistent and producing a fuel that is as close to certified biodiesel as possible. In order for homemade "raw biodiesel" to be called "biodiesel," certain test criteria have to be met in accordance with the ASTM test standard D6751 for B100 biodiesel fuels. (See www.astm.org for further details.) Homemade biodiesel is unlikely to meet this requirement, and using any fuel without this certification rating could damage expensive equipment and might negate automotive fuel system warranties. It is one thing to pour questionable homemade fuel into a 1978 Volkswagen Rabbit that is worth less than the

Figure 16.4-11. Homemade biodiesel quality control is based on quality WVO, accurate measurement of reactants, and careful adherence to procedures including successive washing and pH adjustment of the final product. Dr. André Tremblay, Professor and Chair of the University of Ottawa Chemical Engineering Department, is seen comparing ASTM and homemade biodiesel.

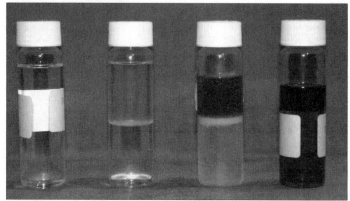

Figure 16.4-12. In this image, ASTM-certified biodiesel is shown at the left and after water washing (to demonstrate its purity) (second from left). Note that the wash water is clear and that there are no soaps at the interface between the biodiesel and the water. The tube shown far right is a sample of homemade biodiesel made from WVO. The sample shown second from right is the same homemade biodiesel after water washing. Notice the residual soaps in the wash water and at the interface layer between the water and "biodiesel." Additionally, the pH level of the biodiesel is a corrosive 10.5, indicating the presence of a considerable amount of unreacted sodium hydroxide.

value of the fuel in the tank; it is quite another to pour it into a shiny new BMW 535d diesel-powered vehicle.

Home brew quality control is based on quality WVO, accurate measurement of reactants, and careful adherence to procedures including successive washing and pH adjustment of the final product. Unfortunately, the ASTM test program assessment is fairly expensive, costing upwards of $1,000 per batch. If you have deep pockets and would like to see how well your process is working, by all means send in a sample for testing. Just remember that successive batches will yield differing results no matter how hard you try.

The major issue in producing quality fuel is having a quality supply of WVO. My personal recommendation is to use only canola, soy, or other non-hydrogenated vegetable oil and to try to find a local restaurant that does not "overwork" its oil. Some restaurants will place their oil in one fryer for "lighter" foods such as French fries and then transfer it to a second fryer for "darker" cooking such as deep-fried ribs or chicken wings. Oils that have been used twice will be high in free fatty acids, requiring more sodium hydroxide and washing to produce a quality fuel.

Canola oil is an excellent choice for colder climates as it does not solidify or reach its gel point until long after animal and hydrogenated fats have turned to sludge.

Figure 16.4-13. My personal recommendation is to use only canola, soy, or non-hydrogenated vegetable oil and to find a local restaurant that does not "overwork" its oil. Oils that have been overused will be high in free fatty acids, requiring more sodium hydroxide and washing to produce quality fuels. Here, Steve Anderson is picking up a delivery of waste vegetable oil from a local restaurant.

Figure 16.4-14. Canola oil, shown left, is an excellent choice of oil for colder climates as it does not solidify or reach its gel point until long after animal fats and hydrogenated oils (right) have turned to sludge. This demonstrates why people should not eat products cooked in animal and hydrogenated oils, as they will congeal like this inside your veins.

Cost and Supply of Raw Materials

WVO should not cost anything other than your time to collect it. Most restaurants have to pay to have their waste oil removed, either by dumping pails in a garbage bin or through the use of a recycling company (Figure 16.4-4).

In addition to the WVO, significant quantities of methanol and sodium hydroxide will be required. For every 26 gallons (100 liters) of WVO, approximately 5 gallons (20 liters) of methanol and 32 ounces (900 grams) of sodium hydroxide will be used in the reaction process, yielding 26 gallons (100 liters) of biodiesel. The cost of methanol is currently U.S.$3.50 per gallon (~$1.00 per liter) and the cost of sodium hydroxide is $6.00 at the calculated concentration. This yields a finished cost of approximately US $0.80 per gallon (US $0.21 per liter) for the completed biodiesel.

In addition to these costs, there are disposable items such as titration chemicals, gloves, rags, and assorted shop materials which will not impact the production cost significantly.

Personal Safety Issues

Would-be biodiesel producers must also consider their own safety while producing this fuel. The two main ingredients, methyl alcohol (methanol) and sodium hydroxide (also known as cleaning lye) are extremely dangerous. Methanol is flammable and burns with an invisible flame.

Many popular "recipes" for biodiesel suggest that methanol is a common, almost household product. This is simply not true. Methanol use in the home is confined to tiny amounts that are used very infrequently, such as charcoal lighter fluid and model airplane fuel. The production of biodiesel requires enormous amounts of methanol (typically 5 gallons/20 liters per batch) to be used in the reaction process. Residual methanol is also present in both the biodiesel and residual glycerin.

"Methanol is a very dangerous substance, one that people simply take for granted," explains Dr. André Tremblay, Professor and Chair of the University of Ottawa Chemical Engineering Department. "The Material Data Safety Sheets (MSDS) states that ingestion may be fatal or cause blindness, while inhalation of high concentrations of vapor may cause headache, convulsions, visual impairment and possible permanent blindness, and possible death."

Dr. Marc Dubé directs the advanced biodiesel research program at the University and concurs about of the safety concerns with the various substances used in the production of biodiesel. "In a laboratory setting we perhaps have higher safety levels than most people would follow. However, I cannot stress enough that the lack of safety with which some people treat these materials is very dangerous. For example, the air space formed in a storage unit containing unwashed biodiesel will exude methanol vapour. Opening the container will induce fumes into the local area and they can be inhaled by an unsuspecting person. Methanol can also be absorbed through the skin which increases an unsuspecting person's access to additional exposure."

Strong words of caution to be sure, and using caution is the recommended approach to working with any chemical. All chemicals and dangerous products are required by law to carry a Material Safety Data Sheet (MSDS) which explains the dangers, characteristics, handling procedures, fire prevention regulations, and recommended safety measures related to each product. It is recommended that these sheets be obtained and studied for each material you will have in your shop. Be sure to speak with a local safety supply dealer to ensure that you have the necessary first aid and fire fighting materials ready and in working order, and make sure you know how to use these safety devices.

Methanol and sodium hydroxide produce an exothermic (heat from) the chemical reaction that will drive off a considerable amount of vapor. Ensure that the reactor tanks are properly sealed and that the work area is either outside or in a very well ventilated location. Also ensure that the WVO is not overheated to prevent the production of excess methanol vapor during the reaction process.

Lastly, ensure that all chemicals are stored in sealed, locked containers that are appropriate for their use. Storing methanol or any other chemical in clear pop bottles is the surest way to disaster.

Winter Driving and Cold Flow Issues

Automotive petrodiesel is supplied as No. 1 grade or No. 2 grade. No. 2 grade is the preferred fuel as it has higher energy content per unit volume, giving better fuel economy. Unfortunately, when temperatures drop No. 2 is subject to an increase in viscosity known as "gelling," which plugs fuel lines and filters.

The cold weather performance of biodiesel is similar to that of petrodiesel No. 2, although its pour point temperature (the temperature at which it gels) is slightly better than No. 2. Animal fats, lard, and hydrogenated oils become solids at much higher temperatures, making them poor choices for cold weather fuel feedstock (as well as for your veins and arteries).

Since No.1 has a pour point of approximately -40°F (-40°C), fuel suppliers blend No. 1 and No. 2. diesel in ratios appropriate for the geographical location and time of year to improve cold flow capabilities. It is tempting to simply add biodiesel to this cold weather petrodiesel; unfortunately, this can be a difficult game to play. The fuel suppliers have already blended the fuel for the coldest expected temperature in your local area. Adding biodiesel to the mix will raise the gel-point temperature by diluting the amount of No. 1 diesel present in the mix.

Figure 16.4-15. The cold weather performance of biodiesel is similar to that of petrodiesel No. 2. In this image B100 canola-based ASTM-certified biodiesel has started to warm up after been subjected to a low temperature below its cloud point. The effects of clouding are clearly visible and should this condition persist fuel filter clogging may occur.

In very cold environments such as the northern United States and most of Canada, blending may need to be restricted to B10 until sufficient experience can be gained with local temperature conditions and their effect on fuel.

Biodiesel can be splash-blended by putting room-temperature biodiesel into the tank of the vehicle and then adding sufficient petrodiesel to achieve the desired blend level.

Although additives can be used, there is considerable argument as to their effectiveness as well as concern about their toxicity. A car that won't start or freezes up during a cold snap can be thawed out by towing it to a heated garage. Once bitten by the frozen fuel line bug, keep an eye on blending ratios and back off on biodiesel concentrations during extremely cold weather conditions.

Summary

We've covered some of the pros and cons of biodiesel. If you're still not sure whether making your own fuel is right for you, let's take a tour of a small brewery that employs environmentally sound business practices, including the production of biodiesel.

Figures 16.4-16 through 16.4-18. This sequence of pictures compares ASTM-certified B100 biodiesel (light color) and homemade B50 canola-based biodiesel blended with winter petrodiesel (dark color) at varying temperatures.

Figure 16.4-16 was taken at 32°F (0°C) and both fuels are becoming cloudy but remain liquid and usable.

Figure 16.4-17 shows the fuels at 0.5°F (-17.5°C). The canola-based B100 biodiesel is about the same consistency as peanut butter.

The B50 blended fuel is still liquid, as shown in Figure 16.4-18 (right).

16.5 A Quick Stop at Church

Perhaps one of the most interesting aspects of researching this book has been discovering the diversity of people who enjoy bucking The System, all the while managing to make a living and help the environment at the same time. On a beautiful day in early spring, I decided to visit one such group of friends who fit into this category and stop by church at the same time.

Figure 16.5-1. After a wonderful drive through central Ontario farm country, you will come across the tiny hamlet of Pithwicks Corners, just east of Campbellford.

Figure 16.5-2. Here in an 1878 Zion church, John Graham and his wife Cherie decided to set up their business, The Church-Key Brewing Company, producing a high-quality product and selling into the premium beer market.

After a wonderful drive through central Ontario farm country, you will come across the tiny hamlet of Pithwicks Corners, just east of Campbellford. It is here in an 1878 Zion church that John Graham and his wife decided to set up their business, the Church-Key Brewing Company.

John explained that his facility is pretty small compared to the commercial "factory" brewers; Church-Key focuses on specialty beers and a quality product. "There is no way on earth that we can compete with the big boys. Our cost base is much higher, so we have to work within the niche area of the market. We have an excellent team producing a very high quality beer, which allows us to make a living; however, I knew we could do more." I immediately assumed that an expansion or takeover by a competitor was in the works, but John just smiled and brought me in for a tour.

Figure 16.5-3. To the untrained eye, the brewery looks pretty much exactly the way one would expect. The big tanks dominate the room and the pungent smell of roasted hops and yeast fills the air, while refrigeration and heating equipment broods over the workers. It all seems rather normal.

To the untrained eye, the brewery looks pretty much exactly the way one would expect. The big tanks dominate the room and the pungent smell of roasted hops and yeast fills the air, while refrigeration and heating equipment broods over the workers. In an industrial sort of way, it all seems rather normal. "Since we have to run a tight ship to stay in business, we decided to take it all the way," John continued. "We have adopted energy efficiency and environmental sustainability as key goals throughout the brewing process and in the operation of the building. We started with the obvious things first, such as compact fluorescent lighting, and moved on to more advanced ideas. Water must be heated as part of the brewing process, which requires fossil fuels, but we also realized that the commercial fridges in the building gave off a lot of heat, so we installed a heat recovery unit that captures this waste heat and transfers it to water used for brewing and cleaning. The propane-fired water boiler requires less energy as a result of being supplied with preheated water. It probably took less than one week to pay back the cost of the unit and we now have a source of clean, "free" energy."

John went on to explain that the reverse problem occurs at the end of the brewing cycle. "Once the brewing process is completed, the next step is to pump the hot (151°F/66°C) wort into a refrigeration unit. With all of the cold water flowing into the building, we are able to use the incoming cold water to "pre-cool" the wort before it enters the fridge. At the same time, this cold water is warmed up a bit before it enters the heat recovery unit I mentioned earlier. With a bit of thinking like this, you can reuse what is typically waste energy just by understanding your processes and using a bit

Figure 15.4-4. John realized that the commercial fridges in the building gave off a lot of heat, so he installed a heat recovery unit that captures this waste heat and transfers it to the brewing water before it enters the boiler unit. It took less than one week to pay back the cost of the unit, and the brewery now has a source of clean, "free" energy.

of creative plumbing." The resulting energy savings help to make the micro-brewery viable, while at the same time reducing the greenhouse gases and other atmospheric emissions. Saving a kilowatt of electricity is always cheaper and more environmentally sustainable than using a kilowatt, regardless of where the energy comes from.

There is more. John pointed out that all of the trucks at the brewery are diesels, and judging by the GMC 7500 sitting out front they're pretty big as well. "Aside from our retail outlet here in the church, we deliver and sell our beer only in draft kegs to bars and restaurants over a wide area of Ontario," John explained. "Fuel costs for beer delivery are a big part of our operating budget, so we decided to take a look at the feasibility of biodiesel production from waste vegetable oil (WVO)."

John and some of his adventurous friends did some research and decided that the easiest way to get into the biodiesel production game was simply to

Figure 16.5-5. "Fuel costs for beer delivery are a big part of our operating budget, so we decided to take a look into the feasibility of biodiesel production from waste vegetable oil (WVO). We quickly realized that we have a zero-cost base for collecting the WVO, as we are delivering beer to the same clients that have this waste product to get rid of," John Graham, brewmaster explained.

roll up their sleeves and give it a try. "Together we pooled our money and purchased a Fuel Meister system from Biodiesel Solutions, as well as assorted pumps and storage tanks to help with the collection and handling of the fuel. We quickly realized that we have a zero-cost base for collecting the WVO, as we are delivering beer to the same clients that have this waste product to get rid of. It works out well for both of us, because we are helping them get rid of a product that costs money to dispose of. The restaurants return the favor by keeping the oil clean, eliminating dish water and bacon grease for example, and having it ready for pickup. Everyone wins."

After loading the WVO into the storage tanks on board the trucks, they return to a small rented facility in the area and convert the WVO into biodiesel. "We have made well over thirty 40-gallon (150-liter) batches of mostly good biodiesel, as well as a couple of barrels of gravy!" John exclaimed. "We are pretty happy with the process and have learned a few tricks along the way to make the work easier, but all in all, it has been a very positive experience and has certainly helped to reduce our operating costs even further."

As we toured the facility, John went on to explain that this past winter they started to test biodiesel with home heating. They are pleased to report having no problems whatsoever with a B50 (50% biodiesel with petrodiesel) blend in a standard oil furnace. "We will move cautiously in making the transition, but because my buddies and I are quite handy, I can't see this being any problem in the future."

Figure 16.5-6. The old church is a bit tough to heat, but John plans to convert the boiler to run on biodiesel, further reducing operating costs.

Speaking of problems, I asked John if they had any trouble using the homemade biodiesel in the delivery trucks. "No, not really. We did push it one winter with B50 and the weather got too cold and we plugged the fuel filter, but that was just being cocky."

In this part of Ontario, winter temperatures can get pretty darn cold, requiring low concentrations (less than B33 is recommended) of biodiesel blend. "Once you can plant your tomatoes, it's time to switch back to B100," John mused.

Figure 16.5-7. Because the brewery requires so much more refrigeration room in the summer, John added a Thermo King diesel-powered trailer unit that will run on B100 biodiesel in the summer months, with the indoor coolers taking care of off-peak demand.

"The old church is a bit tough to heat, but we have a solution for that as well," John explained. "Later this year we intend to convert the oil boiler to run on biodiesel, which is just a natural extension of what we are already doing. The limiting factor may be our biodiesel production capacity."

John also went on to explain that the brewing business has a huge swing in production, with beer sales going through the roof during the hot summer months. The ever-resourceful team picked up an old diesel-powered Thermo King refrigeration unit that would normally be mounted on a rail car. "Because we require so much more refrigeration room in the summer, we added the Thermo King and will run it on B100 biodiesel in the summer months. Then when our sales drop in the fall, our indoor coolers can handle the demand. This also has the effect of leveling out our biodiesel consumption and production." John has clearly thought through his business model, all the while keeping environmental sustainability front and center in his planning.

Of course the last stop on our tour is to drop by the retail store upstairs to pick up a sample of Church-Key's excellent product—in the name of research, of course.

Figure 16.5- 8. Of course the last stop on our tour is to drop by the retail store upstairs to pick up a sample of Church-Key's excellent product—in the name of research, of course.

Figure 16.5-9. John Graham (right) and his biodiesel brewing friends Eric Dickinson (center) and Helmut Klein (left) (Paul Moring is missing) are smiling. They save money and the environment and literally get to drink the fruits of their labor.

16.6 Biodiesel 101

DISCLAIMER

The following is not intended to provide an exhaustive methodology on the production of ASTM-quality biodiesel fuel. Rather, it offers an examination of the basic steps necessary to make the fuel and acts as a guide in determining if this process is suitable for you.

Further information on biodiesel production will be available in William Kemp's forthcoming book on biodiesel production for home and farm, which outlines methods for producing biodiesel fuel from waste oils as well as oil seed crops. Watch for it in the spring of 2006 at www.newsociety.com.

CAUTION

The manufacture and storage of biodiesel fuel as well as its component chemicals are dangerous and involve a degree of risk. Ensure that you are familiar with all necessary material safety data sheets (MSDS) and that you follow the proper procedures for handling and storing chemicals. Ensure that everyone involved in the process is familiar with proper first aid and fire protection procedures.

The following story and photo collection depict the production of biodiesel using an "open processor" method which is known to emit dangerous chemical vapors. This method has been chosen for discussion as it is the only method that allows direct photography of the production and reaction process. However, because the various chemicals are highly dangerous, you should consider only a closed processing system for producing your own biofuel.

The author, publisher, and contributors assume no liability for possible errors and omissions in the following discussion or for any personal injury, property damage, and consequential damage or loss which might occur as a result of using the information in this book, however caused.

"I'm Just a Simple Home Brewer"

Steve Anderson likes to refer to himself as a simple "home brewer." He has no pretensions of putting Big Oil out of business or getting rich making biodiesel. He just wants to save a few bucks and, for good measure, maybe even help the earth to live just a bit longer.

Over the last few years, Steve has produced a couple of thousand gallons of biodiesel in his workshop/garage. It was a cool day in late winter day when I visited Steve and found him filling up a small bucket from a batch of biodiesel fuel that feeds the furnace. "I have a great deal with the owner of this garage," Steve explained. "All I have to do is keep the building warm all winter and I get to use half of the place for biodiesel production. It's a pretty good deal considering how little it costs to make the heating fuel."

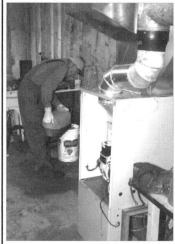

Figure 16.6-1. On a cool day in late winter I visited Steve Anderson and found him filling up a small bucket from a batch of biodiesel fuel that feeds the oil furnace.

Steve uses a very simple open processing unit fabricated from an old oil drum to produce his biodiesel. Without counting the "tinker factor," Steve estimates that he has invested about $500 in his setup. He explained that he has to invest some money this year to upgrade his production equipment.

"I realize that I shouldn't be using an open processing system, because methanol gives off poisonous vapors. However, I have been careful to make sure the garage is open and to cover the reactor and methoxide drums when they are in use. Upgrading to a closed system is one of my next projects." Based on Dr. Tremblay's earlier comments regarding the chronic effects of methanol, this is no doubt a smart move.

Figure 16.6-2. Steve uses a very simple open processing unit fabricated from an old plastic oil drum to produce his biodiesel. Without counting the "tinker factor," Steve estimates that he has invested about $500 in his setup.

Figure 16.6-3. A close-up view of Steve's laboratory indicates that very little complex machinery or equipment is required to produce biodiesel.

A close-up view of the laboratory indicates that very little complex machinery or equipment is required to produce biodiesel. As Steve works through the process, it becomes clear that cleanliness and careful attention to detail are the main ingredients in successfully producing fuel. "I have made a couple of batches of glop," Steve informed me. "But most of the time the process works beautifully because I have worked with my waste oil suppliers to try to get a consistent product from them. They know I will be there twice a week to pick up the oil, in turn saving them thousands of dollars a year in recycling costs. In return, they ensure that the oil is not contaminated coming out of the fryer and waiting for pickup. I also check my "suppliers" to ensure that they are not using animal or hydrogenated fats, as any oil that is solid at room temperature will make a biodiesel that will gel when the weather gets cool." Feedstock oils such as these may be fine for someone living in the southern, more temperate areas of the United States, but pose considerable trouble for northerners.

Figure 16.6-4. Steve is careful not to take any hydrogenated or animal fats. These oils tend to be solid at room temperature and will quickly turn into a butter-like substance when the outside temperature begins to cool. This may not be a problem in the sunbelt of the United States, but it poses serious problems for northerners.

The Production Stages of Biodiesel

The production of biodiesel may appear more difficult than it actually is. However, it is a smart person who produces the first batch or two based on a smaller scale. After all, if you are going to make a mistake and produce a batch of glop, it might as well be a small batch.

Many beginning home brewers start out producing a 1-litre (approximately 1 quart) batch of biodiesel using an old blender and reducing the amount of WVO, methanol, and sodium hydroxide by a factor of 100 based on the standard recipe that follows. For example, using virgin vegetable oil, a mini batch requires 1 liter WVO, 0.2 liter (200 ml) methanol, and 3.5 grams of sodium hydroxide. All other procedures remain the same as for a large 100-liter (26.4-gallon) batch. In fact, this recipe may be scaled up or down to suit your particular needs and processing equipment requirements.

Waste Oil Preparation

Figure 16.6-5 (above). The first step in producing biodiesel is to filter the WVO using a filtration tank comprising an open-ended tank fitted with a pair of pantyhose. This unit filters out food particles and other debris that may be present in the oil.

Figure 16.6-6 (left). Steve is shown pouring WVO from one of his "suppliers" into the oil filtration unit.

Figure 16.6-7. While the oil is filtering through the stocking material, Steve cleans out the reactor vessel by carefully eliminating any traces of sodium hydroxide (lye) and glycerin from previous reactions. Leftover materials, especially water, increase the risk of the reaction making gravy rather than useful biodiesel.

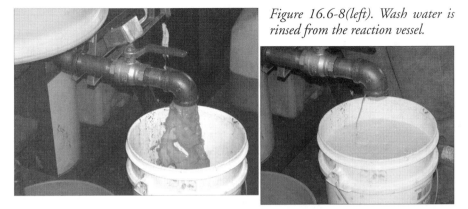

Figure 16.6-8(left). Wash water is rinsed from the reaction vessel.

Figure 16.6-9 (above right). Soaps and glycerin contaminate the wash water. Unreacted sodium hydroxide may cause wash or waste water from the biodiesel process to be slightly "basic." Mixing small amounts of household vinegar (acetic acid) with the water will neutralize the base, bringing the pH level of the water to 7, or neutral. When the water has been neutralized, it may be composted or disposed of in a leeching bed. Do not pour waste water into a drain connected to a septic system.

Figure 16.6-10. The filtered WVO is now transferred to buckets...

Figure 16.6-11. ...and placed in the reaction vessel. 100 liters (26.4 gallons) of oil are processed into biodiesel at one time. (Calculations for mass and volume are greatly simplified using the metric system.)

Figure 16.6-12. Steve attaches a standard electric paint-mixing paddle to a variable speed drill to mix the oil and other chemicals. (As methanol fumes are very flammable, I explained to Steve that sparks produced by the drill motor commutator could ignite the vapors. Use Totally Enclosed Fan Cooled (TEFC) motors or preferably explosion-proof models available at any motor supply store or www.grainger.com.

Figure 16.6-13. The mixing drill is attached to a bracket, with the paddle immersed in the oil.

Figure 16.6-14. The drill is started and a slow, thorough mixing of the oil begins.

Figure 16.6-15. Steve has attached a 120-Volt water heater element to a bracket and placed it in the reaction tank oil. Heaters of this type must never be activated unless they are submersed in fluid. The WVO will require less time and energy to heat in an insulated and closed reaction vessel.

Figure 16.6-16. The heater and mixer are submersed in the WVO. Note the small thermal "bulb" just to the right of the drill. This device measures the temperature of the oil and sends a signal to a control unit.

Figure 16.6-17. The heater control unit receives a signal from the temperature sensing bulb, ensuring that the oil temperature reaches and does not exceed 131°F (55°C).

Figure 16.6-18. Methanol will be added to the WVO in subsequent steps. It gives off large amounts of explosive fumes when it boils (148.5°F/65°C), so it is important to control the temperature accurately.

Titration

If you think back to high school chemistry, you will recall the pH scale which is used to grade the strength of acids and their opposite cousin, "bases." The scale is numbered 0 through 14, with 7 being the halfway or "neutral" point on the scale. Pure water is neutral and therefore has a pH reading of 7. As the number drops below 7, the substance becomes a stronger acid. Likewise, as the reading increases beyond 7, the substance becomes more basic. Virgin vegetable oil is by nature acidic, and used fryer oils become increasingly acidic the longer and "harder" they are used.

The biodiesel transesterification process requires a base material, sodium hydroxide (lye), to act as a catalyst to cause the "exchange" of glycerin for methanol molecules. The titration process will determine the level of Free Fatty Acids (FFA) contained in the oil as well as the amount of base material required to ensure that the transesterification process is "pushed" to completion, thereby producing biodiesel and not "glop."

An acidic substance can become neutralized with the addition of a basic substance of equal but opposite strength. Using this rule and the information derived from the titration process, it is possible to calculate the amount of sodium hydroxide necessary to neutralize the FFAs and ensure that the biodiesel conversion process proceeds correctly.

In order to perform a titration process, make sure that you have a well-ventilated area. Then assemble the necessary chemicals, laboratory glassware, mass scale, and supplies:

- pH indicator known as phenolphthalein or the indicator from a swimming pool test kit known as "phenol red." Phenolphthalein is a laboratory grade (more expensive) indicator and is slightly more accurate than swimming pool phenol red, but either indicator can be used.
- distilled water
- 100% pure sodium hydroxide or lye. Be sure to keep the lye dry, even preventing exposure to humid air.
- 99% isopropyl alcohol (Use care when purchasing alcohol as there are numerous concentrations available. If in doubt, consult with a pharmacist.)
- syringes graduated in 1 ml increments
- graduated cylinder or other accurate liquid measuring devices
- assorted small jars or beakers. At least one jar should have a sealable top to store 1 liter of fluid as a reference solution.
- triple-beam mass balance
- latex rubber gloves
- eye protection

Procedure

Note: Ensure that all beakers, containers, and syringes are clean and used for only one chemical. Syringes and glassware may be reused each time you make a new batch of biodiesel, provided they are used for only one of the following steps:

1. Measure out 1 gram of sodium hydroxide on the mass balance. One gram is a very small amount; be very careful with this measurement. Alternatively, measure out 4 grams of lye and make a larger (4 liter)

batch of reference solution. Measuring a larger quantity will divide any measurement error into a larger sample, improving the accuracy of the solution.

2. Measure 1 liter of the distilled water using a graduated cylinder. Place the water in the sealable bottle and add to it 1 gram of sodium hydroxide. Swirl the sealed bottle to ensure the sodium hydroxide is fully dissolved. This solution should be labeled "Reference Solution."

3. Draw 10 ml of isopropyl alcohol and place it in a small beaker.

4. Add 2 drops of phenolphthalein indicator into the alcohol and swirl until well mixed.

5. Draw a sample of WVO and add 1 ml to the phenolphthalein/alcohol solution. Use extreme care in ensuring that exactly 1 ml is dispensed. It is best to fill the syringe with, for example, 5 ml of oil and then dispense down to 4 ml rather than trying to measure a single milliliter of oil.

6. Vigorously swirl the beaker to ensure complete mixture. If you are using a sealing jar, shaking will work as well.

7. Draw 10 ml of the reference solution into a syringe. Swirling the beaker of phenolphthalein/alcohol/WVO in one hand, carefully drip ½ (CAM one-half) ml amounts of reference solution into the beaker at a time. The solution will be a musty yellow at the beginning of this test and will turn a uniform bright pink when the titration is complete. (The solution will turn pink when the reference solution is first injected but will return to yellow as you continue to swirl the beaker. The solution must be uniformly pink for at least 30 seconds before the titration is considered finished.)

8. Record the amount of reference solution required to complete the titration process. This will be the amount of reference solution left in the syringe subtracted from 10. For example, if 7.5 ml remain in the syringe, then 2.5 ml were used (10-7.5 ml=2.5 ml used).

9. Add 3.5 to the amount of reference solution used in Step 8 and then multiply this number by 100. This equals the mass in grams of sodium hydroxide that will be required to complete the transesterification process for 100 litres of WVO.

In the example in Step 8, 3.5 ml would be added to 2.5 yielding 6, multiplied by 100 to give a result of 600 grams of sodium hydroxide required for every 100 litres of WVO.

Use the balance beam scales to accurately measure the required amount of sodium hydroxide.

10. It is recommended that you clean up the lab and do the test at least one more time to ensure accurate results. It is very easy to make a mistake, and accuracy is what produces a high-quality fuel.

11. Store the measured sodium hydroxide in an airtight jar and put it aside in preparation for the next stage.

Figure 16.6-19. **Steps 1 and 2**. *Measure 1 liter of the distilled water using a graduated cylinder. Place the water in the sealable bottle and add to it the 1 gram of sodium hydroxide. Swirl the sealed bottle to ensure that the lye is fully dissolved. This solution should be labeled "Reference Solution."*

Figure 16.6-20. **Step 3**. *Steve is lecturing on the importance of using 99% pure isopropyl alcohol for the titration process. If you are unsure of the type of alcohol, consult a local pharmacy.*

Figure 16.6-21. **Step 3**. *Draw 10 ml of isopropyl alcohol and place it in a small beaker.*

Figure 16.6-22. **Step 4**. *Add 2 drops of phenolphthalein indicator into the alcohol and swirl until well mixed.*

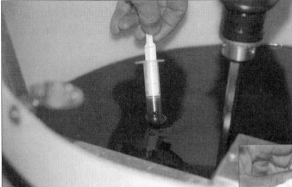

Figure 16.6-23. **Step 5**. *Draw a sample of WVO.*

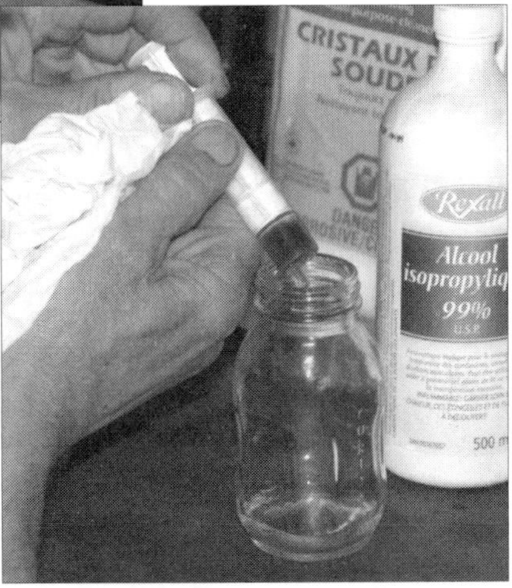

Figure 16.6-24. **Step 5**. *Add 1 ml of WVO to the phenolphthalein/alcohol solution. Use extreme care in ensuring that exactly 1 ml is dispensed. It is best to fill the syringe with, for example, 5 ml of oil and then dispense down to 4 ml rather than trying to measure a single milliliter of oil.*

Figure 16.6-25. **Step 7**. *Draw 10 ml of the reference solution into a syringe.*

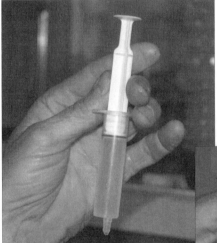

Figure 16.6-26. **Step 7**. *Swirl the beaker of phenolphthalein/alcohol/ WVO in one hand, while carefully dripping ½ ml amounts of reference solution into the beaker at a time. The solution will be a musty yellow color at the beginning of this test.*

Figure 16.6-27. **Step 7.** *The solution will turn a uniform bright pink (which is a bit hard to see in a black and white photograph) when the titration is complete. (The solution will turn pink when the reference solution is first injected but will return to yellow as you continue to swirl the beaker. The solution must be uniformly pink for at least 30 seconds before the titration is considered finished.)*

Figure 16.6-28. **Step 9.** *Steve is shown using a triple balance beam scale to accurately measure the amount of sodium hydroxide calculated in Step 8. It is set aside ready for the sodium methoxide steps to follow.*

Figure 16.6-29. **Step 9.** *The accuracy of the titration process and measurement of the sodium hydroxide catalyst will determine the success or failure of the biodiesel transesterification process.*

Preparation of the Sodium Methoxide Solution

Sodium methoxide is a very dangerous, flammable, and noxious solution. Given the health and safety issues involved, I do not recommend the "open vessel" method of handling of this material.

Procedure

1. Draw 20 liters of methanol (20% by volume of the WVO to be reacted) into a mixing vessel.
2. Carefully add the premeasured sodium hydroxide to the methanol. The mixing of these two components will generate a considerable amount

of heat in an "exothermic" process. The fumes from this solution are deadly. Gas masks and other organic filters are useless against the sodium methoxide vapor.

3. Mechanically stir the mixture to ensure complete solution, since sodium hydroxide does not actively dissolve in methanol.

Figure 16.6-31. A cordless electric drill is fitted to the stirring paddle and the trigger is fitted with a lock, allowing the drill to operate unattended.

Figure 16.6-30. Steve's sodium methoxide reaction vessel is made from a simple high-density polyethylene (HDPE #2) plastic 22-liter pail. A hole is drilled into the removable top to accept a paint-stirring paddle.

Figure 16.6-32. A drum of methanol "racing fuel" is fitted with a siphon pump.

Figure 16.6-33. Steve measures 20 litres of methanol (20% by volume of the WVO to be reacted) into the pail.

Figure   16.6-34.   The premeasured sodium hydroxide from the titration stage is added to the methanol. The exothermic or "heat-producing" reaction of the two chemicals will produce heat and sodium methoxide.

Figure 16.6-35. The sodium methoxide reaction vessel is allowed to sit or is mechanically stirred until all of the sodium hydroxide dissolves in the methanol.

Figure 16.6-36. Steve removes the electric heating element from the WVO reaction tank in preparation for the addition of the sodium methoxide solution.

Figure 16.6-37. The WVO mixer speed is increased to ensure complete agitation of the WVO and sodium methoxide solution.

Reaction of the Waste Vegetable Oil

Once the sodium hydroxide has been fully dissolved into the methanol, the sodium methoxide solution is ready to be added to the heated WVO to begin the transesterification process.

Transesterification involves two phases: a byproduct phase that produces glycerin and a second phase that produces "raw biodiesel". Glycerin may be composted or converted into soap as desired.

While some use the raw biodiesel as soon as it has been reacted and separated from the glycerin, others advocate further processing. Raw biodiesel is 100% incompatible with modern automotive engines and will cause premature failure. If you are attempting to produce a fuel with the intention of saving money, premature engine failure will quickly reverse any potential financial savings.

Procedure

1. Carefully add the sodium methoxide solution to the heated and stirred WVO.
2. Mechanically stir the solution of WVO and sodium methoxide continuously for one hour. During this time, the solution will begin to turn a "chocolate milk" color as the reaction proceeds, then a low-viscosity (thin) clear brown color. This is a good indication that the reaction has proceeded correctly.
3. Stop the mixer after one hour and allow the solution to sit for 24 hours. During this time the solution will separate into two distinct layers or "phases." The denser phase is glycerin and unreacted sodium hydroxide which, being heavier than biodiesel, will sink to the bottom. The lighter solution floating on top of the glycerin is the "raw" biodiesel or Fatty Acid Methyl Ester (FAME) phase.
4. Remove the dark brown, very viscous glycerin layer from the bottom of the reaction vessel by opening the bottom valve and allowing the glycerin to flow into a bucket. As soon as it is drained from the tank, the raw biodiesel will follow. Close the drain valve at this point.
5. You may decant the raw biodiesel into storage pails or leave it in the reaction vessel, depending on your desire to produce a high- or low-quality fuel.

Figure 16.6-38. Steve carefully pours the 20 liters of sodium methoxide into the reaction vessel, which is filled with 100 liters of heated WVO.

Figure 16.6-39. As soon as the sodium methoxide touches the WVO, the reaction process begins. Note the light and dark tinges in the photograph.

Figure 16.6-40. The transesterification process is well underway. The WVO/sodium methoxide solution has turned the color of chocolate milk.

Figure 16.6-41. The reaction is nearly complete after one hour of continuous stirring. The WVO has become very clear and its viscosity (thickness) has dropped dramatically. After stirring, the solution is allowed to sit for 24 hours to allow the glycerin and raw biodiesel (FAME) to separate into two distinct layers or phases.

Figure 16.6-42. After a 24-hour settling period, the heavier, denser glycerin and unreacted sodium hydroxide are drained from the reaction vessel. The lighter, less dense raw biodiesel remains.

Figure 16.6-43. In the "simple" production method, raw biodiesel is drained from the reaction vessel and placed in buckets and allowed to sit for a further three days to one week. Although the majority of the transesterification process is completed in one hour, the reaction will continue for a longer period, allowing additional glycerin to separate from the raw biodiesel. Soap, a byproduct of the reaction process, can be seen floating on top of the raw biodiesel.

Figure 16.6-44. After allowing the raw biodiesel to sit one week, Steve wipes the floating soap layer away.

Figure 16.6-45. The raw biodiesel is decanted into storage buckets.

Figure 16.6-46. The clear, low-viscosity raw biodiesel is visible in this photograph.

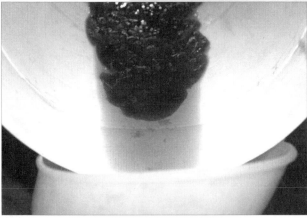

Figure 16.6-47. As the raw biodiesel is decanted, glycerin and unreacted sodium hydroxide remain behind, the result of slow, low-level reactions taking place during the seven-day settling period.

Biodiesel: Raw or Cleaned?

Some people advocate using the raw biodiesel directly. Steve Anderson estimates that he and his friends have traveled over a couple of hundred thousand miles in many different vehicles on the raw fuel without any problems. The same is true for John Graham and countless other people who use the fuel at this stage of the overall biodiesel production process. Why all the fuss?

Raw biodiesel fuel is just that: raw. The reaction process will continue for a long period of time, causing glycerin to drop out of the fuel. In addition, unreacted sodium hydroxide is present in the raw fuel, causing the pH to remain very high compared to the ideal neutral pH 7 reading. High pH level is an indication that the fuel is corrosive and will, over time, damage expensive engine components.

Although the number of people who use homemade biodiesel in its raw form is probably equal to the number who use cleaned biodiesel, raw fuel will not come close to meeting ASTM standards for quality. To skip the washing, pH adjustment, and quality control steps is simply gambling with the longevity of your engine or furnace. Older, less sensitive diesel equipment may operate on raw biodiesel for many years with apparently little problem. For newer engines with computer-controlled management systems, burning raw fuel constitutes a death sentence.

Raw biodiesel has a higher viscosity than washed biodiesel because of suspended glycerin, which affects the fuel spray and atomization pattern. Being a heavier fuel with lower vapor pressure, raw biodiesel does not burn completely, increasing atmospheric pollutants. At the same time, fuel spray condenses on the relatively cool cylinder walls and washes past the piston rings, fouling engine oil with insoluble material.

Corrosive, unreacted sodium hydroxide will attack fuel system components, particularly in late-model, high-pressure common rail direct injection (CDI) engines. Excess methanol is a solvent which will destroy polymer components used in engine fuel-system seals and hoses.

In short, using unwashed, raw biodiesel is akin to gambling. You may save money by making homemade fuel, but will the savings be depleted by expensive engine repairs? Taking the time to perform the remaining few steps will help to ensure that fuel quality issues are minimized.

Washing and pH Adjustment

Washing biodiesel is a process that works exactly the way it sounds. Washing is a water-rinsing process that "draws" glycerin, soaps, unreacted sodium hydroxide, and excess methanol from the raw fuel stock. None of these chemical compounds is meant to be part of commercial diesel fuel, and automotive manufacturers cannot ascertain how these materials will affect their engines and fuel systems. While these compounds might have little or no effect on short-term engine life, diesels are designed for the long haul. Shortening engine life with cheap fuel will never provide the economics that motivate the move towards homemade fuel in the first place.

Methanol, as we have already discussed, can cause long-term health problems as a result of inhalation or skin contact. Some pioneers suggest that excess methanol in the biodiesel fuel provides "extra" energy and better gas mileage. Maybe so, but at what cost? Methanol will vaporize at low temperatures, accumulating in the fuel tank and in storage containers used to hold the biodiesel. Opening any of these containers may expose you to methanol vapors, contributing to chronic (long-term) poisoning. In addition, methanol lowers the "flash point" of biodiesel fuel, increase the risk of accidental ignition of the fuel.

Although the raw fuel shown in Figure 16.6-46 appears to be very clear, almost commercial in quality, the dissolved materials in the fuel make it incompatible with ASTM standards.

Figure 16.6-48. Raw biodiesel fuel is just that: raw. The reaction process continues for a long period of time, causing glycerin to drop out of the fuel. In addition, unreacted sodium hydroxide is present in the raw fuel, causing the pH and corrosiveness to remain very high. In this view the left bottle contains raw biodiesel. The second bottle from left shows the "first wash" fuel, with milky residues of glycerin, soaps, and sodium hydroxide. The third bottle from left shows biodiesel after a second washing stage. The sample to the right is clear, washed biodiesel with a proper non-corrosive pH of 7.

Figure 16.6-48 demonstrates the need for washing raw biodiesel fuel. The bottle at the left contains a sample of the raw biodiesel fuel shown in Figure 16.6-46. Although it looks fine in the photograph, the pH level is extremely high, measuring a very corrosive 10.5. The second bottle from left shows equal parts of raw biodiesel and water after they have been mixed together, shaken, and allowed to settle over an eight-hour period. After this simple washing process, a white milky substance forms which is a combination of glycerin, soaps, and unreacted sodium hydroxide that has been extracted from the raw biodiesel.

The third bottle from the left shows a sample of raw biodiesel that has been through a first washing step and then subjected to a second wash. The results show a marked improvement, with a very small layer of soap and glycerin at the interface between the biodiesel and the water. Note that the second-stage rinse water is almost perfectly clear and may be recovered and used for the first wash water of another batch. This is known as counter-current rising and saves a considerable amount of water in the washing process.

The bottle on the right is biodiesel that has been washed three times, pH corrected, and then subjected to a drying process to remove excess moisture left in suspension during the washing stage. The end result is a clean, clear fuel that might meet ASTM standards.

There are as many ways of washing biodiesel as there are of making a perfect cup of coffee. The most common methods used by home brewers are bubble washing and mist washing, with bubble washing offering one of the simplest means of getting the job done.

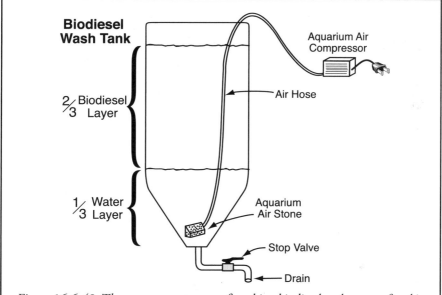

Figure 16.6-49. There are as many ways of washing biodiesel as there are of making a perfect cup of coffee. The most common methods used by home brewers are bubble washing and mist washing, with bubble washing offering one of the simplest means of getting the job done. A thin vertical tank (or the reaction vessel itself) is equipped with a stop valve and drain equipped to conveniently draw the contents from the tank. For each 100 liters of raw biodiesel, approximately 30 to 40 litres of clean tap water and 50ml of acetic acid (household vinegar) are added to the tank.

Procedure

1. Equip a thin vertical tank (or the reaction vessel itself) with a stop valve and drain equipped to conveniently draw the contents from the tank. For each 100 liters of raw biodiesel, add approximately 30 to 40 litres of clean tap water.
2. Add 50 ml of acetic acid (household vinegar) to the biodiesel/water solution.

3. Place a large aquarium air stone in the tank and connect it to an air hose and standard aquarium air compressor which is mounted above the maximum fluid level of the tank. Plug in the compressor and let the solution bubble for eight hours. Air bubbles will carry water through the biodiesel, capturing the unreacted sodium hydroxide, soap, and glycerin.

4. Stop the air compressor (or let the timer expire) and allow the biodiesel to settle for two hours or more. Then drain off the majority of the wash water without discarding any biodiesel. Discard the first wash water.

5. Test the pH of the biodiesel, adding additional acetic acid as needed to lower the pH to a neutral reading of 7. Measuring pH may be done using a litmus paper test kit (see Figure 16.6-50) or an electronic measuring meter available at any laboratory supply store.

6. Repeat the wash step three times. (The amount of acetic acid will be decreased or eliminated for subsequent washes as the pH of the biodiesel approaches 7.)

7. After the last wash, drain off a small amount of biodiesel to ensure that all of the wash water has been removed. Biodiesel from the final wash is now ready for drying.

Note: Using a counter-current washing system will allow the wash water from the third wash to be reused for the second wash of the next batch of biodiesel. Likewise water used in the second wash may be used in the first wash of the next batch. The first wash water is always discarded. Provided you have the necessary water storage room, this process will reduce the amount of water consumed in the washing process.

Drying Biodiesel

Biodiesel that has just been washed will be hazy as a result of microscopic water molecules in suspension. In order to bring the water level below the limit set by the ASTM standard, mechanical drying is required.

As with washing, there are a number of different methods for drying biodiesel. Although heating the fuel slightly above the boiling point of water is perhaps the simplest method, it requires a considerable amount of energy and there is the potential for fire. Current home-built systems rely on a combination of low-level biodiesel heating, agitation of the biodiesel, and blowing heated air over the biodiesel to aid in water evaporation.

The method outlined below uses the same washing tank (after removal of the last wash water) depicted in Figure 16.6-49. Heated air is blown into the raw biodiesel, causing simultaneous agitation and evaporation. The process will work considerably faster if the ambient air is dry, with a relative humidity of less than 50%.

Procedure
1. Connect a heated air blower from a bubble-style whirlpool bath (available at any spa or hot tub dealer) to a length of air hose. Insert the open end of the air hose into the wash tank.
2. Connect the air blower to an industrial lamp dimmer of sufficient electrical capacity to operate and vary the speed of the blower motor.
3. Activate the blower, starting at low speed and air pressure and gradually increasing the speed until the biodiesel is fully agitated but not splashing out of the container. Operate the blower in this manner for a period of one hour.
4. At the end of one hour, turn off the blower and draw a sample of the biodiesel into a beaker. Allow the biodiesel to return to room temperature. The fuel should be clean and clear when observed with a bright light source. Haziness indicates that further drying is required.

Quality Control
It is not possible for the home brewer to produce a fuel that meets ASTM standard D6751. Many of the ASTM tests require access to a specialized gas chromatograph to determine the presence of impurities in biodiesel. Alternatively, it is possible to have a private laboratory test a sample of homemade fuel, although the cost will be in excess of $1,000 and the results will be valid only for a single sample. Batch-to-batch processing will yield considerable variation in quality, requiring retesting each time. This is not a reasonable proposition. The following table demonstrates the properties necessary for certified biodiesel, indicating the difficulty of performing the tests at home.

However, even though a home brewer does not have access to the specialized laboratory equipment required to meet the ASTM standard, quality control cannot be omitted.

Visual Inspection
During the reaction phase, WVO and sodium methoxide combine to create a dark glycerin layer which sinks to the bottom of the vessel and a lighter-colored biodiesel or FAME layer which floats on top. Occasionally, a third

Table 1 Detailed Requirements for Biodiesel (B100)[A]

Property	Test Method[B]	Limits	Units
Flash point (closed cup)	D 93	130.0 min	°C
Water and sediment	D 2709	0.050 max	% volume
Kinematic viscosity, 40°C	D 445	1.9-6.0[C]	mm^2/s
Sulfated ash	D 874	0.020 max	% mass
Sulfur[D]	D 5453	0.05 max	% mass
Copper strip corrosion	D 130	No. 3 max	
Cetane number	D 613	47 min.	
Cloud point	D 2500	Report[E]	°C
Carbon residue[F]	D 4530	0.050 max	% mass
Acid number	D 664	0.80 max	mg KOH/g
Free glycerin	D 6584	0.020	% mass
Total glycerin	D 6584	0.240	% mass
Phosphorus content	D 4951	0.001 max	% mass
Distillation temperature	D 1160	360 max	°C
Atmospheric equivalent temperature.			
90% recovered			

[A] To meet special operating conditions, modifications of individual limiting requirements may be agreed upon between purchaser, seller and manufacturer.

[B] The test methods indicated are approved referee methods. Other acceptable methods are indicated in 5.1.

[C] See X1.3.1. The 6.0 mm^2/s upper viscosity limit is higher than petrodiesel and should be taken into consideration when blending.

[D] Other sulfur limits can apply in selected areas in the United States and in other countries.

[E] The cloud point of biodiesel is generally higher than petrodiesel and should be taken into consideration when blending.

[F] Carbon residue shall be run on the 100% sample (see 5.1.10).

Table 1 Reprinted, with permission, from D 6751-03a Standard Specification for Biodiesel Fuel (B100) Blend Stock for Distillate Fuels, copyright ASTM International, 100 Barr Harbor Drive, West Conshohocken, PA   19428.

soapy-white layer may form at the interface between the glycerin and FAME layers. Either two or three distinctive layers indicate a successful reaction. Glop or a gelatinous mass of "gravy" is a pretty clear indication that the reaction has failed.

Following the washing and drying procedure the biodiesel should be clear, without any sign of "haze" when observed at room temperature (70°F/20°C). The color of biodiesel changes depending on the feedstock oil and degree of "use" it received in the fryer process; color is not an indication of quality.

Haziness results from excess water in the oil, typically above the ASTM

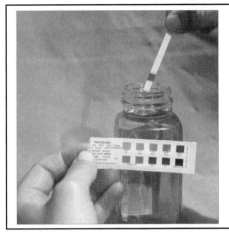

Figure 16.6-50. Accurate pH measurement indicates the amount of unreacted sodium hydroxide and the acidity of the biodiesel fuel. In this view, washed biodiesel is tested using a litmus paper test kit. Electronic, direct reading pH meters are also available.

limit of 0.05% by volume. Additional drying time usually resolves this problem.

pH Measurement

Accurate pH measurement indicates the amount of unreacted sodium hydroxide and the acidity of the biodiesel fuel. The ideal measurement is pH 7 or neutral. Either a litmus test kit (see Figure 16.6-50) or a commercial electronic meter can be used.

Specific Gravity

Specific gravity compares the density of an unknown fluid in relation to distilled water, which has a reading of 1.000. Petrodiesel No. 2 has a reading of 0.830 at 60°F (15°C), WVO from canola is 0.925, and biodiesel is between

Figure 16.6-51. Specific gravity compares the density of an unknown fluid in relation to distilled water, which has a reading of 1.000. If the specific gravity for biodiesel is above 0.900, it is an indication that the reaction is not complete and that excess glycerin remains in the fuel.

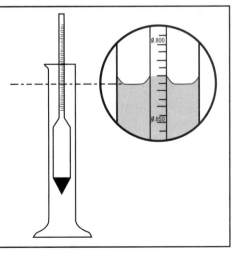

0.850 and 0.900, with an average near the midpoint of 0.875-0.880. If the specific gravity for biodiesel is above this typical range, it is an indication that the reaction is not complete and that excess glycerin remains in the fuel.

Specific gravity is read using a calibrated float such as the model shown in Figure 16.6-51. A sample of biodiesel at the temperature required by the specific gravity device is placed in the cylinder and the float is immersed in it. A reading is taken by sighting along the lower side of the fluid meniscus as shown.

Viscosity

Viscosity is the measure of fluid flow or "thickness." Excess viscosity will lead to fuel spray and atomization problems, poor combustion, increased exhaust emissions, and possible fuel injector or pump failure.

The simplest way to measure viscosity is to use a pipette and stop watch. A known sample quantity of biodiesel is placed in the pipette. The stop watch is started, the fuel is allowed to flow out, and the time required to drain the pipette is recorded. Samples of petrodiesel, WVO, and ASTM-certified biodiesel are subjected to the same test and the recorded times are compared. Biodiesel has a slightly higher viscosity than petrodiesel and will take longer to drain. Likewise, WVO has a higher viscosity that biodiesel and will take longer to drain. Comparing the times of commercial and homemade biodiesel relative to the reference standards of petrodiesel and WVO will indicate whether the biodiesel has the required viscosity. Biodiesel that is too viscous, approaching WVO, may indicate a poor reaction.

Cloud and Gel Points

Cloud and gel points may be determined by placing the biodiesel in a fridge or freezer and subjecting it to periodic agitation and temperature measurement. While this is taking place, look for cloudy formations (cloud point temperature) and thickening (gel point temperature).

Animal fats and hydrogenated vegetable oils have cloud and gel point temperatures higher than those of pure vegetable oils such as soy or canola (rapeseed). Knowledge of these setpoints can prevent fuel filter plugging in cold weather; if weather extremes are a concern, the setpoints can be adjusted by lowering the concentration of biodiesel. In the northern reaches of the United States and Canada B20 can be used without concern for cold weather problems.

Total Glycerin

Measuring total glycerin is a difficult task. Some experiments have been conducted where a sample of homemade biodiesel is resubjected to the reaction process. The test can be completed by taking a one-liter sample of the washed and dried biodiesel and reacting it with sodium methoxide made from 3.5 grams of sodium hydroxide and 200 ml of methanol. Careful attention to material measurement is required to yield useful results.

In theory, any unreacted glycerin should remain after the transesterification process, although the results can be difficult to verify.

Storage

Biodiesel can be stored in high density polyethylene (HDPE #2) barrels or steel fuel drums, preferably in a cool, darkened area. Biodiesel (as well as petrodiesel) should be used a soon as possible and not stored for extended periods. Fill storage vessels to the maximum, eliminating as much air space as possible to reduce condensation and the potential for methanol vapor from forming.

If fuel must be stored for an extended period, consider adding a biocide additive, which prevents the formation of bacterial grown, and is available at any diesel fuel distributor.

Figure 16.6-52. Biodiesel can be stored in high density polyethylene (HDPE #2) barrels or steel fuel drums, preferably in a cool, darkened area. Biodiesel (as well as petrodiesel) should be used as soon as possible and not stored for extended periods. Sean McAdam from The Veggie Gas Company (www.veggiegas.ca) is shown filling a storage tank after producing a bath of biodiesel.

Fuel may be dispensed using a low-cost electric fuel-dispensing pump such as the model available from Biodiesel Solutions, or it may be gravity fed. In either case, ensure that a 5-micron or small marine-type fuel filter with a water extraction feature or agglomerator is fitted in the fuel supply line.

Sean McAdam of Veggie Gas Company notes that his Hummer H1 running on biodiesel will emit fewer greenhouse gases than a 2005 Honda Civic Hybrid. I guess this is why Sean is fond of saying, "Mother Nature doesn't care what you drive; she cares what you burn."

Figure 16.6-53. Fuel may be dispensed using a low-cost electric fuel-dispensing pump such as the model available from Biodiesel Solutions, or it may be gravity fed. In either case, ensure that a 5-micron or small marine-type fuel filter with a water extraction feature or agglomerator (glass bowl to the right of the fuel supply meter) is fitted in the fuel supply line. Peter Schneider of VeggieGas Company is shown filling up a Hummer model H1 running on B100 biodiesel. The Hummer will produce fewer greenhouse gases than a 2005 Honda Civic Hybrid.

17
Living with Renewable Energy

Solar Thermal Systems

Once a solar thermal system has been installed and commissioned, very little maintenance is required. For the majority of homeowners, sweeping snow from the collectors (assuming they are safely accessible) is the limit of day-to-day maintenance.

The majority of systems installed throughout North America use non-toxic antifreeze known as propylene glycol to transfer heat from the collector panels to the storage tank in the house. Propylene glycol is rated for a maximum operating temperature of 325°F (163°C), which is well above the normal operating temperature of the system. During the summer months, particularly when people are on vacation, stagnation of the propylene glycol may occur as a result of the low heating load of the home. Should stagnation continue for extended periods of time, the glycol solution will become "sticky" and gum up the works.

It is recommended that a trained service technician analyze the glycol solution every second year. During this test, all system components should also be checked to ensure peak operating efficiency. You may also examine the glycol solution for pH and freeze protection using a home test kit as outlined in Chapter 5.3.

Solar Electric Systems

Well, let's get ready to break out the champagne and "throw the switch." All the wiring is done, the circuits have been checked and rechecked, and the electrical inspector has given the OK. It's time to get things running. So where do you start? It can't really be that simple. After all, there's a lot of equipment involved.

Getting Started

The best place to start is with safety and also with some background on operating a system. Each source of renewable energy, whether it's a wood stove or PV panels, has an operating guide supplied with it. All manufacturers provide a list of the safety tools and supplies you will need when working with their products. This equipment has been covered in the previous chapters, so go back and review the material as necessary. The most obvious requirements are common sense and the ability to learn as you work with the various components.

Once all of the systems are checked out in accordance with the manufacturers' requirements, we can start things moving. Obviously, the first thing to do is to energize the system as discussed in the following paragraphs.

Figure 17-1. The coal-fired grid may become a thing of the past if sufficient renewable energy and efficiency measures are adopted by society at large. (Courtesy Ontario Clean Air Alliance)

Step 1. Record the specific gravity for each battery cell prior to initial operation unless you are using maintenance-free or NiCad batteries, in which case this step is not required. Recording specific gravity will provide a basis for determining the state of the batteries upon receipt and may be required if you have any warranty issues. Use a felt pen or marking plate to give each individual cell a number. If your system is 12 V, you will have 6 cells. Likewise 24 V and 48 V systems will have 12 and 24 cells respectively per battery bank. If you have multiple sets of batteries wired in parallel, such as those shown in Figure 10-19, you will have double the number of cells in your bank.

Step 2. Energize the main circuit breaker connecting the batteries to the inverter. At this point, the inverter will be ready for startup. Follow the manufacturer's suggested startup procedure to get the unit online. You will need to set up the following basic parameters:
- battery charger rate;
- battery bank capacity;
- battery type;
- over-discharge protection;
- search mode sensitivity.

More advanced inverters require you to set many more options, and this can be done initially or on an as-needed basis. For first-time operators, this would be a good time to have your system dealer available to do a bit of "hand holding."

Step 3. If your inverter is equipped with a "search mode" option, set the search sensitivity. This setting adjusts the current threshold required to bring the inverter out of the low-power "sleep" mode and into full-power operation. As we learned in Chapter 1.3, the search mode works by sending a brief pulse of power into the power lines. When all circuits are turned off, the flow of electricity is stopped. At the instant the circuit is closed (when a light switch is turned on, for example), the current pulse travels through the load and back to the inverter, signaling it to activate. Sleep mode saves considerable electrical energy when the inverter is not required, for example when everyone is sleeping.

Search mode can be tricky to adjust, as the sensitivity differs depending on which load is activated. For example, an 8-Watt CF lamp uses far less power than a toaster. Following the inverter manual instructions, perform this adjustment using a variety of electrical loads in the house.

Step 4.   Program the energy meter according to the manufacturer's instructions. Typical data required include the battery bank voltage and capacity (in amp-hours). With all renewable source supplies turned off, the inverter will be drawing power of varying amounts depending on the electrical loads activated at the time. Try turning a light on or off to see the amount of current draw. Each load will require more or less current (in amps), causing the meter to register accordingly.

Step 5.   Close the circuit breaker for the first renewable-source supply. The input power will vary depending on whether it is sunny and/or windy outside. The energy meter should now start to record the energy going into the battery. A common problem at startup is the current shunt wiring (Figure 14-22) that is connected to the meter backwards. A quick test is to turn off all electrical loads and monitor the meter when energy is being produced. The meter should indicate a "battery charging" condition, with the capacity moving slowly towards the full state. Repeat this step for each renewable source.

Step 6.   Close the circuit breaker for the generator feed inputs. Start the generator (if one is installed) either manually or by activating the autostart function in accordance with the inverter instruction manual. Once the generator has started and stabilized for about twenty seconds, the inverter will "click over" to battery-charge mode.

When this occurs, the batteries should start to charge at the inverter's maximum rate (or as programmed). The energy meter will record the resulting power input and charging will progress.

While generator charging is in progress, verify that the house electrical power is active. All lights and appliances should be able to run as if they were running on the inverter. Note that the generator power may add a "flicker" condition to lights. This is completely normal operation.

Step 7.   If the battery specific gravity is lower than 75% full upon receipt, as determined in Step 1, allow the generator to operate for one charging cycle as directed by the inverter manual. This will ensure that the battery electrolyte is up to 100% capacity as you begin to work with your new system.

Figure 17-2. Saving energy is always more economical than generating additional energy. This is true whether you are talking about an urban homeowner or the electrical utility.

Step 8. Take time to read the energy meter manual to understand the relationship between the meter and batteries. Remember that the meter is only a "guesstimate" of actual battery state of charge.

Up and Running

The real beauty of a renewable energy system is its ease of operation once everything is commissioned and running. Correct operation and proper attention to the system will prolong its life. The most neglected part of a system is the batteries, as those unfamiliar with renewable energy systems often do not realize how much energy they are using or understand the relationship between personal energy use and battery depth of discharge.

If there is any point that needs to be made clear from the beginning it is that you do not have a line of credit with your battery bank. There is only so much "cash" in the savings account. Use it up and that's it; the lights go out. Keep draining the bank continually and the bank will close permanently.

Read the battery manual to ensure that you understand how your energy usage, energy meter, and battery specific gravity work together. **CHECK THE SPECIFIC GRAVITY EVERY DAY UNTIL YOU UNDERSTAND THIS RELATIONSHIP.** Once you fully understand the energy meter/specific gravity interaction, you need to check the battery bank only once a month. But until then, use care and follow these steps to the letter:

- Battery specific gravity will read abnormally low immediately after a full day of charging. This is because of tiny gas bubbles suspended in the electrolyte, causing the density of the fluid to decrease. (Gas bubbles are lighter than water or sulfuric acid.) Wait at least one or two hours after charging before taking the specific gravity reading.

- Likewise, specific gravity reading should not be taken when the batteries are under a full load. Wait until the fridge stops or the microwave is not being used before taking readings.

- Battery electrolyte specific gravity readings that are outside room temperature range will need to be "corrected" in accordance with the manufacturer's data.

- Use a quality hydrometer to take the readings. Look at the scale from straight on to ensure that the reading is correct.

- Compare the specific gravity readings with the battery depth-of-discharge chart in Chapter 9, Table 9-1. If the battery bank indicates that it is 90% full (i.e. 10% depth of discharge) and the energy meter agrees, great! If not, the meter may need some fine-tuning if the readings are too far out of whack. You may wish to look at a figure called the *charge efficiency factor* setting. This is a fancy term for the fudge factor to help calibrate the meter-to-battery setting. Decrease the efficiency factor if the meter thinks the batteries are charged more fully than they really are. Likewise, increase the efficiency factor if the meter reads lower than the specific gravity reading.

- As time goes by, the readings of even the best energy meter will fall out of step with the batteries. At this point, read the meter manual to determine how to reset the meter to agree with the battery capacity. This step is normally accomplished when the batteries are fully charged immediately after an equalization charge, approximately four times per year.

- Speaking of equalization, watch the specific gravity between cells. The readings should all be approximately the same. If the cells are starting to get out of balance by a reading of more than 0.010 (ten points on the scale), it's time to equalize.

- Equalization may be accomplished using grid power, a generator, or better yet, the PV panel or the wind turbine on sunny/windy days. Whatever method you choose, it is necessary to "program" the charge controller and/or inverter to start the equalization process. Many charge controllers have a single button which increases the battery maximum charge voltage, thus enabling equalization mode. The inverter will often have a

Figure 17-3. Saving energy is not limited to your home. Hybrid vehicles such as this model from Toyota will reduce your automotive energy bill by 50% and smog-forming emissions by 90%.

similar button or setting which is activated when equalization is started from a generator.

- If you use Hydrocaps with your battery bank you **MUST** remove them during equalization.

What Else?

There really isn't much else to think about. Equipment manufacturers will give you a list of yearly maintenance steps to follow. Some are absurdly simple, such as spring-cleaning the PV cells to wash off any dirt on the glass. Wind turbines require an inspection, although even this can be pretty simple. Bergey WindPower suggests checking to see if the unit is turning once every year....Honestly, it's right there in the manual!

Of course there are some tricks you will learn over time that help make living with renewable energy easier:

- Watch the battery specific gravity or energy meter when you have a long-term blackout of one day or longer. This is where many newcomers to renewable energy get into trouble with their batteries. Recognize when the batteries are depleted enough to warrant shutting the system down

or running a generator.
- During severe lightning storms, consider furling your wind turbine to limit mechanical stress during the storm. Consult the manufacturer's manual for additional details.
- Make seasonal adjustments to a manual PV array: latitude +15E towards the vertical for winter; latitude -15E for summer.
- Remember never to smash ice and snow off the PV panels. Simply brush off the top layer of snow with a squeegee or brush. The sun will quickly take care of the rest.
- Rest easy knowing that you are not the only one living lightly on the planet. Support abounds from dealers, like-minded neighbors, the Web, and publications such as *Private Power Magazine* and *Home Power* (www.homepower.com). Renewable energy is here to stay.

Most importantly, enjoy your handiwork and marvel at the elegant simplicity of it all.

Figure 17-4. Heating your pool or hot tub with solar energy eliminates fossil-fuel heating, extending the swimming season without increasing the operating cost.

18
Conclusion

Renewable energy works. Yes, there is a lot of technology to understand before embarking on this path, but the rewards make every bit of effort worthwhile. Lorraine tells me she feels more "connected" in her actions and her interaction with the environment.

Once you starting harvesting your own energy and seeing what it takes to heat a house or run a light bulb, endless, mindless waste becomes intolerable. As we approach the end of 2005, hundreds of thousands of people are moving toward a more sustainable life style. This movement started with the hippies in the '60s, and after the psychedelic haze left that era environmental concerns and technology started to enter the mainstream. Renewable energy has now arrived.

Companies such as BP, Kyocera, Siemens, and Xantrex are not in the business to help out ageing hippies; they are here because renewable energy is business—big business. Governments, NGOs, industry, and grassroots people like you and me are making the transition to a cleaner, more sustainable future for the sake of our children and the planet. Let's hope that the rest of society gets the message before it's too late.

"To see a world in a grain of sand, and heaven in a wildflower; to hold infinity in the palm of your hand and eternity in an hour - is inspiration."

William Blake

Appendix 1
Cross Reference Chart of Various Fuel Energy Ratings

Fossil Fuels and Electricity

Heating Fuel	BTU per Unit
Heating Oil	142,000 BTU/gallon (38,700 kj/L)
Natural Gas	46,660 BTU/cubic-yard (37,700 kj/m3)
Propane	91,500 BTU/gallon (26,900 kj/L)
Electricity (resistance heating)	3413 BTU/kWh (3600 kj/kWh)
Coal (air dry average)	12,000 BTU/LB (27,900 kJ/kg)

Renewable Energy Heating Fuels

Heating Fuel	BTU per Unit
Shelled Corn	7000 BTU/lb (16,200 kJ/kg)
	14,000,000 BTU/ton (12,700 MJ/tonne)
Firewood by weight (all types) *	8000 BTU/lb (18,500 kJ/kg)
Hardwood Firewood by volume: * Ash	25,800,000 BTU/cord (27,200 MJ/cord)
Beech	28,900,000 BTU/cord (30,500 MJ/cord)
Red Maple	22,300,000 BTU/cord (23,500 MJ/cord)
Red Oak	27,200,000 BTU/cord (28,700 MJ/cord)
Hybrid Poplar	18,500,000 BTU/cord (19,500 MJ/cord)
Mixed Hardwood (average)	27,000,000 BTU/cord (30,000 MJ/cord)
Mixed Softwood (average)	17,500,000 BTU/cord (18,700 MJ/cord)
Wood Pellets	20,700,000 BTU/ton (19,800 MJ/tonne)
Biodiesel	128,000 BTU/gallon (35,500 kJ/L)

* All firewood has the same heating or carbon content per pound or kilogram of mass. However, the density of softwoods is much lower owing to increased air and moisture content, resulting in lower BTU content per unit mass.

Appendix 2
Typical Power and Electrical Ratings of Appliances and Tools

Appliance Type	Power Rating (Watts)	Energy Usage per Hour, Day or Cycle
Large Appliances:		
Gas clothes dryer	600	500 Wh per dry cycle
Electric clothes dryer	6,000	5 kWh per dry cycle
High efficiency clothes washer	300	250 Wh per wash
Ten-year-old vertical axis clothes washer	1,200	720 Wh per wash
Ten-year-old refrigerator	720	5 kWh per day
New energy-efficient refrigerator	150	1.2 kWh per day
Ten-year-old chest freezer	400	3 kWh per day
New energy-efficient chest freezer	140	0.9 kWh per day
Dishwasher "normal cycle"	1,500	800 Wh per cycle
Dishwasher "eco-dry cycle"	600	300 Wh per cycle
Portable vacuum cleaner	600	600 Wh per hour
Central vacuum cleaner	1,400	1.4 kWh per hour
Air conditioner 12,000 BTU (window)	1,200	1.2 kWh per hour
AC submersible well pump (1/2 hp)	1,150	200 Wh per cycle
DC submersible well pump	80	160 Wh per day
DC slow pump (includes booster pump)	80	160 Wh per day

Appliance Type	Power Rating (Watts)	Energy Usage per Hour, Day or Cycle
Small Appliances:		
Microwave oven (0.5 cubic foot)	900	0.9 kWh per hour
Microwave oven (1.5 cubic foot)	1,500	1.5 kWh per hour
Drip style coffee maker (brew cycle)	1,200	1.2 kWh per hour
Drip style coffee maker (warming cycle)	300	0.3 kWh per hour
Espresso/cappuccino maker	1,200	300 Wh per cycle
Food processor	300	50 Wh per cycle
Coffee grinder	100	10 Wh per cycle
Toaster	1,200	150 Wh per cycle
Blender	300	50 Wh per cycle
Hand mixer	100	10 Wh per cycle
Hair dryer	1,500	200 Wh per cycle
Curling iron	600	100 Wh per cycle
Electric toothbrush	2	50 Wh per day
Electric iron	1,000	1 kWh per hour

Appendix 2 Continued

Appliance Type	Power Rating (Watts)	Energy Usage per Hour, Day or Cycle
Electronics:		
Television –12 inch B&W	20	20 Wh per hour
Television –32 inch color	140	140 Wh per hour
Television –50 inch high definition	160	160 Wh per hour
Satellite dish and receiver	25	25 Wh per hour
Stereo system	50	50 Wh per hour
Home theater system (movie)	400	1 kWh per movie
Cordless phone	3	72 Wh per day
Cell phone in charger base	3	72 Wh per day
VCR/DVD/CD component	25	25 Wh per hour
Clock radio (not including inverter waste)	5	120 Wh per day
Computer (Desktop)	60	60 Wh per hour
Computer (Laptop)	20	20 Wh per hour
Monitor - 15 inch	100	100 Wh per hour
Monitor - 15 inch flat screen	30	30 Wh per hour
Laser printer (standby mode average)	50	50 Wh per hour standby
Laser printer (print mode)	600	600 Wh per hour printing
Inkjet printer (all modes)	30	30 Wh per hour
Fax machine	5	120 Wh per day
PDA charging	3	72 Wh per day
Fluorescent desk lamp	10	10 Wh per hour

Appendix 3
Resource Guide

Energy Efficiency Councils and Societies:

American Council for an Energy Efficient Economy
Website: www.aceee.org
Phone: 202-429-8873
Publishes guides' comparing the energy efficiency of appliances

American Solar Energy Society
Website: www.ases.org
Phone: 303-443-3130
ASES is the United States chapter of the world Solar Energy Society. They promote the advancement of solar energy technologies.

The American Wind Energy Association
Website: www.awea.org
Phone: 202-383-2500
The AWEA is the trade association for developers and manufacturers of wind turbine and associated equipment and infrastructure.

California Energy Commission
Website: www.energy.ca.gov
Phone: 916-654-4058
The CEC is the strongest supporter of grid inter-connected renewable energy systems in North America. Their website explores what is happening in California in this regard.

Canadian Standards International
Website: www.csa-international.org
Phone: 416-747-4000
Develops standards for the Canadian marketplace. Tests and administers safety certification work in North America.

Canadian Wind Energy Association
Website: www.canwea.ca
Phone: 800-992-6932
The CAWEA vision is to have 10,000 MW of wind power systems installed in Canada by 2010. They promote all aspects of wind energy and related systems to the industry.

David Suzuki Foundation

Website: www.davidsuzuki.org
Phone: 614-732-4228

Dr. Suzuki is a lecturer and TV broadcaster promoting energy efficiency, global climate change and ocean sustainability. The website provides links and publications on all manner of environmental sciences.

Electro Federation of Canada

Website: www.micropower-connect.org
Phone: 905-602-8877

The Electro Federation is a consortium of electrical manufacturers working in many disciplines of electrical engineering and sales. The micropower-connect division is dealing with small (<50kW) distributed energy producers interconnecting to the grid in Canada.

Energy Star

Website: www.energystar.gov
Phone: 888-782-7937

Their website reviews energy efficient computers and electronics.

The Green Power Network

Website: www.eren.doe.gov/greenpower

This website describes the status of utility interconnection guidelines on a state-by-state basis. Also provides information on where to purchase green electricity when connected to the grid.

National Renewable Energies Laboratory (NREL)

Website: www.nrel.gov
Phone: 303-275-3000

The NREL is the national renewable energy research laboratory in the United States.

Natural Resources Canada

Website: www.nrcan.gc.ca
Phone: N/A

The Government of Canada hosts this website which includes the office of energy efficiency. Many resources are presented in this fact filled site.

Rocky Mountain Institute

Website: www.rmi.org
Phone: 970-927-3851

The Rocky Mountain Institute is a think tank regarding all energy efficiency issues. Their website contains a great deal of source information for books and applied research.

Underwriters Laboratories Inc.

Website: www.ul.com
Phone: 847-272-8800

Develops standards for the United States marketplace. Tests and administers safety certification work in North America.

Trade Publications and Magazines:

Home Power Magazine

Website: www.homepower.com
Phone: 800-707-6585

This magazine bills itself as "The hands-on journal of home-made power". Based in Oregon, the magazine deals primarily with a south and west coast flavor. Extensive details related to producing alternate energy.

Private Power Magazine

Website: www.privatepower.ca
Phone: 800-668-7788

Private Power is a new comer to the Canadian market place. Focuses on hybrid installations required in the North.

Alternative Power Magazine

Website: www.altpowermag.com
Phone: N/A

This web-based magazine offers general news-like stories related to all aspects of alterative power for home, automobiles and industry.

Manufacturers:

Photovoltaic Panels and Equipment

Photovoltaic panel manufacturers do not supply directly to end consumers, as they rely on their large distribution networks throughout the world. For information purposes, here are some of the major suppliers in this field. Contact the manufacturer or visit their website for distributors in your local area.

AstroPower

Website: www.astropower.com
Phone: 302-366-0400

BP Solar

Website: www.bpsolar.com
Phone: 410-981-0240

Evergreen Solar Inc.

Website: www.evergreensolar.com
Phone: 508-357-2221

Kyocera Solar Inc.

Website: www.kyocerasolar.com
Phone: 800-544-6466

Matrix Solar Technologies

Website: www.matrixsolar.com
Phone: 505-833-0100

RWE Schott Solar

Website: www.asepv.com
Phone: 800-977-0777

Schott Applied Power Corporation

Website: www.schottappliedpower.com
Phone: 888-457-6527

Sharp USA

Website: www.sharpusa.com
Phone: 800-BE-SHARP

Siemens Solar Inc.

Website: www.siemenssolar.com
Phone: 877-360-1789

Solardyne Corporation

Website: www.solardyne.com
Phone: 503-244-5815
Manufacturer of small solar modules for charging laptop computers, cell phones, etc.

Solarex

Website: www.solarex.com
Phone: 301-698-4200
Solarex was recently acquired by BP Solar

PV Module Mounts

Array Technologies Inc.

Website: www.wattsun.com
Phone: 505-881-7567
Manufacturer of the Wattsun active tracking system

Sun-Link Solar Tracker

Website: www.northernlightsenergy.com
Phone: 705-246-2073
Manufacturer of Sun-Link active tracking system.

Two Seas Metalworks

Website: www.2seas.com
Phone: 877-952-9523
Manufacturer of fixed PV racks. Also supply battery racks.

UniRac Inc.

Website: www.unirac.com
Phone: 505-242-6411

Zomeworks Corporation

Website: www.zomeworks.com
Phone: 800-279-6342
Manufacturer of fixed and passive tracking mount systems

Wind Turbines:

Atlantic Orient Corporation

Website: www.aocwind.net
Phone: 802-333-9400

Bergey Windpower Inc.

Website: www.bergey.com
Phone: 405-364-4214

Southwest Windpower

Website: www.windenergy.com
Phone: 520-779-9463

Aeromax Corporation
Website: www.aeromaxwindenergy.com
Phone: 888-407-9463

Bornay Windturbines
Website: www.bornay.com
Phone: +34-965-560-025
Manufacturer of wind turbines from Spain

Lake Michigan Wind and Sun
Website: www.windandsun.com
Phone: 920-743-0456
New and re-built turbines

Jack Rabbit Energy Systems
Website: www.jackrabbitmarine.com
Phone: 203-961-8133

Windstream Power Systems Inc.
Website: www.windstreampower.com
Phone: 802-658-0075

Wind Turbine Industries Corporation
Website: www.windturbine.net
Phone: 952-447-6064

True North Power Systems
Website: www.truenorthpower.com
Phone: 519-793-3290

Windturbine.ca
Website: www.windturbine.ca
Phone: 886-778-5069

Micro Hydro Turbine Systems
Canyon Industries Inc.
Website: www.canyonindustriesinc.com
Phone: 360-592-5552

Energy Systems and Design
Website: www.microhydropower.com
Phone: 506-433-3151

Harris Hydroelectric
Website: www.harrishydro.com
Phone: 831-425-7652

HydroScreen Co. LLC
Website: www.hydroscreen.com
Phone: 303-333-6071
Manufacturer of intake screen for micro-hydro systems

Jack Rabbit Energy Systems
Website: www.jackrabbitmarine.com
Phone: 203-961-8133
Home Power Systems
Website: www.homepower.ca
Phone: 604-465-0927

Battery Manufacturers

Dyno Battery Inc.
Website: www.dynobattery.com
Phone: 206-283-7450

HuP Solar-One Battery
Website: www.hupsolarone.com
Phone: 208-267-6409

IBE Battery
Website: www.ibe-inc.com
Phone: 818-767-7067

Rolls Battery Engineering (USA)
Surrette Battery Company (Canada)
Website: www.surrette.com
Phone: 800-681-9914

Trojan Battery Company
Website: www.trojanbattery.com
Phone: 800-423-6569

U.S. Battery Manufacturing Company
Website: www.usbattery.com
Phone: 800-695-0945

Hydrogen Recombining Caps
Hydrocap Catalyst Battery Caps
Website: N/A
Phone: 305-696-2504

D.C. Voltage Regulators
Morningstar Corporation
Website: www.morningstarcorp.com
Phone: 215-321-4457

RV Power Products
Website: www.rvpowerproducts.com
Phone: 800-493-7877

Steca Gmbh
Website: www.stecasolar.com
Phone: N/A

Xantrex Technology Inc.
Website: www.xantrex.com
Phone: 360-435-2220

Inverters
ExelTech Inc.
Website: www.exeltech.com
Phone: 800-886-4683

Out Back Power Systems
Website: www.outbackpower.com
Phone: 360-435-6030

SMA America Inc.
Website: www.sma-america.com
Phone: 530-273-4895

Xantrex Technology Inc.
Website: www.xantrex.com
Phone: 360-435-2220

Backup Power Gensets

Epower
Website:
www.epowerchargerboosters.com
Phone: 423-253-6984

Generac Power Systems Inc.
Website: www.generac.com
Phone: N/A (sold through Home Depot)

Hardy Diesel & Equipment Inc.
Website: www.hardydiesel.com
Phone: 800-341-7027

Kohler Power Systems
Website: www.kohlerpowersystems.com
Phone: 800-544-2444

Energy Meters

Bogart Engineering
Website: www.borartengineering.com
Phone: 831-338-0616

Brand Electronics
Website: www.brandelectronics.com
Phone: 207-549-3401

Xantrex Technology Inc.
Website: www.xantrex.com
Phone: 360-435-2220

Miscellaneous

Bussmann
Website: www.bussmann.com
Phone: 314-527-3877
Fuses and electrical safety components

Delta Lightning Arrestors Inc.
Website: www.deltala.com
Phone: 915-267-1000

Digi-Key (Canada and USA)
Website: www.digikey.com
Phone: 800-DIGI-KEY
Supplier of many electrical wiring components

Electro Sonic Inc. (Canada)
Website: www.e-sonic.com
Phone: 800-56-SONIC
Supplier of many electrical wiring components

Real Goods
Website: www.realgoods.com
Phone: 800-919-2400
Suppliers specializing in renewable energy systems

Siemens Energy and Automation Inc.
Website: www.siemens.com
Phone: 404-751-2000
Fused disconnect switches and circuit breakers

Xantrex Technology Inc.
Website: www.xantrex.com
Phone: 360-435-2220
D.C. circuit breakers, battery cables, power centers, metering shunts, fuses

Appendix 4
Magnetic Declination Map for North America

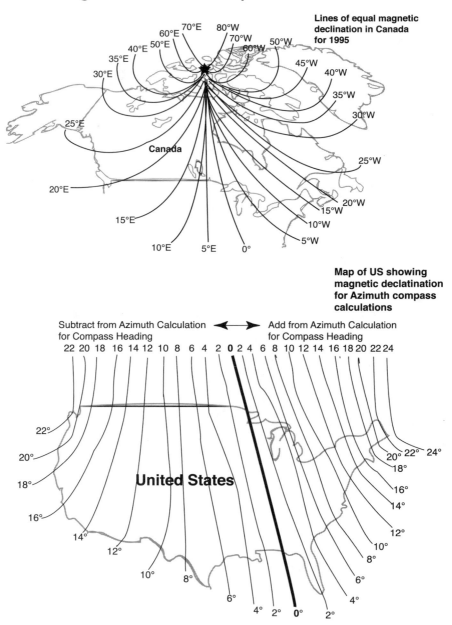

Map indicates the correction heading required to locate true north (and solar south). If you live along the line running from Manitoba through to southern Texas, your heading will be due north when the compass is pointing approximately 8 degrees east.

Appendix 5
Map of Winter Average Sun Hours per Day
for North America

"North American Sun Hours per Day (Worst Month)"

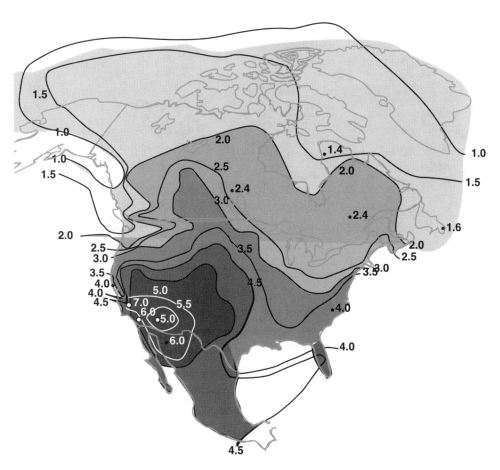

Map indicates the average winter sun hours per day. Use this map when calculating the average energy output of a PV system used all year round.

Appendix 6
Map of Yearly Average Sun Hours per Day for North America

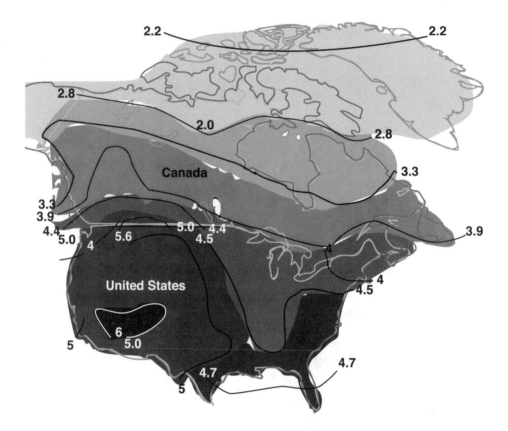

Map indicates the yearly average of sun hours per day. Use this map when calculating the average energy output of a PV system used seasonally during the spring/summer months.

Appendix 7
Electrical Energy Consumption Worksheet

Appliance Type	Appliance Wattage (Volts x Amps)	X	Hours of Daily Use	=	Average Watt-hours Per Day

Total Watt-hours Per Day for all Appliances =

Start by checking your major electrical loads in the house. Anything that plugs into a regular wall socket will have a label that tells you the voltage (usually 120 or 240 for larger appliances) and the current or wattage for that item. Although these labels can over state energy usage, they are a good guide for calculating energy consumption. Enter the wattage data from the labels on the *Energy Consumption Worksheet*. (If the appliance label does not show wattage, multiply amps x volts to calculate watts.)

The next step is to see if you can estimate how many hours you use the device per day. If you use a device only occasionally, try to estimate how long it is used per week and divide this time by 7. Your calculator will quickly give you your "daily average energy usage".

Appendix 8
Average Annual Wind Speed Map for North America

Wind Class / Speed Chart
Class 1:   3.8 m/s (8.5 mph)
Class 2:   4.8 m/s (10.8 mph)
Class 3:   5.4 m/s (12.1 mph)
Class 4:   5.8 m/s (13.0 mph)
Class 5:   6.2 m/s (13.9 mph)
Class 6:   6.7 m/s (15.0 mph)
Class 7:   7.5 m/s (16.8 mph)

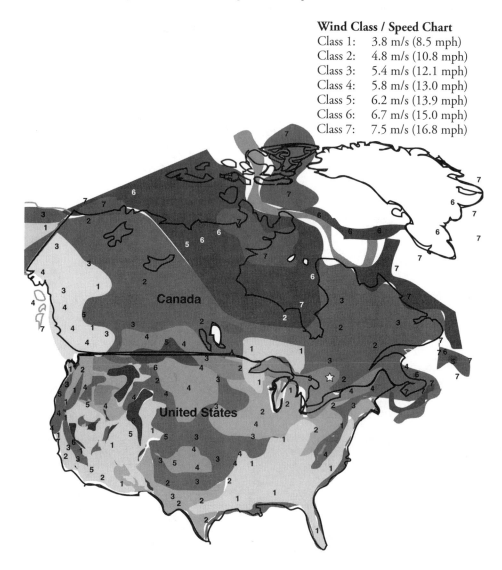

Map indicates the average annual wind speed in meters per second averaged over a ten-year period. Elevation of anemometer is 33 feet (10 m). Use this map with care as local wind speed levels will vary. Refer to resource guide for further details on wind mapping for your specific area.

Appendix 9a
Polyethylene SDR Pipe Friction Losses
Pressure Loss from Friction in Feet of Head per 100 Feet (30 m) of Pipe

(Table Courtesy of Energy Systems and Design)

Flow in US GPM	Pipe Diameter in Inches							
	0.5	0.75	1	1.25	1.5	2	2.5	3
1	1.13	0.28	0.09	0.02				
2	4.05	1.04	0.32	0.09	0.04			
2	8.60	2.19	0.67	0.19	0.09	0.02		
4	14.6	3.73	1.15	0.30	0.14	0.05		
5	22.1	5.61	1.75	0.46	0.21	0.07		
6	31.0	7.89	2.44	0.65	0.30	0.09	0.05	
7	41.2	10.5	3.24	0.85	0.42	0.12	0.06	
8	53.1	13.4	4.14	1.08	0.51	0.16	0.07	
9		16.7	5.15	1.36	0.65	0.18	0.08	
10		20.3	6.28	1.66	0.78	0.23	0.09	0.02
12		28.5	8.79	2.32	1.11	0.32	0.14	0.05
14		37.9	11.7	3.10	1.45	0.44	0.18	0.07
16			15.0	3.93	1.87	0.55	0.23	0.08
18			18.6	4.90	2.32	0.69	0.30	0.09
20			22.6	5.96	2.81	0.83	0.35	0.12
22			27.0	7.11	3.36	1.00	0.42	0.14
24			31.7	8.35	3.96	1.17	0.49	0.16
26			36.8	9.68	4.58	1.36	0.58	0.21
28				11.1	5.25	1.56	0.67	0.23
30				12.6	5.96	1.77	0.74	0.25
35				16.8	7.94	2.35	1.00	0.35
40				21.5	10.2	3.02	1.27	0.44
45				26.8	12.7	3.75	1.59	0.55
50				32.5	15.4	4.55	1.91	0.67
55					18.3	5.43	1.96	0.81
60					21.5	6.40	2.70	0.94
65					23.8	7.41	3.13	1.08
70					28.7	8.49	3.59	1.24
75					32.6	9.67	4.07	1.4
80						10.9	4.58	1.59
85						12.2	5.13	1.77
90						13.5	5.71	1.98
95						15.0	6.31	2.19
100						16.5	6.92	2.42
150						34.5	14.7	5.11
200							25.0	8.70
300								18.4

Appendix 9b
PVC Pressure Class 160 PSI Pipe Friction Losses
Pressure Loss from Friction in Feet of Head per 100 Feet (30 m) of Pipe

(Table Courtesy of Energy Systems and Design)

Flow in US GPM	Pipe Diameter in Inches							
1	0.05	0.02						
2	0.14	0.05	0.02					
2	0.32	0.09	0.04					
4	0.53	0.16	0.09	0.02				
5	0.80	0.25	0.12	0.04				
6	1.13	0.35	0.18	0.07	0.02			
7	1.52	0.46	0.23	0.08	0.02			
8	1.93	0.58	0.30	0.10	0.04			
9	2.42	0.71	0.37	0.12	0.05			
10	2.92	0.87	0.46	0.16	0.07	0.02		
11	3.50	1.04	0.53	0.18	0.07	0.02		
12	4.09	1.22	0.64	0.20	0.09	0.02		
14	5.45	1.63	0.85	0.28	0.12	0.04		
16	7.00	2.09	1.08	0.37	0.14	0.04		
18	8.69	2.60	1.33	0.46	0.18	0.07		
20	10.6	3.15	1.63	0.55	0.21	0.09	0.02	
22	12.6	3.77	1.96	0.67	0.25	0.10	0.02	
24	14.8	4.42	2.32	0.78	0.30	0.12	0.04	
26	17.2	5.13	2.65	0.90	0.35	0.14	0.05	
28	19.7	5.89	3.04	1.04	0.41	0.16	0.05	
30	22.4	6.70	3.45	1.17	0.43	0.18	0.05	
35		8.90	4.64	1.56	0.62	0.23	0.07	
40		11.4	5.89	1.98	0.78	0.30	0.09	0.02
45		14.2	7.34	2.48	0.97	0.37	0.12	0.04
50		17.2	8.92	3.01	1.20	0.46	0.14	0.04
55		20.5	10.6	3.59	1.43	0.55	0.16	0.05
60		24.1	12.5	4.21	1.66	0.64	0.18	0.07
70			16.6	5.61	2.21	0.85	0.25	0.09
80			21.3	7.18	2.83	1.08	0.32	0.12
90				8.92	3.52	1.36	0.39	0.14
100				10.9	4.28	1.66	0.48	0.18
150				23.2	9.06	3.5	1.04	0.37
200					15.5	5.96	1.75	0.62
250					23.4	9.05	2.65	0.94
300						12.6	3.73	1.34
350						16.8	4.95	1.78
400						21.5	6.33	2.25
450							7.87	

Appendix 10
Voltage, Current and Distance (in feet) Charts for Wiring (AWG gauge)
(1% Voltage drop shown – Increase distance proportionally if greater losses allowed)

12 Volt Circuit

Amps	WIRE GAUGE									
	10	8	6	4	2	1	1/0	2/0	3/0	4/0
1	48	74	118	187	299	375	472	594	753	948
5	10	15	24	37	60	75	94	119	151	190
10	5	7	12	19	30	38	47	59	75	94
20	2	4	6	9	15	19	24	30	38	48
40	1	2	3	5	7	9	12	15	19	24

24 Volt Circuit

Amps	WIRE GAUGE									
	10	8	6	4	2	1	1/0	2/0	3/0	4/0
10	10	15	24	37	60	75	94	119	151	190
20	5	7	12	19	30	38	47	59	75	94
30	3	5	8	12	20	25	31	40	50	63
40	2	4	6	9	15	19	24	30	38	48
50	2	2	4	6	10	13	16	20	25	31
100	1	1	2	4	6	8	9	12	15	19
125	1	1	2	3	5	6	8	10	13	16
150	1	1	2	2	4	5	6	8	10	13

48 Volt Circuit

Amps	WIRE GAUGE									
	10	8	6	4	2	1	1/0	2/0	3/0	4/0
10	20	30	48	74	120	150	188	238	302	380
20	10	15	24	38	60	76	94	118	150	188
30	6	10	16	24	40	50	62	80	100	126
40	4	8	12	18	30	38	48	60	76	96
50	4	4	8	12	20	26	32	40	50	62
100	2	2	4	8	12	16	18	24	30	48
125	2	2	4	6	10	12	16	20	26	32
150	2	2	4	4	8	10	12	16	20	26

Appendix 11
Wire Sizes Versus Current Carrying Capacity
(Size shown is for copper conductor only)

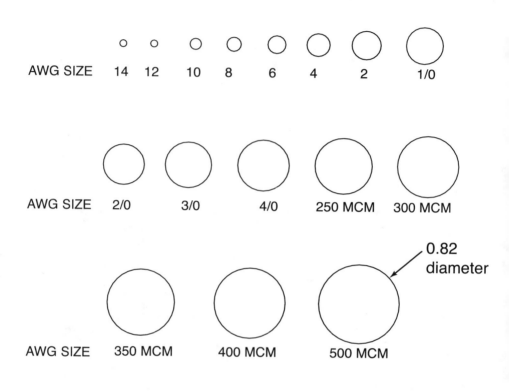

AWG	Maximum Current
14	15
12	20
10	30
8	55
6	75
4	95
2	130
0	170
2/0	195
3/0	225
4/0	260

Appendix 12

A Few Words on Toilets, Greywater, and Waste Management

Like everything else in the city, potable water and waste management are invisible services provided to citizens at an impossibly low price. The average North American takes for granted that clean water magically flows from a tap and that human wastes simply disappear (which of course, is a good thing). Yet we pay a mere dollar or so a day for the privilege, some would say the right, of having these essential services. I placed a phone call to the municipal department of a major city and asked if they could provide me with the cost breakdown for city water. This list included items such as alum, chlorine, filtration, and pumping. I then asked, "So how much do you charge for the actual water?" His reply: "Oh, the water doesn't cost anything. We just pump it. It's the chemicals and purification that cost money." I remember his words every time I go to the developing world and watch people wash food and clothes in fetid puddles, if any are to be found. Southern California and parts of the US Midwest are not too far behind, as wasteful agricultural irrigation methods and household water practices drain the mighty river and aquifer systems millions of people depend on for their survival.

Because humans are so disconnected from the ecosystems that support their very lives, few people consider what they would do if water and sewer were not available and therefore don't make an effort to reduce their current water consumption and waste effluents. Clean water is competing with fossil fuel as the natural commodity headed for extinction.

When you decide to take the plunge and embark on an off-grid life style, these questions become front and center issues, not just academic questions.

We discussed the harvesting and efficient use of water in Chapter 2.5. This appendix will provide a brief overview of wastewater treatment options for an energy-efficient, off-grid, environmentally sustainable home.

For most people, country living, wastewater, and "septic system" are synonymous. True, septic systems work well and are well understood by rural plumbers and tradespeople. However, before we delve into the mechanics of such a system, let's take a moment to review some of the theory behind what we are trying to accomplish.

Defining Wastewater

Wastewater can be divided into two broad categories:

- Greywater: wastewater that has a low pathogen density but contains a combination of organic and inorganic waste materials from everyday activities such as laundry, bathing, and dishwashing;
- Blackwater: water that contains a high-density pathogen load from human organic wastes from toilets.

Wastewater Treatment

The disposal of these waste streams can be treated as two completely different processes, although a typical septic system combines their disposal as one integrated system. The average North American consumes approximately 100 gallons of water per day which breaks down as:

Greywater:	57 gallons/216 liters
Blackwater:	38 gallons/144 liters
Potable water:	5 gallons/19 liters

Provided the greywater is not laden with large amounts of inorganic chemicals, it may be disposed of rather easily by suitable leaching through the earth. The use of biodegradable laundry detergent, bath soap and shampoo and home cleaning products will reduce inorganic compounds dramatically, placing very little stress on the ecosystem as it absorbs and breaks down the waste stream materials.

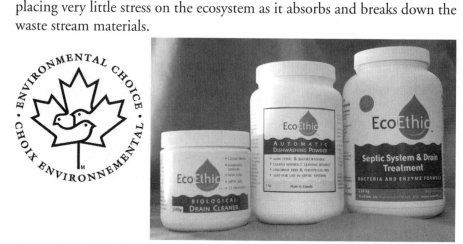

Figure A12-1. The first step in developing any wastewater treatment system is to use water as efficiently as possible and to use biodegradable cleaning products. EcoEthic Inc. (www.ecoethic.ca) provides a full line of home cleaning products that are free of phosphates, petroleum solvents, chlorine, dyes, perfumes, and other environmentally harmful ingredients. EcoEthic products, and others like them, work effectively and can be used even in sensitive wetland areas.

Blackwater, on the other hand, is loaded with pathogens that are dangerous to human life and must be eliminated. Treatment of human wastes present in blackwater is based on the nutrient cycle of all living organisms. Every organism consumes nutrients (food) and produces a waste by-product. Bacteria decompose these waste materials, turning them back into nutrients and leaving the water in the waste stream to be returned to the ground water table.

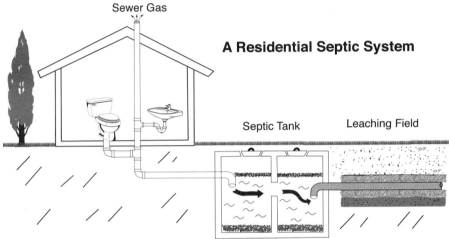

Figure A12-2 The residential septic system is a full sewage treatment plant that processes and purifies both the blackwater and greywater waste streams simultaneously.

Blackwater (toilet waste) is over 99% water, which can be evaporated or clarified and allowed to mix with the greywater stream. The residential septic system is a full sewage treatment plant that processes and purifies both the blackwater and the greywater waste streams simultaneously. The septic system shown in Figure A12-2 allows blackwater and greywater to mix in the main sewer "stack" pipe of the home and feeds this waste stream to the septic tank, where baffles cause the solids to drop to the bottom while turbid water is allowed to flow across the tank baffles and out to the leaching field.

Figure A12-3 shows the connection of household appliances to the main waste stack. Note that the waste stream flows down to the septic tank under the influence of gravity, while "sewer gases" flow upwards to the roof vent. To prevent sewer gases from flowing into the home, plumbing appliances are equipped with either a wet or a dry "gas trap." The wet trap is used in sinks, showers, and washing machines and works by allowing a small amount of water to remain in a U-shaped section of pipe, preventing sewer gas from

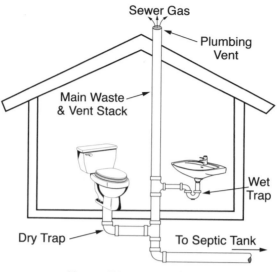

House Waste Plumbing

Figure A12-3. This detailed view shows the connection of household appliances to the main waste stack.

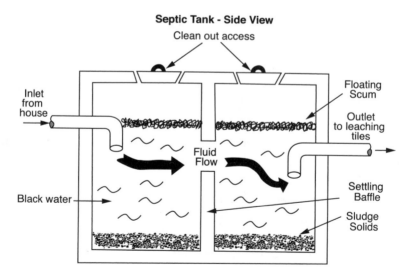

Figure A12-4. Septic tanks may be solid concrete or plastic and are buried just below the surface of the earth. Cleanout access doors are provided for periodic pumping of residual solids that are not broken down during the digestion stage. Septic tanks must be pumped out every three to five years or solids will flow out to the leaching bed, plugging the system and requiring extensive and costly repairs.

flowing into the drain and out into the room. A toilet uses a dry trap arrangement, which provides a wet U-trap within the toilet structure itself.

Figure A12-4 shows a detailed view of a septic tank. They may be either concrete or plastic and are buried just below the surface of the earth. Provided the home is used continuously, the tank will not freeze in winter as a result of the heat generated by the waste digestion process and warm waste water flowing into the unit. Cleanout access doors are provided for periodic pumping of residual solids that are not broken down during the digestion stage. Septic tanks must be pumped out every three to five years or solids will flow out to the leaching bed, plugging the system and requiring extensive and costly repairs.

As wastewater flows into the septic tank, internal baffles slow the flow of water and direct solids to the bottom of the tank, forming a sludge layer, while greases and lighter wastes form a floating scum layer. Turbid water remains in the central area of the tank and will periodically flow out to the leaching bed whenever additional contents are added to the tank.

Innumerable bacteria work in all three layers of the septic tank, digesting and breaking down solid effluent matter. About 95% of the solids making up the sludge and scum layers can be broken down by bacteria; the remaining 5% form a solid layer that must be pumped out as discussed earlier.

The leaching bed shown in Figure A12-5 is part of the sewage treatment and is not simply a means of ridding the system of excess water. Partially treated effluent flows from the septic tank into the leaching field where it flows into perforated pipe buried in the soil. Effluent is further broken down by ground-based bacteria and the earth's natural filtration capability as effluent and waste water percolate through the soil. The soil acts as a biological filter, removing harmful materials before they reach the groundwater table which feeds the house water supply.

A properly functioning septic system is a natural process in which ground water is recycled over and over again.

Disadvantages of Septic Systems

You would think that the septic system is the perfect answer to household waste treatment. Why consider anything else? There are a number of reasons:
* a building lot that is very close to fragile wetland ecosystems;
* the cost of a septic system;
* occasional or infrequent use of recreational property;
* a lot that is too small for a full septic system.

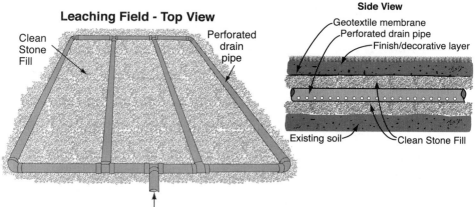

Leaching Field - Top View

Clean Stone Fill

Perforated drain pipe

From Septic Tank

Side View

Geotextile membrane

Perforated drain pipe

Finish/decorative layer

Existing soil

Clean Stone Fill

Figure A12-5. The leaching bed shown in this schematic drawing and photograph is part of the sewage treatment system and is not simply a means of ridding the septic tank of excess water.

In these cases, alternative systems to consider include:
- a full-size effluent holding tank that stores both blackwater and greywater;
- a partial holding tank and leaching bed;
- a composting toilet and leaching bed.

Holding tanks are exactly that: tanks with sufficient capacity to hold several months' worth of total effluent. Holding tanks may be also be downsized if they can be equipped with leaching beds to dispose of excess waste fluids.

Composting toilets have the advantage of decomposing waste solids directly and eliminating the need for periodic pumping. They are also less costly than full septic tanks, although a leaching bed is still required to eliminate greywater. Composting toilets can also be used in boats and RVs where waste management can be very complicated.

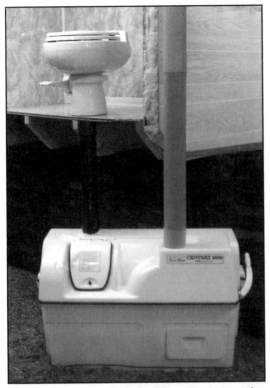

Figure A12-6. Composting toilets have the advantage of decomposing waste solids directly and eliminating the need for periodic pumping. They are also less costly than full septic tanks, although a leaching bed is still required to eliminate greywater. (Courtesy Sun-Mar Corporation)

A Sun-Mar Centrex composting toilet is shown in Figure A12-6. In this system an ultra-low flush toilet dispenses approximately 1 pint (0.5 liter) of water per flush into a central chamber located in the basement or crawlspace area. A natural "bulking" material similar to peat moss is added to the chamber every two or three days and a handle is rotated, activating the composting material and exposing it to the correct amount of oxygen and moisture. Excess liquid is drained into an evaporation chamber, allowing moisture to escape through the vent system. (Some composting systems allow for the connection of a moisture drain pit, similar to a miniature leaching bed, for the elimination of excess fluids.) Most models are available as electric or non-electric, allowing them to operate in the off-grid home without any electricity demand.

Periodically, the drum of dry, clean compost is removed from the system and may be added to potting soils or spread on flower gardens just like regular compost.

Composting toilets are also available as self-contained units, having the toilet and composting chamber fitted as one component. These units require very little space and can even be installed in a closet.

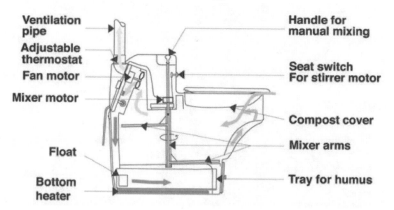

Ventilation pipe

Adjustable thermostat

Fan motor

Mixer motor

Float

Bottom heater

Handle for manual mixing

Seat switch For stirrer motor

Compost cover

Mixer arms

Tray for humus

Figure A12-7. A natural "bulking" material similar to peat moss is added to the chamber every two or three days and a handle is activated, rotating the composting material to expose it to the correct amount of oxygen and moisture. Excess liquid is drained into an evaporation chamber, allowing moisture to escape through the vent system. Most models are available as electric or non-electric, allowing them to operate in the off-grid home without any electricity demand. (Courtesy EcoEthic Inc.)

Wastewater Gardens

Figure A12-8. The wastewater garden is an innovative technology that takes advantage of the fact that plants need water and nutrients. The system "grows away" wastewater from residences, eliminating effluent discharge into the ground. (Courtesy Centre for Sustainable Watersheds)

Nutrients in wastewater are fertilizers. The wastewater garden is an innovative technology that takes advantage of the fact that plants need water and nutrients. The system "grows away" wastewater from residences, eliminating effluent discharge into the ground. The technology is particularly ideal for seasonal locations in proximity to watercourses, areas sensitive to groundwater contamination, or locations where the bedrock is close to the surface.

The system works by accepting the effluent from septic tanks, composting toilet leachate (excess moisture), and greywater and feeding it into a subsurface fluid distribution system or bed, where naturally occurring microorganisms convert the chemical constituents of the wastewater into nutrients for plant growth. Plants use the nutrients as they grow and also take up water and release it into the atmosphere. In areas where freezing occurs, the wastewater garden may be enclosed in a greenhouse or solarium.

The benefits of wastewater gardens include:

- zero discharge and no environmental contamination;
- elimination of holding tank pumpouts;
- suitability for sites with bedrock, shallow soil, poor drainage, or high water table;
- horticultural appeal.

As with all "non-standard" plumbing systems, contact your local public health unit for detailed assistance and regulations regarding waste management in your area.

Figure A12-9. In areas where freezing occurs, the wastewater garden may be enclosed in a greenhouse or solarium, offering free fertilizer to sustain beautiful plants. (Courtesy David Del Porto, Ecological Engineering Group)

Appendix 13
Kicking the Propane Habit

I would like to add a few words at the very end of The Renewable Energy Handbook to discuss an interesting anomaly related to off-grid homes. Many people like to "ooh and aah" and gush eloquently about how energy-efficient or environmentally sustainable off-grid living is. While there is no doubt that some off-gridders live very efficient lifestyles, most simply trade limited electrical energy for another form; propane or wood.

With an off-grid system it is imperative to reduce electrical or "plug loads" to the point where household appliances can be supported by the size of renewable energy system you have chosen or can afford to build. Every other energy-consuming device normally moves to propane (or a combination of propane and wood heat).

If you consider this for a moment, you will realize that it is the most energy-intensive devices that are moved to propane in an effort to reduce the capital cost of the electrical generation system. The cook stove, clothes dryer, water heater, and backup space heating all end up connected to the big gas bottle out back.

Now consider that most people in the country will add to this list of fossil fuel-consuming devices one or more cars (or more often pickup trucks) to commute to a distant town, gas lawn mowers or tractors, weed trimmers, snow blowers, gas barbeques, chain saws, ATVs, and snowmobiles.

No, off-grid living **is not** sustainable living unless you as a homeowner understand that you are a living being interacting with a natural ecosystem. Consider that the couple who live downtown, walk to work, and have no yard to maintain will win the environmentalist award over most off-gridders hands down.

In order to avoid this dirty little secret, it is necessary to take complete stock of your life, tools, habits, and interactions with the ecosystem to determine if you are living sustainably. It is fine to grow a lovely organic garden, but your efforts are completely negated if you are using a '72 GMC pickup truck that belches oily smoke to purchase supplies in a distant town.

Most people who live off-grid and try to confront this issue will find that it is almost impossible to reduce their net impact on the environment to the same level as that of a couple living downtown. This does not mean throwing away the shovel, purchasing an Armani suit, and moving to the city to sell insurance, but it does mean that you have to consider every aspect of your life style and separate "wants" from "needs and ego" when making purchasing decisions.

The purpose of these few words is not to be condescending to rural off-grid people (after all, I have been one for over 12 years), but rather to make you aware that sustainable living is not the "in thing" of the day. It is a process that requires awareness and careful attention to everyday decisions. Hopefully, one day, all of society will understand this message.

INDEX

About the Author

William Kemp

As a consulting electronics/software designer, Bill develops high performance embedded control systems for low environmental impact hydroelectric utilities worldwide. He is actively involved in the development and advocacy of green-technologies including; renewable energy heating, energy efficiency, sustainable development as well as photovoltaic, micro-hydro and wind electric systems. Bill advises on grid-intertie and off-grid domestic systems and is a leading expert in small and mid-scale (<20MW) renewable energy technologies. He is the author of the best selling book The Renewable Energy Handbook and has just completed his second book titled $mart Power; an urban guide to renewable energy and efficiency. In addition he has published numerous articles on small-scale private power and is the chairman of an electrical safety standards committee with the Canadian Standards Association. He and his wife Lorraine, live off the electrical grid on their hobby/horse farm in eastern Ontario. Visit Bill's website at <u>www.balancetoday.ca</u>.

ΛZTEXT PRESS

Environmental Stewartship

In our private lives, the principals in Aztext Press try to minimize our impact on the planet, living off the grid and producing our electricity through renewable energy sources and looking at all facets of our lives, including our personal transportation, heating, cooling, food choices, etc.

We continue to evaluate our books to try minimize their impact, and pick the right mix of recycled stock that still allow photographs to be clear and crisp and offer a quality product to motivate our readers.

In the last 10 years we have planted over 2,000 trees, and each year make a commitment to plant more and nurture existing trees so they thrive and maximize their potential to act like the lungs of the planet and clean the air.